I0486930

FAMILY PLATFORM

Wellbeing of Families in Future Europe - Challenges for Research and Policy

FAMILYPLATFORM
Families in Europe Volume 1

Edited by Uwe Uhlendorff, Marina Rupp & Matthias Euteneuer

This publication was produced by FAMILYPLATFORM, 2011.

ISBN 978-1-4477-4149-7.

FAMILYPLATFORM (SSH-2009-3.2.2 Social platform on research for families and family policies) is funded by the EU's 7th Framework Programme (€1,400,000) for 18 months (October 2009-March 2011).

The Consortium consists of the following 12 organisations:

1. Technical University Dortmund (Co-ordinators)
2. State Institute for Family Research, University of Bamberg
3. Family Research Centre, University of Jyväskylä
4. Austrian Institute for Family Studies, University of Vienna
5. Demographic Research Institute, Budapest
6. Institute of Social Sciences, University of Lisbon
7. Department of Sociology and Social Research, University of Milan-Bicocca
8. Institute of International and Social Studies, Tallinn University
9. London School of Economics
10. Confederation of Family Organisations in the European Union (COFACE), Brussels
11. Forum Delle Associazioni Familiari, Italy
12. MMMEurope (Mouvement Mondial des Mères-Europe), Brussels

Contact *info@familyplatform.eu* or visit *http://www.familyplatform.eu* for more information.

Thanks to ILGA-Europe and Sandro Weltin/Council of Europe for use of photos on the cover.

Typesetting and cover design by Lila Hunnisett (*http://lilahunnisett.com/*).

EUROPEAN COMMISSION
European Research Area

SEVENTH FRAMEWORK PROGRAMME

Funded under Socio-economic Sciences & Humanities

Unless otherwise stated, the views expressed in this publication do not necessarily reflect the views of the European Commission.

Contents

Introduction
Uwe Uhlendorff, Marina Rupp & Matthias Euteneuer 7

Chapter 1: Research on Families and Family Policies in Europe - Major Trends
Edited by Kimmo Jokinen & Marjo Kuronen 13

1.1 **Introduction** 13

1.2 **Changing family structures and demographic processes** 14
1.2.1 Fertility and demographic development 14
1.2.2 Change of family forms 18
1.2.3 Reconstituted families 23
1.2.4 Rainbow families 23
1.2.5 Conclusions 25

1.3 **Gender and generations** 26
1.3.1 Transition to adulthood 27
1.3.2 Gender, parenthood, paid and unpaid work 30
1.3.3 Intergenerational relations in families 38
1.3.4 Gender, generations and family violence 41
1.3.5 Conclusions 43

1.4 **Social inequalities and living environments** 45
1.4.1 Social inequalities, diversity and wellbeing of families 45
1.4.2 Families and poverty 50
1.4.3 Physical living environment and housing 54
1.4.4 Conclusions 57

1.5 **Social conditions of migrant families** 59
1.5.1 Demographic impact of migration 59
1.5.2 Families, gender, generation and migration 61
1.5.3 Conclusions 63

1.6 **Media, communication and information technologies** 64
1.6.1 The changing place of media in the European home 64
1.6.2 Media technologies and associated risks 67
1.6.3 Parenting, media, everyday life and socialisation 68
1.6.4 Conclusions 70

1.7 **Family policies and social care policies** 70
1.7.1 State family policies in Europe 71
1.7.2 Childcare policies 75

1.7.3 Cash and tax benefits for families 79
1.7.4 Social care for older people 82
1.7.5 Local family policies 84
1.7.6 Conclusions 85

1.8 **Gaps in existing research** 87

1.9 **References** 94

**Chapter 2: Critical Review of Research on Families and Family
Policies in Europe**
Karin Wall, Mafalda Leitão & Vasco Ramos 119

2.1 **Focus group sessions** 122
2.1.1 Existential Field 1 - Family structures and family forms 122
2.1.2 Existential Field 2 - Development processes in the family 128
2.1.3 Existential Field 3 - State family policies 135
2.1.4 Existential Field 4 - Family, living environments and local policies 145
2.1.5 Existential Field 5 - Family management 152
2.1.6 Existential Field 6 - Social care and social services 160
2.1.7 Existential Field 7 - Social inequality and diversity of families 167
        Topic 1 - Social inequalities and families 169
        Topic 2 - Migration 172
        Topic 3 - Poverty 175
        Topic 4 - Family violence 176
2.1.8 Existential Field 8 - Family, media, family education and participation 177

2.2 **Workshops on key policy issues** 183
2.2.1 Workshop 1 - Transitions to adulthood 183
2.2.2 Workshop 2 - Motherhood and fatherhood in Europe 187
2.2.3 Workshop 3 - Ageing, families and social policy 193
2.2.4 Workshop 4 - Changes in conjugal life 197
2.2.5 Workshop 5 - Family relationships and wellbeing 203
2.2.6 Workshop 6 - Gender equality and families 206
2.2.7 Workshop 7 - Reconciling work and care for young children:
        parental leaves 211
2.2.8 Workshop 8 - Reaching out to families: the role of family
        associations and other institutions 218

2.3 **Methodological issues identified in focus groups and
      workshop sessions** 222

2.4 **Final comments: selected elements on the research and
      policy agenda** 225

**Chapter 3: Facets and Preconditions of Wellbeing of Families -
Results of Future Scenarios**
Olaf Kapella, Anne-Claire de Liedekerke & Julie de Bergeyck 237

3.1 **Introduction** 237
3.1.1 Scientific background for the work 239
3.1.2 Major trends in the Existential Fields 239
3.1.3 Key aspects of the wellbeing of the family 245

3.2 **Methodological approach** 248
3.2.1 The Foresight Approach 248
3.2.2 Methodological to constructing Future Scenarios 249

3.3 **Possible family and living forms in 2035** 252
3.3.1 Scenario 1: Equal opportunities, open migration, diverse
    education and values, mix of private and public care systems 252
3.3.2 Scenario 2: Increasing inequalities, no migration (or very select),
    private education and extreme values, privatisation of care systems ... 256
3.3.3 Scenario 3: Increasing inequality, open limited migration, private
    education, accepted diverse values, privatisation of care systems ... 262
3.3.4 Scenario 4: Equal opportunities at a low level, restricted migration,
    rigid public education with very specific curricula, accepted
    diverse values, public care systems 264

3.4 **Key policy issues and research questions** 269
3.4.1 Importance of intergenerational solidarity and communities 270
3.4.2 Importance of sufficient time for families 273
3.4.3 Unpaid work and care arrangements 276
3.4.4 Children's perspectives: rights, best interests, and impact on wellbeing 278
3.4.5 Family transitions 279
3.4.6 Family mainstreaming and individualisation 280
3.4.7 Impact of technological advance on families 282

3.5 **Summary and conclusions** 284

3.6 **Annex - Living arrangements and family forms** 285
3.6.1 Scenario 1 285
3.6.2 Scenario 2 292
3.6.3 Scenario 3 300
3.6.4 Scenario 4 305

3.7 **References** 315

Chapter 4: Research Agenda on Families and Family Wellbeing for Europe
Marina Rupp, Loreen Beier, Anna Dechant & Christian Haag
(with the support of Dirk Hofäcker and Lena Friedrich) 317

4.1   **Introduction** 317
4.1.1 Main societal trends 317
4.1.2 Key recent policy issues 318

4.2   **Main research areas and methodological issues** 322

4.3   **General methodological remarks** 323

4.4   **Family policies** 328

4.5   **Care** 336

4.6   **Life-course and transitions** 339

4.7   **'Doing family'** 343

4.8   **Migration and mobility** 347

4.9   **Inequalities and insecurities** 350

4.10 **Media** 354

4.11 **Summary** 354

4.12 **References** 356

Annex 1 – Participants in FAMILYPLATFORM 365
Annex 2 – Author Biographies 373
Annex 3 – FAMILYPLATFORM Reports and Publications 379

Introduction

Uwe Uhlendorff, Marina Rupp & Matthias Euteneuer

European societies have undergone profound changes in family life over the last few decades. Putting it very simply, these changes have involved a diversification of family forms over the life-course of family members. As an integral part of this process, families are developing different ways of dealing with parenthood, child rearing, and work-life balance. One result of this is that there is a lack of suitable models for how to best reconcile work and family life. Establishing a fulfilling family life is therefore much more of an individualistic challenge for every family member and for the family unit as a whole.

Despite this, there are considerable cross-national differences between European societies regarding the living conditions of families. Legal systems, welfare structures, educational systems, health-care provision and economic policies vary from country to country, and the structures of families and trends in these areas are therefore quite diverse.

Social innovations and evidence-based policies are needed to cope with the new plurality of family life. In doing so, they should also tackle the decrease in fertility rates all over Europe, increases in rates of divorce and separation of families and changes in gender roles. Family-related issues are an important factor in the formulation of national social policies. Family policy is not an explicit area of competence of the European Union, although many family-related issues are on the European agenda. These are dealt with using the open method of co-ordination by EU Member States. They include gender equality, reconciliation of work and family life, intergenerational solidarity, life-long learning, and the expansion of day-care systems for children.

The European Union took an important step towards strengthening family-related policy issues with the establishment of the European Alliance for Families in 2007. Although this has given greater prominence to family-related issues, there is a continuing need for further research on family issues to enhance policy strategies and improve the wellbeing of families. A first step in this direction was taken by the European Foundation for the Improvement of Living and Working Conditions: the Second European Quality of Life Survey[1], which focused on the theme of family life and work, looking mainly at how to achieve a better balance between work and family life across Europe. Furthermore, several research projects on

---

[1] See *http://www.eurofound.europa.eu/publications/htmlfiles/ef0852.htm*.

family issues have been initiated and funded by the European Commission, many of them within the Seventh Framework Programme, which includes FAMILYPLATFORM in its roster.

The main purpose of FAMILYPLATFORM as a co-ordination and support action for the European Commission was to build up a social platform involving a wide range of stakeholder representatives, including policy makers and family and welfare organisations, grass root initiatives and researchers. The idea was to match different stakeholder groups and their perspectives, to identify vital societal challenges regarding the future wellbeing of families, and to derive key policy questions from interactions between stakeholders. The final objective of FAMILYPLATFORM was to launch a European agenda for research on the family, to enable policy makers and others to cope with the challenges facing families in Europe.

This book summarises the main results of FAMILYPLATFORM, focussing on four areas:

1. State of the art of existing research on family life and family policies;
2. Critical review of existing research;
3. Key policy questions and research issues focused on the wellbeing of families;
4. Research Agenda on Families and Family Wellbeing for Europe.

Overall, more than 120 civil society representatives, policy makers, and scientific experts were involved in the work of FAMILYPLATFORM. Encouraging diverse societal groups to share and negotiate their sometimes quite contradictory perspectives and thoughts and ensuring an effective working process in managing all of the tasks of the project was an undeniable challenge. But overall there has been very fruitful and productive co-operation between these diverse groups, resulting in a great deal of shared learning for everyone involved.

State of the art of existing research on family life and family policies

The state of our knowledge on families has only partially kept up with changes in society, family life and its global frameworks. In general this is due to the great variety of family life and its legal and social contexts. In addition, European policies and research are currently confronted with a situation in which some aspects of family life are thoroughly researched, while others (such as rare family types) remain largely unexplored. In addition, the intensity of research covering specific themes varies between European countries and regions. For these reasons, the first objective of FAMILYPLATFORM was to establish an empirical foundation for further discussion and decisions, by working out the current state of family research and bringing recent and relevant research findings together. An overview of policies and social systems was also compiled, to help give shape to the contextual framework of family life.

As family is related to nearly every area of society, FAMILYPLATFORM had to define specific areas of major concern in order to have a concrete starting point. The following (so-called) "Existential Fields" were taken into account when outlining the current state of family research, identifying significant trends and differences between countries, discovering research gaps, and analysing methodological problems:

1. Family structures and family forms in the European Union;
2. Family developmental processes;
3. State family policies;
4. Family living environments;
5. Family management;
6. Social care and social services;
7. Social inequality and diversity of families;
8. Media, communication and information technologies.

Different expert groups worked on the Existential Field reports, summarising the state of the art of European research in each field. Each report provides an overview of the focal points of research over the last few decades, highlights trends (in family life, as well as family policies) and points out gaps in existing research. These reports were the basis for intense discussion in workshops in Jyväskylä (Finland) in February 2010. Chapter One *"Research on Families and Family Policies in Europe: Major Trends"* by Marjo Kuronen and Kimmo Jokinen is based on this work, and provides an in-depth overview of existing family research in Europe.

Focused critical review of existing research

One of the special characteristics of FAMILYPLATFORM, which made it a social platform rather than a 'simple' research project, was involvement of a wide range of stakeholder representatives. For the critical review on the state of the art it was essential to include the views of representatives of family associations as well as policy makers and social partners. Participants in the critical review process worked out key policy questions and appropriate research perspectives. This was a very fruitful step in the work of the platform, as these groups seldom meet up to engage with each other's thoughts, understandings and agendas. By critically reviewing the current state of research from different perspectives, future challenges for family research and important research gaps were highlighted, and key policy questions for future Europe identified.

To encourage critical comments and statements from a wide range of experts and stakeholders, two discussion forums were established. First, a conference took place in Lisbon in the spring of 2010. This conference was not only an opportunity for participants to hear statements on the state of the art reports, but also saw eight focussed discussion groups and eight workshops take place. More than 120 participants engaged in lively and open discussions, providing the platform with recommendations for future research and key policy questions, each discussion being documented by a rapporteur. The conference in Lisbon was thus a milestone in the work of FAMILYPLATFORM[2]. In addition, an internet platform opened up further possibilities for discussion and involvement of stakeholders who were unable to come to the conference. Its design provided an opportunity to ask questions, get in contact with researchers, and most importantly, to add critical statements or new ideas online. In Chapter Two, Karin Wall, Mafalda Leitão and Vasco Ramos present the major findings of this stage of the work.

Key policy questions and research issues focused on the wellbeing of families

One of the main findings of FAMILYPLATFORM is that the concept of "wellbeing of families" should be considered an important long-term compass when implementing research and developing policy.

To help achieve this, the Foresight Approach was used. It enabled a group of experts and stakeholder representatives to generate common visions of the future, and to explore possible strategies for dealing with their possible consequences. In the spring and summer of 2010, more than 35 researchers, policy makers and representatives of civil society organisations met to discuss and develop four future scenarios using this approach. The participants worked out the preconditions and facets of wellbeing for families, described factors that may have a strong impact on families in the future, and tried to forecast future developments that challenge the wellbeing of families. Based on these assumptions, four future welfare societies and 16 family narratives were sketched out. By elucidating these scenarios, policies to support the wellbeing of families were defined, and areas for future research to support such policies were highlighted.

The method and the results of this procedure are summarised in Chapter Three, "Facets and Preconditions of Wellbeing of Families: Results of Future

---

[2] All of the statements and rapporteur reports are currently available to download from the FAMILYPLATFORM website (http://www.familyplatform.eu).

Scenarios" by Olaf Kapella and Anne-Claire de Liedekerke. It has attracted the attention of scientists and stakeholders, evoking vibrant discussion.

The European Research Agenda

As shown in Diagram 1, the European Research Agenda brings together all of the previous steps, distilling the key findings and concerns of stakeholders into an agenda for research on families for the European Union and its Member States. Taking all of the prior stages of the work into account, it outlines major societal trends, challenges for policy and main areas for future research, and considers methodological issues. It can be seen as a roadmap for future research on families, providing not only smaller topics for research, but also societal challenges that need to be tackled using a multidisciplinary and multi-research method approach.

To enable the involvement of stakeholders in this final stage of the work, a conference in Brussels took place where over 100 representatives from civil society organisations, policy and scientific backgrounds were able to give their input on a preliminary outline of the agenda. Loreen Beier, Anna Dechant, Christian Haag and Marina Rupp present a shortened version of the Research Agenda in Chapter Four.

Diagram 1. The road to the European Research Agenda

This book is the result of the encounters of many experts from all over Europe and beyond, creating a lively think tank on family issues. We want to thank everybody involved in the process for sharing their thoughts and ideas, and for their commitment and their contribution to the project. Special thanks go to all of the members of the Advisory Board, and also to the external experts for their valuable input at every stage of the project. This volume is based on the scientific work of all members of the Consortium, who carefully compiled all of the results.

In addition, we would like to thank Linden Farrer for doing a great job co-ordinating production of this book, and Elie Faroult, whose experience helped guide the work on the Future Scenarios. Finally, we would like to thank Pierre Valette and Marc Goffart from the European Commission (Directorate-General Research & Innovation) for their extensive advice at every important stage of our work, and also Ralf Jacob (Directorate-General Employment, Social Affairs & Equal Opportunities) for his support throughout the project.

FAMILYPLATFORM was funded by the European Union's 7th Framework Programme (Socio-economic Sciences and Humanities 2009) for 18 months (October 2009-March 2011).

Chapter 1: Research on Families and Family Policies in Europe: Major Trends

Edited by Kimmo Jokinen & Marjo Kuronen

1.1 Introduction

The work of FAMILYPLATFORM encompasses four key steps, the first of which is to chart and review the major trends of comparative family research within the EU. This first step consists of eight Existential Field Reports, two additional Expert Reports and WP1 Final Report, *State of the Art of Research on Families and Family Policies in Europe.* The partners involved have conducted extensive and systematic literature reviews on European comparative research published since the mid-1990s using existing scientific and statistical databases, reports from previous and ongoing EU-funded research projects, and other relevant publications, which are occasionally supplemented with own analyses of data available[1].

The Family Research Centre and the Unit of Social Work at the University of Jyväskylä, Finland, was responsible for co-ordinating the first stage of FAMILYPLATFORM. This chapter draws together the main results, conclusions and major trends identified in the more extensive Existential Field reports. It is a newer and shorter version of the Final Report that was edited by Marjo Kuronen with contributions from Kimmo Jokinen and Teppo Kröger; Johanna Hyväluoma assisted in technical editing and proofreading. Despite this, it is still very much the result of joint effort by the whole of the FAMILYPLATFORM Consortium.

There are significant cross-national differences in the living conditions of families between different European Union Member States. Legal systems, welfare structures, education systems, health and social care service systems and economic systems and conditions vary from country to country. Consequently, European family structures and family forms, as well as respective trends and developments are quite diverse. It is therefore crucial to provide a comprehensive overview of various fields of family life and family policies, in order to derive conclusions for political practice and further research. To this end, this chapter is organised into eight sections: 1) Introduction; 2) Changing family structures and

---

[1] This chapter is based on the reports of the eight Existential Field Reports written by Consortium partners, the two Expert reports, and the WP1 Final Report edited by Marjo Kuronen: All of these reports, and other outcomes of FAMILYPLATFORM are available at *http://hdl.handle. net/2003/27684.*

demographic processes; 3) Gender and generations; 4) Social inequalities and living environments; 5) Social conditions of migrant families; 6) Media, communication and information technologies; 7) Family policies and social care policies; and 8) Gaps in existing research.

1.2 Changing family structures and demographic processes

It is a well-known and documented fact that family structures and family forms have changed considerably throughout Europe since the 1960s and 1970s. A review of existing research and statistics shows there has been comparatively high growth in the number of family forms within European countries over recent decades.

The degree to which these transformations have materialised varies considerably between European countries. There is still a large variety of different, nationally or regionally specific patterns, often strongly connected to different cultural backgrounds or family policy models. Therefore, it is too simple to speak about "the European family". The Nordic countries represent one end of the scale, with late marriages, modest marriage rates and a high proportion of out-of-wedlock births. These countries have moved considerably far from the "traditional" family model. At the other end of the scale are the Southern European countries, where family patterns are still much in line with the 'traditional model', with a central importance placed on marriage, low divorce rates, and low incidence of out-of-wedlock births, with new family forms not being widespread.

These developments largely rely on long-term trends. Most recent data suggest that there may be some signs of a "flattening out" of previous highly dynamic processes, in the move away from the "traditional" family model. However, data indicating this is often very recent, and it is hard to say whether it is indicative of a more general future trend. Even if the trend towards "new family forms" comes to a halt, a return to a "nuclear family model" is unlikely.

1.2.1 Fertility and demographic development

Knowledge on fertility points to considerable shifts in demographic behaviour throughout recent decades. The decision for both marriage and family formation has shifted to ever later ages in virtually every European country. The medium age of women giving their first birth is lowest in Eastern Europe, i.e. in the former post-socialist countries, with average ages ranging between 25-27 years. In contrast, highest average ages, around 30, are observed in the UK and Switzerland. Southern European and German-speaking countries show similarly high ages, while only Portugal with a comparatively early age appears to deviate from the Southern European pattern. Increases over time appear to have been most pronounced in Central European and Nordic countries.

The medium age of first marriage of women in Europe shows a very distinct country-specific pattern, with women in the Nordic countries displaying the highest average age. Women in Central and Eastern European countries (CEE) (except for Slovenia), display comparatively low average ages of marriage. The pattern in the postponement of first marriages of men is not as clear: whereas men in Eastern European countries are the youngest, and men in Sweden and Denmark are the oldest to marry, the pattern in the centre of Europe appears to be more mixed (OECD, 2009a).

Figure 1. Average age of women at first childbirth, 1970-2005, by country

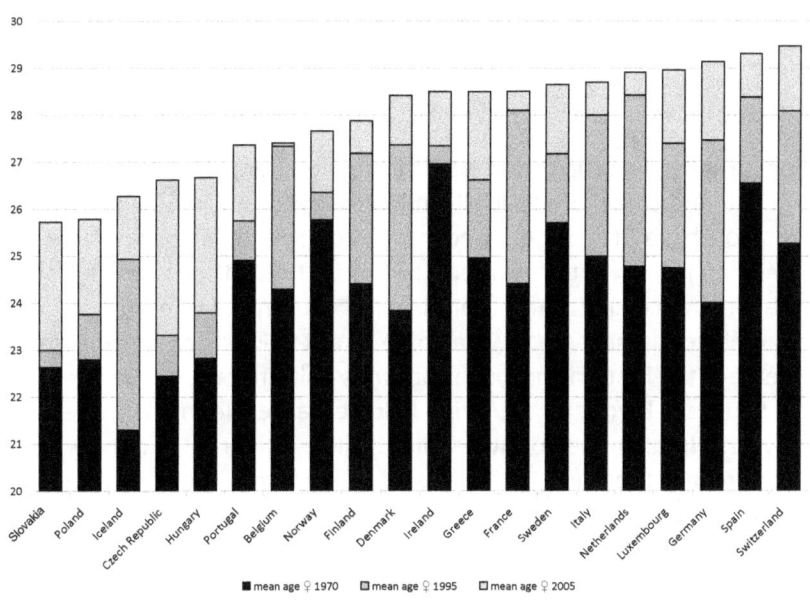

Source: *OECD, 2009a*

Figure 2. Average age of women at first marriage, 1970-2004, by country

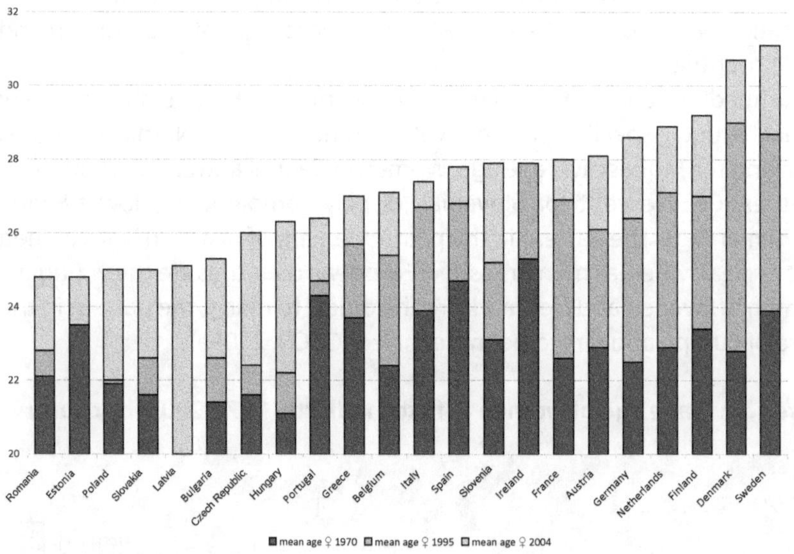

mean age ♀ 1970 mean age ♀ 1995 mean age ♀ 2004

Source: *OECD, 2009a*

Looking at both marriage and family formation trends simultaneously, data appear to indicate that especially in Northern Europe, marriage and family formation have increasingly become decoupled, as a considerable share of children is born out-of-wedlock. Since the 1970s, their share first started to rise in France and in the Scandinavian countries. Since then, the trend has remained largely stable. The share of out-of-wedlock births started to increase in the Eastern European countries only after the 1990s. The Central and Southern European countries followed a development somewhat in between. Notably, the countries with a high incidence of out-of-wedlock births are also those with the highest fertility levels. The postponement or denial of marriage thus cannot be seen as a major driver of declining fertility in modern European societies.

What emerges today is a change in the very concept of marriage. Marriage has increasingly come to be a subjective experience: choosing to marry or not to marry has become an individual decision. The French sociologist Théry (1993) has called this phenomenon "démariage". The traits of instability and uncertainty, which distinguish the transformations in the contemporary family, are in line with the atmosphere that characterises society at large, marked by a climate of uncertainty as far as work and social stability are concerned, aggravated by a high level of economic instability.

In this context, there is a perception that even creating a family has become an individual - even a risky - enterprise (Beck/Beck-Gernsheim, 1994: 29).

Due to the increasing postponement of family formation decisions, period-specific fertility rates[2] in all European countries have declined throughout recent decades. In Northern and Central Europe fertility fell from around 3 children per woman in 1965 to less than 1.8 in the mid-1990s. Southern European countries appeared to follow this general trend with a ten-year time lag. In Eastern Europe, fertility levels started to decline after the transition from state socialism to market economies in the 1990s. In Southern and Eastern European countries recent declines have resulted in very low fertility levels of less than 1.2 children per woman, that have led demographers to describe these countries as displaying "lowest-low fertility" (Kohler/Billari/Ortega, 2006). In recent years, the lowering trend in fertility levels has "flattened", with only marginal changes since the 1990s. Some researchers even point to partial recovery in period-specific fertility levels since the turn of the century, especially in Northern and Western Europe.

However, regarding long-term fertility developments, the period-specific fertility rates may be partially misleading. While women indeed are increasingly postponing family formation and the birth of their first child to ever later ages, it could, in principle, be assumed that women nonetheless are not generally reducing their overall lifetime fertility, but simply shifting their "family phase" to later stages in their life-course. Cohort-specific fertility rates appear to indicate that postponement of first childbirth might be partially compensated for by "recuperation" behaviour later (Frejka et al., 2008: 6). Given the fact that reliable data on cohort-specific fertility are available only up to the birth cohort of 1965, it is too early to judge whether this recuperation effect will "balance out" fertility rates in the long run.

A possible indicator of future fertility trends is childbearing preferences, as reflected in the perceived ideal number of children in a family, and individual intentions to have (more) children in the future. In almost all European countries, the general ideal and the personally favoured number of children are well above the actually realised fertility figure; a finding that recent sociological research (e.g. Blossfeld et al., 2005) interprets as reflecting a personally perceived inability to start a family, for example, due to rising individual uncertainties. Alternative explanations have stressed the role of a general value change towards more "post material values" such as self-fulfilment, which have contributed to a decline in the importance of more "collectivist" family values (Inglehart, 1990; Lesthaeghe/van de Kaa, 1986).

[2] Defined as the average number of children that would be born per woman, if all women lived to the end of their childbearing years and bore children according to a given fertility rate at each age.

The traditional view that low fertility trends are an outcome of increased female labour force participation can be dismissed, even if this relationship is rather complex. Today, countries with a high share of employed women simultaneously display highest fertility rates. It is not employment as such but the way in which the reconciliation between work and family is facilitated that drive women's childbearing considerations (D'addio/D'Ercole, 2005; also Ahn/Mira, 2002: 669-670; Rindfuss *et al.*, 2003: 411; Philipov *et al.*, 2009: 26). Family and gender policies, as well as work-related institutions, may contribute to explaining the extent of these differences (Engelhardt/Prskawetz, 2004: 55-56). A macro-level comparison shows that both higher fertility and female employment rates are simultaneously found in countries where institutional support for working parents is fairly comprehensive (Philipov *et al.*, 2009: 27-28). Therefore, one of the most important future challenges will be to enable parents to fulfil their fertility aspirations, which may well be achieved through well-designed family policy packages.

1.2.2 Change of family forms

Family forms have become more diverse in nearly all European countries in recent decades. The idea of a standard "nuclear family model", i.e. a household with married heterosexual couple and their biological children, has been replaced by a variety of different alternative family forms and lifestyles (Kapella *et al.*, 2009). Especially in Northern and Western European countries, the recent decline of the "golden age of marriage" with high fertility and marriage rates, low divorce rates and an early start to family formation (Peuckert, 2008: 341), has been accompanied by an increase in less institutionalised relationships.

Scholars have identified many different factors that have contributed to the 'crisis of matrimony' in contemporary Europe. At the socio-economic level, the transformation of the labour market, with increasing labour force participation of women, has made a tie of marriage a choice rather than "destiny", a rite of passage into adulthood. At the cultural level, the process of secularisation (Norris/Inglehart, 2004) has contributed to the gradual spread and affirmation of cohabitation. Universal education and the emergence of collective movements such as feminism have played a key role in undermining the model of the traditional, patriarchal family. The marital tie in itself is no longer crucial; rather, marriage is induced by the individual sentiments of each of the partners to seek a union (e.g. Weigel, 2003). Paradoxically, the tendency to place love at the basis of contemporary marriage constitutes one of the elements of its fragility and instability.

Decreasing marriage rates and increasing divorce rates

Since the mid-1960s, marriage rates in Europe have declined, and have only recently stabilised. While the marriage rate was 7.64 marriages per 1.000 persons in 1965, it has fallen to as low as 4.87[3] in 2007 (Eurostat, 2010).

When comparing European nations, some Northern European countries display high marriage rates (e.g. Denmark with 6.81 in 2008), following modest increases since 2003. Eastern European countries are rather heterogeneous concerning marriage patterns (Eurostat, 2010). A major reason might be that in some Eastern European countries the influence of the Catholic and Orthodox Church is still significant, i.e. in Poland and in Romania. Since these religions advocate a more traditional family model, it is not surprising that their citizens show the highest marriage rates within Eastern Europe. Despite this, marriage rates have declined strongly between 1990 and 1992 in all Eastern European countries (as well as in the Eastern part of Germany; see Eurostat, 2010), most likely a repercussion of both rising insecurities following the breakdown of the socialist regime, but also the discontinuation of political support for the "nuclear family" model (see Peuckert, 2008: 358). In most Central European countries marriage rates have fallen since the early-1960s, and are now slightly below the European average (*ibid.*). In most Southern European countries marriage rates have also fallen continuously and are either well below the European mean (e.g. Italy, Spain or Portugal), or just above the average.

At the same time, the proportion of cohabiting families has increased. Generally speaking, cohabiting couples with children are most common in Northern Europe and in France and very rare in Southern Europe (Kiernan, 2004). Still, cohabitation often makes up a "preliminary" form of partnership before getting married. This indicates that overall, getting or being married is still very important for most Europeans (Kiernan, 2003; Spéder, 2005). Thus, the number of (long-term) cohabiting couples with children is still low, but recently has been increasing. In most Northern and Western European countries, except Western Germany and the Benelux countries, over 40% of cohabiting couples already have children. Still, the percentage of first-born children of cohabiting parents is much higher than for second- or later-born children.

While marriage rates have decreased in Europe, divorce rates have constantly risen – more than doubling from 0.8 (divorces per 1000 persons) in 1965 to 2.0 in 2005[4]. The highest rates are observed in Lithuania, the

---

[3]   Figure is estimated.
[4]   No actual data available.

Czech Republic, Belgium, Denmark, and Latvia. In Germany, Sweden and Slovakia, rates are rather moderate, whereas in Greece, Italy and Ireland, divorce rates are very low (Eurostat, 2010). Taken together, there appear to be only small regional differences. The most obvious pattern is that lowest rates are observed in countries with a high proportion of Catholics.

The impetus for divorce is increasingly coming from women, and is often explained by women's increased financial independence. However, recent research shows that the relationship between women's employment and the increase in the divorce rate varies according to socio-cultural context. In countries with greater gender equality, like the Netherlands and the UK, the financial independence of women has a positive effect on marital stability, while in countries where equality is still far from being achieved, like Italy, the increase in the presence of women in the workforce is accompanied by increased instability. Furthermore, it is not so much women's employment as the nature of the relationship itself that generates instability in a marriage (Saraceno/Naldini, 2007; MacRae, 2003).

A large number of studies, both economic and social, have examined the consequences of divorce on men and women. As far as the economic consequences of divorce are concerned, those mainly fall upon women (Mckeever/Wolfinger, 2001; Aassve et al., 2006). Separation and divorce also influence men who, while suffering less financially, seem to suffer other negative effects, such as deterioration in the quality of their life style, housing and general consumption, as well as deterioration in the quality of their relationships with family and friends. Obviously, a crucial role in regard to the economic and social consequences of divorce is played by the social welfare system and the services it offers, which differ from one European country to another (Kalmijn/Rigt-Poortman, 2006; Uunk, 2004).

Re-marriage and non-traditional styles of living

As divorce rates have risen, the relative incidence of re-marriages has also risen. In nearly all European countries, the percentage of first matrimonies as a share of marriages in total decreased through 1960 and 2006[5]. Cross-national comparisons show that the Eastern European countries had the lowest increase in re-marriages. In contrast, the Northern European states show increases of about 10%. Central European countries are more hetero-geneous: whereas Belgium, the UK and Luxembourg display a high increase in re-marriage, Germany and France show few differences over time.

---

[5] No overall data for these two points of time is available. Most post-socialistic countries just offer data since 1995 or later.

Southern countries (except Portugal with a moderate increase) as well as Ireland show almost no differences at all (Eurostat, 2010).

Despite the developments outlined above, the "nuclear family model" with married parents clearly remains dominant in all European regions. In this context, Peuckert (2008) differentiates three regional types with relative homogenous characteristics: the Northern European states, where non-traditional styles of living are more widespread; the Western European states with a dominance of the "modern nuclear family model" (even though it is decreasing); and the Southern states (including Ireland), which are still traditionally oriented (*ibid.*: 368). As mentioned above, Eastern European countries are more heterogeneous regarding the dominance of a specific family type, but in general seem to lose their inclination towards the traditional model.

Figure 3. Share of family-types in the EU27 countries, 2007

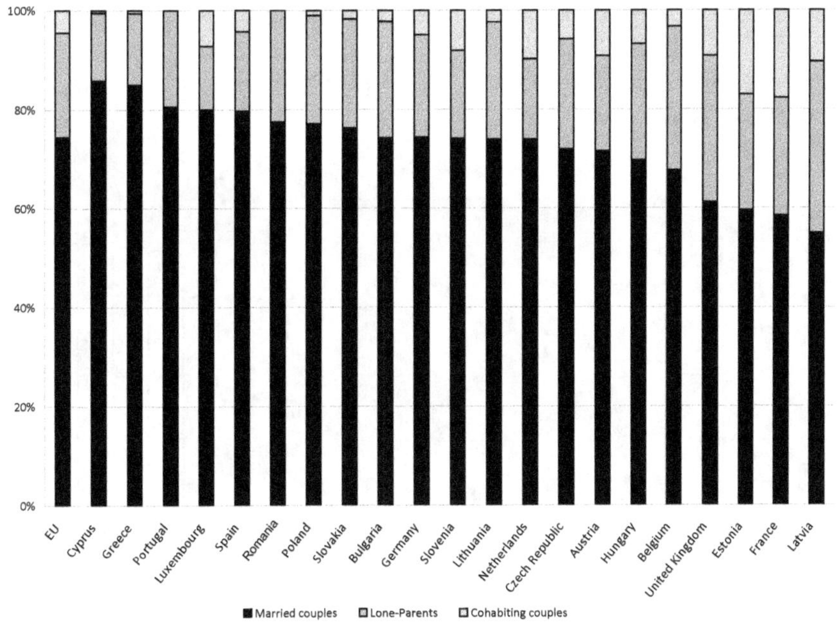

Source: *Labour Force Survey microdata, 2007, ifb-calculations (unweighted data)*

Lone-parent families

The decline in institutionalised relationships is accompanied by increases in other, previously less widespread forms of family life, such as lone parent-hood, reconstituted and cohabiting families. During the dominance of the "nuclear family model", lone parenthood often resulted from the death of a partner. Today there is a comparatively high percentage of unmarried as well as divorced (or separated) single parents. They are mainly mothers who live alone with their children (European Commission, 2007a: 13). Since the 1980s, the share of lone-parent families rose from 10 to 27 percent in the EU15[6] in 1999 and was at about 21% in the EU27 in 2008[7].

As shown in Figure 4, a large number of lone-parent families are found in the UK, Central European countries, and in Eastern European countries like Estonia, Lithuania and Latvia. Very low rates can be observed in Southern Europe and Luxembourg (Rost, 2009: 13). The composition of lone-parent families, especially in Southern European countries, shows a very high share of divorced and widowed mothers. In contrast, there are only a few unwed lone parents. In the Central and Western European countries, there is a dominance of divorced lone mothers and a moderate share of single unmarried ones. The highest proportion of this group can be observed in Denmark, the United Kingdom, Ireland, and in Eastern part of Germany (European Commission, 2007a: 18ff.).

Figure 4. Share of lone parents in all family-households in the EU27 countries[8], 2007

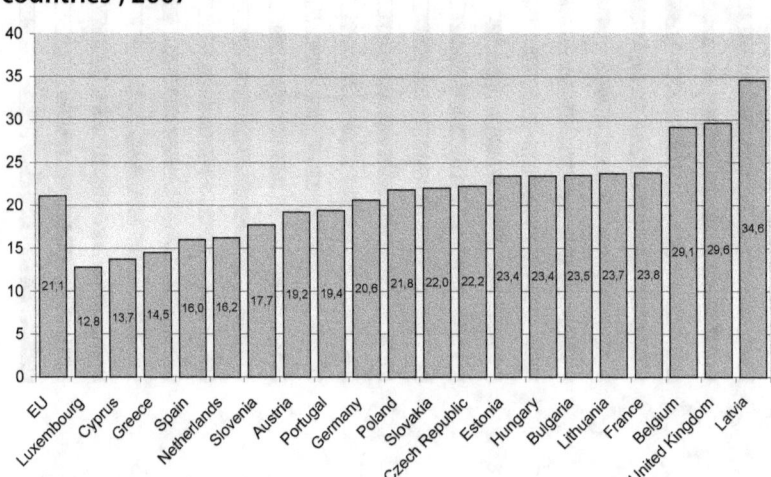

Source: *Labour Force Survey microdata, 2007, ifb-calculations (unweighted data)*

---

[6] The 15 EU Member States prior to enlargement in 2004:Austria, Belgium, Denmark, Finland, France, Germany, Greece, Ireland, Italy, Luxembourg, the Netherlands, Portugal, Spain, Sweden and the United Kingdom.
[7] The above time series, however, need to be treated with care, as data is scarce. Furthermore, occasional evidence points to substantial variations in data on lone parenthood between different data sources.
[8] For Denmark, Finland, Ireland, Italy, Malta and Sweden there was no data available. So the figure for the EU27 depends on the other available countries.

1.2.3 Reconstituted families

Another important issue with regard to the rising diversity of family forms is the increase in reconstituted families. Most recent literature distinguishes different types of these: simple stepfamilies with children from just one side, complex stepfamilies with children from both sides or even with shared children, and multi-fragmented families with more diffuse family formations (see Peuckert, 2008; Steinbach, 2008). Complex stepfamilies as well as multi-fragmented families are also called patchwork families (Nave-Herz, 2004: 33).

As comparative data show, it is not so much the sheer number of reconstituted families and lone-parent families that is new, but rather the substantial increase in their numbers over time. Comparing the different countries with available data in the Generation and Gender Survey (United Nations, 2005), Germany today has the highest percentage of reconstituted families (around 14%). At the other end of the scale, Bulgaria and the Netherlands have the lowest rates at around 4%.

Notably, reconstituted families play an important role in the context of fertility. They are more likely to have additional children, because, on the one hand, they mostly intend to start a new family and, on the other, the motivation of the childless partner to fulfil his or her wish to have a child is rather high. This so-called "parenthood effect" positively affects the probability of having shared children (Prskawetz et al., 2003: 108). The magnitude of the effect varies depending on the number of children both spouses have previously had (Thomson, 2004). Women in particular, who bring their own biological children into the new relationship, are generally more willing to have another shared child as compared to men.

1.2.4 Rainbow families

Rainbow families are still a very rare phenomenon in Europe. Rainbow families are defined "by the presence of two or more people who share a same sex orientation (e.g. a couple) or by the presence of at least one lesbian or gay adult rearing a child" (Allen/Demo, 1995: 113). There are two main types of rainbow families: those where the child(ren) stem(s) from previous heterosexual relationships and those where same sex couples realise the desire for a child via reproductive medicine, adoption or fostering (Wegener, 2005: 53ff.). Sample data for Germany indicates that the first type is most widespread, but in recent years, the second type of parenthood has become increasingly frequent (Dürnberger et al., 2009: 15; Jansen, 2010).

Legal recognition of same sex couples shows wide variation between European countries. During the 1980s and 1990s, the Nordic countries were forerunners, giving same sex couples the opportunity to legalise their unions. A new legal term, the registered partnership, was introduced and first passed in Denmark in 1989. Other Nordic countries introduced this new civil status in subsequent years (Norway in 1993, Sweden in 1995, Iceland in 1996), followed by a number of other European countries in later years[9]. Registered partnership does not assign all the same legal rights to same sex couples as marriage provides for heterosexual partners. Other countries, such as Belgium (2000), Slovenia, the Czech Republic (2006) and Hungary (2009) chose a more property and inheritance oriented construction to recognise same sex couples (Verschraegen, 2009: 434). In 1999, the French government chose a unique way and installed PACS ("Pacte civil de solidarité") as a new social status and possibility for heterosexual as well homosexual cohabitees[10].

In 2009, Norway and Sweden completed their process of granting same sex couples the same rights to marriage as to heterosexual couples. Norway and Sweden are the only states in Europe where fully gender-neutral marriage legislation is implemented, while other European countries such as Italy, Ireland, and Poland still have no institution at all to legally recognise same sex couples (ILGA, 2010).

The means by which same sex couples realise their childbearing aspirations, such as adoption and access to reproductive medicine, displays huge variation between countries (Verschraegen, 2009: 434). In Norway, the Netherlands, Sweden, England and Wales, Scotland, Belgium, Iceland and Finland, same sex couples have the right to adopt an unknown child, while in other countries, such as Germany, this is not permitted. Additionally in Germany, France and Denmark registered same sex couples have the right to a so called "stepchild adoption" where the (new) partner of the biological parent can "step-adopt" the child, once the other biological parent is known and agrees to this procedure. In Norway, same sex couples additionally have access to reproductive medicine (Verschraegen, 2009: 434).

Existing international research has mainly focused on legal and juridical aspects, such as the recognition process, the legal differences to marriage or the right to adoption (Biele-Woelki/Fuchs, 2002; Verschraegen, 2009; Festy, 2006). Furthermore, official statistics on rainbow families are based on a very small number of cases, so estimation errors are very probable. The

---

[9] The Netherlands and Spain in 1998, Germany and Portugal in 2001, Finland in 2002, England, Wales and Luxemburg in 2004 and Austria in 2010 (Banens, 2010: 10; Biele-Woelki & Fuchs, 2002: 215ff.; Verschraegen, 2009: 433ff.; Bundeskanzleramt, 2009; Festy, 2006: 419).

[10] This intermediate status, which is neither a union nor a contract, neither private nor public, expresses also the "French ambiguity of responding to increasing cohabitation" (Martin & Théry, 2001: 135).

GGP provides data on rainbow families for only five European countries: In Bulgaria 0.1% of all families are rainbow families (with children), in Germany and the Netherlands 0.7%, in France 0.5% and in Hungary only 0.02%[11] (United Nations, 2005, own calculations).

In Germany, as compared to heterosexual partnerships, same sex couples have a higher educational level, and most of the couples share domestic and paid work more equally. These trends are also confirmed by the first representative German national study on registered partnerships with children, "Children in same sex partnerships" (Rupp, 2009). In addition, a further study demonstrates that in Germany most of the rainbow families are made up of same sex orientated women and their children (Eggen/Rupp, 2010).

Families without a common household

Living apart together and commuter families are relatively new research subjects in Europe. By definition, a living apart together relationship is a couple which does not live in the same household. These people define themselves as a couple, and they perceive that their close surrounding personal network does so as well (Levin, 2004: 227ff.). These might be also families with children, where one parent does not live in the same household.

The data on living apart together and commuter families can be described as incomplete. The data of the Gender and Generation Survey (GGS) provide a first database to describe this family form to a limited extent. Available evidence from GGP data shows that living apart together couples with children are a rare phenomenon. Their percentage in Europe varies between 1.4% in Bulgaria and Hungary to 4.1% in France.

1.2.5 Conclusions

Previous analyses have given a concise overview of recent developments in family structures and family forms in Europe. Taken together, the results demonstrate comparatively high growth in the number of family forms within European countries over recent decades. Some major trends can be identified, based on demographic statistics and existing research:

- Postponement of first childbirth and first marriage, generally decreasing number of children, even though fertility aspirations are still at a comparable high level.

---

[11] This accords to only one case in Hungary. In all GGP-observed countries the cases of rainbow families amount to only 27 cases (Germany).

- Being married has lost its central role as a precondition of family formation, and there are increasing numbers of out-of-wedlock births.
- Decreasing marriage rate, increasing divorce rate and increasing rate of re-marriage.
- As a consequence, a decrease in the incidence of the 'middle-class nuclear family', even though this model remains dominant.
- Increasing diversity of family forms and family life.

However, the degree to which these transformations have materialised varies considerably between European countries. Opposing cases are represented by the Nordic countries, where there has been a considerable move away from the "traditional" family model, and Southern European countries, where family patterns still are much in line with traditional patterns (i.e. central importance of marriage, low divorce rates, low numbers of out-of-wedlock births, low incidence of new family forms). It is important to underline that - at least at present - those countries with the highest degree of recent "de-standardisation" display the highest fertility levels.

Outlining of the above developments largely relies on long-term trends. Most of the outlined trends thus refer to developments over the last few decades. Most recent data suggest, however, that there may be some signs of a "flattening out" of previous highly dynamic processes. However, data indicating this is often very recent, and it is hard to say whether it is indicative of a more general future trend. Furthermore, even though for some indicators there has been a tendency to stabilise in some of the countries analysed, it is unlikely that in the future a full reversal of previous developments, for example postponement of first childbirth or increasing divorce rates, could occur. Therefore, even if the trend towards "new family forms" comes to a halt, a return to a "nuclear family model" is unlikely.

1.3 Gender and generations

Gender and age strongly influence the experiences and everyday life of family members in different life phases. In this section, the developmental processes of families are reviewed with particular emphasis on gender and generational relations in families: on the parent-child relationship, the process of transition to adulthood, relationship between partners, women and men, as spouses and parents, and between older generation, their adult children and grandchildren. In the research on family developmental processes, age is an important reference point. In this respect, an approach involving the conception of the life-course, which is founded on age, may well be a useful one.

1.3.1 Transition to adulthood

Age regulates entries into and exits from various life stages, together with forms of action – for example, when it is appropriate to marry, to have or not to have children, or when to enter the world of work or leave it (Elder, 1975; Elder/O'Rand, 1995). Nonetheless, it is necessary to consider these assumptions with a critical eye. The normative strength of age itself has decreased, and in contemporary society, biological age and social age tend to be separated: the former is no longer an obligatory reference point for the definition of the latter.

Transitions that young people go through in contemporary societies are interrelated and intertwined. So-called "yo-yo" transitions (EGRIS, 2001) are potentially reversible transitions that unfold in respect to the multiplicity of interlacing "strands" that constitute the path to adulthood. Today young people find themselves having to negotiate transitional processes that are made up of a highly complex mixture of dependence and autonomy. This includes their transition from school to work (Walther et al., 2006), relations with their family of origin (Biggart/Walther, 2006; Stauber/du Bois-Reymond, 2006) and when to remain in or leave the family (Buber/Neuwirth, 2009), the development of their emotional lives, their life plans and building of a family of their own.

Figure 5. Different transitions into adulthood

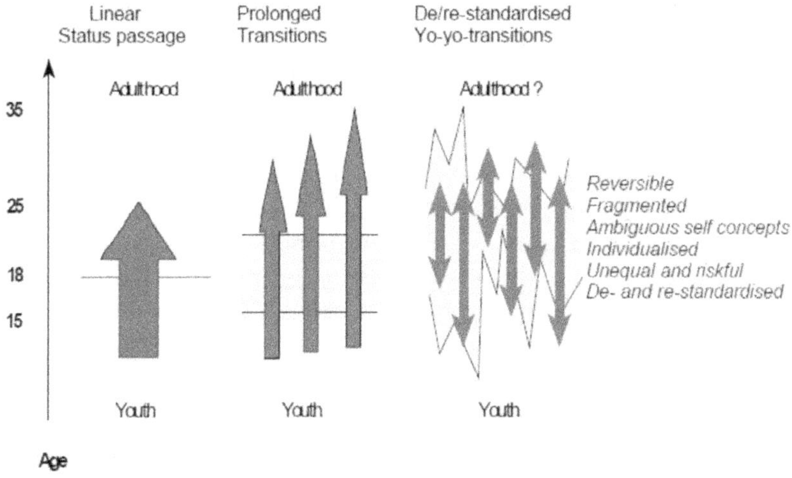

Source: *Walther et al., 2006*

In contemporary society young people do not leave their family of origin at a very early age. Young women leave home earlier than young men. Of particular note is the advanced age at which young people leave their family of origin in Belgium, Italy, Slovakia, Slovenia, and Malta, where on average they continue to live in their parents' home beyond the age of 28 (women leave home on average at the age of 28-30 and men at the age of 31-32). The European country where young people leave home the earliest is Finland, where on average young women and young men become independent at the ages of 22 and 23 respectively (Eurostat, 2009a).

This has a significant impact on the process whereby young people enter into adult life. Two distinct sets of factors and conditions contribute to this phenomenon: on the one hand, the temporal extension of educational/work training paths and the concrete difficulties in entering the workforce, and on the other, the emergence of emotionally closer and more supportive relations between the generations. The family of today is a negotiation and affection-based family, no longer a rule-governed family. This new family type tends to be represented as a place dedicated to caring and protecting, the principle purpose of which is to provide love and security to children, satisfying their economic, social, and affective needs. This can lead to ambivalent consequences: on the one hand, a more open and richer affective relationship between parents and children, but on the other, a more marked and prolonged dependence of children on parents.

Research dealing with transition processes has revealed a marked variability in the life trajectories of different individuals (Arnett, 2004 & 2006; Coté, 2000; Leccardi/Ruspini, 2006). Life trajectories, which for previous generations were more standardised, have become increasingly fragmented, without clearly identifiable connections between one phase and another. Sometimes the phases can even be inverted; this process has been referred to as the de-standardisation of life-courses (see Walther/Stauber, 2002) and manifests itself in what has been called a biography of choice (Furlong/Cartmel, 1997).

Transition to parenthood, in contrast to other transitional processes such as those relating to work, personal relationships, housing, etc., is distinguished by an indisputable irreversibility: becoming a mother or father inevitably involves becoming a parent until the end of one's life. This characteristic plays a fundamental role not just in the construction of identities but also in the construction of representations of reality and societal images of parenthood. Thus, the link between reversibility of life choices and irreversibility of parenthood generates a considerable amount of ambivalence that young people have to cope with in becoming parents.

Becoming a parent involves an extremely profound change not just in the life-course of individuals but also in the nature of the relations within the couple. It is for this reason that couples today, whether married or not, tend to evaluate and weigh up ever more carefully a series of circumstances, both present and future, before committing themselves to bringing a child into the world. Decisions relating to parenthood are influenced by a series of novel considerations (Hobcraft/Kiernan, 1995): first, having a partner. The majority of children are still born to a mother and father who live together. Secondly, completing education and work training. The majority of young Europeans do not become parents before completing their studies. Thirdly, having a job that guarantees an adequate income. Changes in the labour market have led to an increase in and prolongation of the financial dependence of young people on their families. Therefore, young people have to evaluate whether it is economically feasible for them to become parents. Fourthly, having a house of one's own, and fifthly, having a "sense of future". Apart from the concrete factors discussed above, having a child also demands being able to anticipate events at least over the medium term.

An especially important thematic strand for decision-making processes relating to parenthood is timing. After a period of moral panics on the topic of teenage pregnancy, the current discourse - at least on a scientific level - now seems to focus on resources related to an early entry into parenthood (e.g. Phoenix, 1991; Arai, 2009). Teenage parenting may be more of an opportunity than a catastrophe. Recent studies have mostly ignored those young mothers and fathers who intentionally and consciously became parents. Coleman/Cater (2006) show that some of the young fathers and mothers interviewed clearly relate parenthood to an idea of "leading a different life", to ways of fathering and mothering different from those they experienced in their families of origin.

It seems possible to identify certain common traits among the representations of the maternity and paternity of young Europeans today (Walther *et al.*, 2009). A first point to note is the discrepancy between the ideas expressed by young women and young men and the actual practices put into action in family life. While there is a tendency to aspire to more equal and balanced relations within the couple, it seems that in everyday life these aspirations do not find expression in terms of the actual distribution of domestic work, which still penalises the maternal figure, as different national time-budgets show. Alongside traditional visions of parenthood, some new models of parenthood are beginning to emerge in Europe, which make provision for changing gender roles, "doing gender", obligations, and a reallocation of tasks within the family.

One general insight of European research on transitions into adulthood is that young people depend to a large extent on facilitating structures, such as socio-economic resources and opportunity spaces, in order to negotiate, shape and cope with uncertain transitions to family, work and citizenship, especially where they are accompanied by insecurity. However, the success of these facilitating structures in turn cannot secure predictable trajectories. Policies are required, which let young people perceive such structures as accessible, relevant and manageable and in consequence accept and use them.

1.3.2 Gender, parenthood, paid and unpaid work

One of the most studied areas of family research today is labour market participation of women, especially of mothers with young children, and the reconciliation of family and paid work and gendered structures and processes related to it. In this section, we make a distinction between paid work, household duties and childrearing. The three domains are strongly interrelated. This is not only so because the amount of time spent on one of these tasks will inevitably restrict the amount of time available for the others, but also because they are all underpinned by very similar factors: beliefs and values regarding gender roles, the structural environment of families and individual characteristics.

Work-family balance and reconciliation of work and family life have been in the focus of scholars as well as European policy makers for decades, and are attached to a series of policy aims including gender equality, fertility rates, prevention of loss of human capital and economic growth (Knijn/Smit, 2009). The work-family balance approach focuses on state policies as well as employers' measures to facilitate employment of individuals with family commitments.

There is a strong gender aspect involved in this topic. Paid work and unpaid work are unequally distributed between men and women in every European society – although the extent of the differences varies considerably. Cultural traditions and social norms relating to gender roles shape individual attitudes. Despite the efforts of several European societies to create a policy environment of equal treatment for men and women, gender remains a substantial factor in work-distributing behaviour in the labour market as well as in the household.

Main trends and cross-national (dis)similarities in the division of paid work

The most marked change over recent decades in the area of division of labour is the increasing level of female employment that - at the household-level - has led to the expansion of the two-earner model. Female participation in the labour force across the EU is constantly increasing in virtually all

Member States. Consequently, the gender gap in the level of labour market activity is decreasing – falling from 18.6% in 1997 to 13.7% in 2008 in the EU27 countries (Eurostat, 2009a). There was a slight break in this falling tendency and in the rate of decrease when Central and Eastern European (CEE) countries joined the Union. During their socialist decades, most of these countries achieved a female employment rate close to full employment. It dropped radically after the economic collapse in the early 1990s. Only since the mid-2000s has a slightly increasing trend been observed in the CEE countries (Scharle, 2007).

Thus, changing female aspirations have led to increased female labour market participation in many countries. The biggest change in behaviour has taken place among married mothers (OECD, 2007a: 42). Between 2000 and 2006, female employment rates increased in most of the Member States, and strong increases were recorded in a number of Mediterranean countries (Spain, Greece and Italy) and in certain new Member States[12] (notably in Cyprus, Latvia and Estonia) (Eurostat, 2009c: 17). Even so, the proportion of men of working age in employment exceeds that of women throughout Europe (Eurostat, 2008b: 53). In 2007, the employment rate for women was 58.3% in the EU27, significantly higher than that recorded in 2001 (54.3%), although considerably lower than the corresponding rate for men (72.5%). The differences between employment rates for female and male employees are smallest in Sweden and Finland (Kovacheva et al., 2007: 17).

Since men's labour market activity has remained largely stable, it is the change in female employment which has brought about most of the changes in the family. As an obvious consequence, household employment patterns have changed. The male breadwinner model is being increasingly replaced by alternative models. However, the dual full-time earner model is not the dominant one in most of the Member States, and there is little evidence that it will replace the male breadwinner model (Lewis et al., 2008). Instead, a great variety of coexisting models can be observed, with variations not only between but also within countries and within the life cycle of individual families. Overall, the presence of small children greatly increases the prevalence of more traditional gender arrangements.

Various forms of the modified male breadwinner model exist, where one of the partners (mostly the woman) works a more limited number of hours. In the less affluent countries, even the female breadwinner pattern

---

[12] Sometimes referred to as the NMS12. These are the 12 new Member States, ten of which joined the EU in 2004 (Cyprus, the Czech Republic, Estonia, Hungary, Latvia, Lithuania, Malta, Poland, Slovakia and Slovenia) - and are sometimes referred to as the NMS10 - and the remaining two in 2007 (Bulgaria and Romania).

is common, supposedly due to severe labour market difficulties. The dual earner - dual carer model is a minority one even in Scandinavian countries, where citizens are highly supportive of this normative pattern (Aboim, 2010: 101). The dual earner - highly unequal pattern is prevalent in many post-socialist countries. In Switzerland, western Germany and in Spain, the more traditional male earner - female carer model is the prevalent model. The so-called one and half earner model is well above average in Switzerland, Belgium (Flanders), western Germany, the UK, and Sweden, mostly countries where part-time female employment is more common.

Key factors affecting the gender division of paid work

Four main sets of factors can be identified that are closely linked to women's employment rates. These key sets of elements can be labelled as structural, economic, cultural, and individual factors (Haas et al., 2006).

Structural factors include institutional arrangements that support or hinder female employment. Usually they include either the general welfare setting in a given country, or more specifically the impact of some public institutions, such as childcare facilities, parental leave systems, financial support for children, etc. (e.g. Gornick et al., 1997; Szeleva/Polakowski, 2008; Van der Lippe/Van Dijk, 2002). Less attention has been paid to labour market institutions such as availability of part-time work, flexible timetables, or distance work. Among structural factors, economic determinants such as national income and unemployment are sometimes also included (Van der Lippe, 2001; Uunk et al., 2005).

The set of cultural factors include individual attitudes at the micro level and social norms and traditions at the macro level. Cross-national comparisons and country typologies that emphasise structural influences in gendered employment are contrasted with 'culturalist' approaches that prioritise "social values, norms and preferences that go hand in hand with a gender-specific division of labour" (Haas, 2005: 490). At the micro level, a direct causal link is assumed to exist between an individual's attitudes and preferences on the one hand, and his or her behaviour on the other. For example, a woman with a more traditional gender role is probably less likely to re-enter the labour market after her child is born (Uunk et al., 2005).

In a macro-level approach, it is expected that women's behaviour will be influenced by the social norms and values shared in her wider social surroundings. Hakim's Preference Theory (2003) states explicitly that attitudinal factors, such as motivations, aspirations and preferences regarding work and family, are more influential in shaping women's employment behaviour than institutional factors. Hakim differentiates between 'work-centred', 'home-centred' and 'adaptive women'. Work-centred women give preference to work, and home-centred women to family. Adaptive women - the majority - adjust their strategies to the actual situation more flexibly. These women can also be expected to react to (changes in) public policy.

Finally, a range of individual characteristics of the actor has to be taken into consideration. Most relevant are the number and ages of children. Labour supply theory suggests that out of the two parents the one with the higher earning potential - usually the male partner - will specialise in paid work. Consequently, the woman usually reduces the number of hours spent on paid work. Looking at it from a different angle, children raise the value of women's time spent away from paid work and lower her effective market wages, since her decision to take up paid work would imply additional costs to be paid for alternative childcare (e.g. Gornick et al., 1997). The resulting division of paid work between genders is then also reinforced by cultural norms that expect women to take care of the children and men to support the family.

Therefore, policy makers' options for influencing the gender division of paid work are limited by cultural factors as well as economic constraints and individual characteristics. In countries with a high level of policy support for female employment, social norms typically favour less traditional gender roles. That is the reason why it is often not possible to tell whether policy changes would also be effective in a less supportive cultural environment. Economic constraints not only limit the resources available for supporting work-family balance but might also restrict employment opportunities, and therefore negatively influence female employment (Scharle, 2007). At the same time, however, economic necessity might also force women to take up paid work and thereby improve the levels of female employment, contrary to their individual preferences (Van der Lippe, 2001; Uunk et al., 2005). Other individual characteristics also operate independently of policy interventions. Women with higher education attainment levels and better earning potential are more likely to take up paid jobs and to have shorter career breaks when they have children than less well educated women (Vlasblom/ Schippers, 2006; Kangas/Rostgaard, 2007).

The division of unpaid work

Women's increased labour force participation decreases the time they have available to carry out domestic work and puts pressure on men to take on greater responsibilities in the household. The scale of these changes has however remained limited (Margherita/O'Dorchai/Bosch, 2009). Although the gap in the number of hours men and women spend on domestic work has narrowed in recent decades, this is more due to women reducing the number of hours they spend on such activities than to significant changes in men's behaviour (Burchell *et al.*, 2007; Bianchi *et al.*, 2000; Fuwa, 2004; Vannoy *et al.*, 1999).

On average, women in the 18 European countries analysed in the Eurostat report (Aliaga, 2006) perform 66% of all domestic work, although alongside the cross-national similarities in the gender distribution of domestic tasks, there are also cross-national dissimilarities. Employed women do less housework than non-employed women, but they still continue to take the bigger share of domestic work even in dual-earner families. Studies also find cross-national proof of the gender segregation of domestic tasks. Men and women do different housework tasks inside the home, with women usually doing the routine chores (cleaning, laundry, washing) that typically cannot be postponed, and men more intermittent ones (car maintenance or repairs, emptying the trash) (Coltrane, 2000; Gaspar/Klinke, 2009; Eurostat, 2004; Fuwa, 2004).

A substantial part of unpaid work is spent on childcare. There is a clear tendency for an increase in domestic working hours if women have children, particularly when the children are small (Aliaga/Winqvist, 2003; Eurostat, 2003). According to the European Working Conditions Survey (EWCS) data, women do the lion's share of childcare tasks in all countries surveyed (Eurostat, 2004 & 2009b). The division between men and women of time spent on caring for children tends to be most equal in the Netherlands, Nordic countries, and Switzerland, where women spend twice as much time on childcare tasks (16 hours per week) compared to men (7-8 hours per week). The largest gender gap in time spent on caring for children was noted in the Anglo-Saxon countries, with a difference of ten hours per week between women's (14 hours) and men's (4 hours) time spent on childcare tasks (Eurostat, 2009b).

The literature provides evidence that fathers are more involved in childcare when mothers are employed, although mothers still provide more of the care. Furthermore, fathers' involvement in childrearing is increasing slightly – but it reaches varying degrees in the various countries as well as in different types of families (Fisher/McCulloch/Gershuny, 1999; Gauthier/Smeedeng/Furstenberg, 2004). Several studies find that better educated men do more domestic work, while better educated women do less (Batalova/Cohen, 2002; Gaspar/Klinke, 2009; Pittman/Blanchard, 1996). The educational level of the husband is not as important as that of the wife, however, in determining the probability of a more equal gender division of domestic work. The reason for this is that more educated wives spend less time doing housework, not because their husbands participate more in the domestic responsibilities (Ramos, 2005; Work Changes Gender, 2007).

Key factors affecting the gender division of domestic and parenting work

There are diverging views on the reasons why women generally do more housework than men. The time availability argument states that the partner with the most available time will participate most in housework and child-care. This argument is based on the assumption that housework allocation is rationally made in accordance with time commitments of each partner (e.g. Becker, 1981). Accordingly, the partner with a more demanding occu-pation and higher number of paid work hours spends less time on house-hold and on childcare tasks. Empirical results provide mixed support for this argument (Burchell *et al.*, 2007; Gauthier *et al.*, 2004).

The resource-power perspective assumes that women's influence on family decision-making is limited by their usually lower resources. Couples try to negotiate the allocation of time within the household to make the best deal based on self-interest (Brines, 1993). The partner with more resources and higher income and level of education will bargain libera-tion from domestic chore responsibilities (Gaspar/Klinke, 2009) and will spend less time in housework and childcare. Empirical studies show that the division of household labour seems to be more equal when the gap between the relative socio-economic status of spouses narrows. The gender gap in incomes seems to be a contributing factor to the imbalance in the division of domestic labour. Studies find that a smaller gap between wives' and husbands' earnings tends to balance the performance of housework (Gaspar/Klinke, 2009; González *et al.*, 2009).

The socialisation and gender role attitude explanations suggest that husbands and wives perform household labour according to adopted values and beliefs about gender norms (Hiller, 1984; Fenstermaker/West, 2002). Couples with egalitarian gender attitudes are expected to have more equal division of labour, while traditional couples would have a more gendered division of domestic work. According to the gender perspective, domestic work is "a symbolic enactment of gender relations" (Bianchi *et al.*, 2000: 194), rather than a trade-off between time spent in unpaid and paid labour or a rational choice due to the maximisation of family utility. The doing gender approach states that the division of household labour in families involves the production and maintenance of gender itself (Berk, 1985; Ferree, 1990; West/Zimmerman, 1987). Many empirical studies from recent decades have found both men's and women's gender role attitudes as a predictor of the division of domestic labour in various countries (Coltrane, 2000; Davis/Green-stein, 2004; Shelton/John, 1996). Men with less traditional gender ideolo-gies do a greater share of the housework. These findings are confirmed in samples from Germany (Lavee/Katz, 2002), Sweden (Nordenmark/Nyman,

2003), Great Britain (Kan, 2008) as well as in several cross-national studies (Batalova/Cohen, 2002; Davis/Greenstein/Marks, 2007; Fuwa, 2004).

According to the integrative perspective, individual behaviours cannot be separated from the surrounding context. Since contextual variables shape individual behaviours, a holistic approach taking the broader socio-economic and policy context into account, might contribute to explaining patterns of domestic work management. State policies, economic development, the level of gender equality and characteristics of the welfare regime can all influence the division of housework (Batalova/Cohen, 2002; Hook, 2006; Fuwa, 2004; Stier/Lewin-Epstein, 2007).

However, the division of unpaid work seems to be even more resistant to policy intervention. Through the level of female employment, policy might have some influence on the gender distribution of housework. Policy intervention that promotes gender equality in the labour market also increases gender equality within the household. Individual characteristics, however, play a decisive role in this process. Only women with strong individual bargaining power and with modern gender attitudes can benefit from egalitarian welfare policies within their families (Fuwa, 2004).

Country characteristics in gender division of paid and unpaid work

Several studies have shown that the classic trichotomy of Social Democratic, Liberal, and Conservative welfare regimes (Esping-Andersen 1990, see also Section 7) is efficient in explaining some of the between-country variations in the level of female employment as well as in the patterns of division of unpaid work. However, "exceptions" are numerous.

Social Democratic countries in the EU (Sweden, Denmark and Finland) are characterised by a high level of female labour market participation together with a moderate child effect on women's participation rates. In these states, the dual full-time earner model remains the most prevalent form of household strategy, even when there are children in the family (Lewis/Campbell/Huerta, 2008). Widespread support for reconciling work and family life includes a high level of childcare provision and a generous parental leave system (e.g. Gupta/Smith/Verner, 2008; Haas/Steiber/Hartel/Wallace, 2006). Gender equality is integrated into family, social and labour market policy. Empirical studies find the availability of childcare particularly important in boosting female employment in the Nordic countries (e.g. Pettit/Hook, 2005; Uunk et al., 2005). The (relative) gender equality in the labour market is also accompanied by relatively low inequality in the division of unpaid work (Fuwa, 2004; Geist, 2005). This is not only because women in these countries spend less time on domestic work, but also because their male partner does a significant share – especially in childrearing.

The only Liberal country in the EU is the UK. Not surprisingly, the child-effect on women's employment is strong. After childbirth a move not only towards the male breadwinner, but also to the modified breadwinner model can be seen, and this latter effect remains pronounced even when children are of school age (Pettit/Hook, 2005). In the UK, the gender gap in unpaid work conforms to the European average (Fuwa, 2004).

The greatest heterogeneity in household management patterns can be found within Conservative regimes. One would expect a low level of female employment with a marked child-effect, and the dominance of the male breadwinner model combined with an unequal division of household labour. Although these tendencies seem to hold when broad categories are discussed (e.g. Fuwa, 2004; Geist, 2005; Van der Lippe/Van Dijk, 2002), cross-country variations are remarkable. Most importantly, Portugal is marked by a high level of (full-time) female employment, which hardly ties in with the presence of children in the family. Across the EU15, Portugal is the only country outside the Northern region where the dual full-time earner model remains in the majority among parents (Lewis *et al.*, 2008). Uunk *et al.* (2005) suggest that this is likely to be due to the economic pressure on women to have a paid job, while others refer to the existence of a rudimentary welfare state, where female employment is considered to be the norm (Van der Lippe/Van Dijk, 2002: 230). However, the division of unpaid work is more in line with the conservative pattern in Portugal (Fuwa/Cohen, 2007; Voicu *et al.*, 2009).

In other Southern European countries such as Italy, Greece and Spain female employment is lower, and only around one third of couple-parents follow the dual full-time model (Lewis *et al.*, 2008). Findings on the division of domestic tasks also show rather traditional patterns (Aliaga, 2006; Fisher/Robinson, 2009; González/Jurado-Guerrero/Naldini, 2009; Voicu *et al.*, 2009). Still, variations between - but also within - countries are notable. Stier, Lewin-Epstein and Braun (2001) describe Italy as a Conservative country with a high level of support for mothers' employment where both a high level of continuous full-time employment and frequent long-term withdrawal from the job market are present.

In Germany and Austria - Conservative countries with an intermediate support for women's employment (Stier *et al.*, 2001) - female employment is considerably higher, and division of household duties is more equal than in the Southern European Conservative countries, but the child effect is similarly significant. This latter can be put down to the lack of childcare institutions (Jönsson/Letablier, 2005), but also to economic affluence. Extensive parental leave coupled with nearly universal childcare availability in France results in a modest child effect (Pettit/Hook, 2005) typical of a Conservative regime, but not surprising given the high level of support for maternal work. Still, France seems to remain traditional in the division of household labour (Crompton, 2006a).

The Netherlands is remarkably different from other countries in this cluster. Full-time employment of mothers receives little institutional support here but part-time employment (of women and also men) is exceptionally frequent. These situations have clear consequences on female employment in general but also on mothers' employment in particular. After childbirth, not only a move towards part-time employment but also to the traditional male breadwinner model is rather frequent – together resulting in a high level of child effect on female employment (Haas *et al.*, 2006; Lewis *et al.*, 2008; Uunk *et al.*, 2005).

Central and Eastern European countries are not included in the classic typologies of welfare regimes. A widely used strategy in the empirical litera-ture is to include 'Post-Socialist' countries as a separate cluster (Fuwa, 2004; Van der Lippe, 2001). A long tradition of full-time work for men and women, underpinned by the 'official' socialist ideology as well as families' need to rely on two incomes might justify clustering these countries together. There are substantial between-country differences as far as gendered behavioural patterns are concerned, but underlying structural and cultural factors also play a role (Szeleva/Polakowski, 2008). Szeleva and Polakowski (2008) iden-tified four distinct types of childcare policies (implicit familialist, explicit familialist, comprehensive support, female mobilizing) across eight coun-tries in the region. Studies in the division of unpaid work also point towards some dissimilarity within the region. Nevertheless, most studies describe the region with a medium to relatively low level of gender inequality in the field of household work (Fuwa, 2004; Fuwa/Cohen, 2007; Voicu *et al.*, 2009).

Despite a lack of information on some Member States and the limitations on exploring precise trends, we can conclude that there is notable diversity across Europe. No coherent typology of countries that would adequately reflect this diversity has emerged. Furthermore, families do not make their choices for a lifetime but instead, they adapt their actual behaviour to their situation in the various phases of their life-course.

1.3.3 Intergenerational relations in families

Existing comparative research has mostly ignored multigenerational rela-tions in families and concentrated on couple relations and the relationship between parents and (young) children. Concerning generational relations, existing research has focused mainly on intergenerational support patterns.

Most of these studies concentrate on intergenerational transfers of time and money and the existing differences between welfare regimes (e.g. Albertini/Kohli/Vogel, 2007). In social care research, there has been recent interest in studying how care and help are distributed within families between generations. Mainly informal help and care provided by adult children towards their elderly parents has been studied, but also to some extent the role of grandparents in providing care for their grandchildren, and help and financial assistance they provide for their adult children.

Research on multi-generation households in Europe is not extensive, and knowledge on multi-generation households as families is rather sparse. One major finding so far is that European countries differ in the degree to which the "nuclearisation" of the family has occurred, i.e. the degree to which the older people live by themselves either as a couple or alone. The European Quality of Life Survey found that in Italy 25% of all people over 65 still lived in a household with a child, while more than 30% did so in Malta and Poland. In Hungary, Spain, Slovenia, Cyprus and Greece, 20% of the older people still live in these family arrangements. By contrast, in Denmark, Sweden, France and Germany those households represent less than 5% (Saraceno et al., 2005: 17).

Intergenerational family obligations and care relations

Researchers today are more and more interested in the division of care responsibilities and provision between family generations. According to Hagestad and Herlofson (2007) co-longevity has greatly increased the duration of family ties. The parent-child relationship may last 6-7 decades and the grandparent-grandchild bond, 3-4 decades (ibid.: 341). Researchers have recognised that care relations run both ways, i.e. adult children provide care for their parents, grandparents provide care for their grandchildren, and sometimes elderly parents even take care of their adult children. These relations can be mutual and mixed, and related to the provision of formal care services.

According to Saraceno and Keck (2008), a number of studies have found that intergenerational solidarity is alive and strongly reciprocal in all countries, at both the two and the three-generational level, with the middle generation in the "Janus position" (Hagestad/Herlofson, 2007) of redistributing both upwards and downwards. Both long-standing family cultures and welfare state arrangements affect the shape of this solidarity, as well as the overall social care package as a mix of family, volunteer, and public provision. However, Hagestad and Herlofson (2007: 345) note that cases of coinciding responsibilities for older parents and children at the same time are relatively rare. They refer to Dykstra's (1997) overview of 12 European Union countries, showing that only 4% of men and 10% of women had overlapping responsibilities for young children and elderly parents who required care.

SOCCARE project (Kröger, 2004; Kröger/Sipilä, 2005) studied social care arrangements in five different countries representing the variety of European welfare states (Finland, France, Italy, Portugal, and the UK). In relation to multi-generational, "double front carer" families the results showed that the emphasis is on care for older family members. The care of children is generally described as less problematic and more natural (Kröger, 2004: 72-86). Families in Finland, France, and the UK most often used combinations of informal care and publicly provided formal care. Portuguese and Italian families mostly used third sector and private care facilities. The informal non-professional paid sector was found to be wide and varied in Italy, France, and Portugal, offering a range of types of assistance. Concerning the general organisation of the care arrangement, the family and in particular the main care-giver remains the most important resource everywhere. The results affirm the common belief that European social care cultures are diverse, but not completely different.

The role of grandparents

In our societies, where life expectancy is increasing and general health has improved significantly, the figure of the grandparent is becoming more important. Grandmothers and grandfathers are a resource for their children and their children's families (Walther et al., 2009; Kröger, 2004).

Hank and Buber (2009) have investigated cross-national variations in grandparent-provided childcare in ten continental European countries. Across all countries, 58% of grandmothers and 49% of grandfathers provided some kind of care for a grandchild aged 15 or less during the most recent 12-month period. The lowest shares were found in Spain, Italy, and Switzerland, whereas the highest prevalence was in Sweden, France, the Netherlands, and Denmark. However, the order of countries changed remarkably when the researchers made a distinction between regular and occasional care. Sweden, Denmark, and France had below-average levels of regular childcare by grandparents, whereas the respective share in Greece, Italy, and Spain was almost twice as high as in the Scandinavian countries. Austria, Germany, the Netherlands and Switzerland had an average position. Among regular carers, the gender division of carers also changed, with grandmothers having more intensive involvement (ibid.: 60-69).

However, older people are also active subjects in their own lives, deciding autonomously how to spend their time and money. In this new role, which involves social and cultural re-engagement, grandparents are also capable of undertaking new projects (Leccardi, 2009a). Older people's plans often revolve around travel and the possibility of discovering new places and cultures (Pronovost, 1992). That might involve cultural and social interests that were not possible in previous years, for example voluntary or charity work (Verbrugge *et al.*, 1996; Bickel/Lalive d'Epinay, 2001). As pointed out by Facchini/Rampazi (2009), in more extreme cases some of these "young oldies" actually decide to construct a new life for themselves.

1.3.4 Gender, generations and family violence

Violence in families is first of all a gender issue but also a generational issue. Violence within the family has become an important public concern for contemporary societies. Several international bodies (e.g. the European Commission, the United Nations, the World Health Organization) have taken up the fight against violence against children, women and the elderly as one of the priorities of the international political agenda, leading several countries to implement legislation protecting victims of domestic violence and their fundamental rights. Although systematic efforts have been made to identify and analyse violence in the family, it has been difficult to reach a consensus regarding its definition and theoretical boundaries. The conceptual diversity of this field is a significant obstacle to comparisons. Another empirical problem is the under-reporting of violence. Estimates are often based on official reports, which tend to present lower numbers than real values (Knickerbocker/Heyman, 2007).

The majority of existing studies have focused on violence against women but also on violence against children and youth. Much less research has been carried out on violence against men, elderly people, homosexuals and bisexuals, people with disabilities, and immigrant and minority women. Most of the research has been done at the national level. Studies with a comparative European perspective seem to be rather rare, because it is difficult to compare the rates of family violence between different European countries, as existing studies have many important methodological and terminological differences.

The "gender violence" paradigm has dominated research on violence against women. It reduces all violence to two foundations: male abuse is used to maintain power over women, and female violence is defensive and used only for women's own protection. More recently, violence studies against women have mainly been based on two analytical models and measures. The first consists in analysing violence solely perpetrated by males against females. This is the so-called unidirectional model, and has been adopted by important national and international organisations (e.g. the WHO, in countries such as Spain, France and Germany). It is a model which assesses only violence against women. The second (bidirectional) model assesses violence perpetrated by both male against female and female against male. Particularly from the mid-1990s, studies have begun to include other contexts where violence is also perpetrated against women (e.g. public space, workplace), and other types of perpetrators beyond intimate partners (e.g. other relatives, acquaintances) (Martinez/Schröttle, 2006). However, most studies collect information on violence perpetrated by an intimate partner against the woman.

Female violence against men remains a neglected area of study in the field of social sciences. A number of important questions regarding female violence remain unaddressed. Prevalence studies of violence against men in Europe are very scarce. Those that do exist focus on two basic lines of research: sexual violence perpetrated by women against men (Krahé/Scheinberger/Bieneck, 2003); and the way violence against men is socially represented, perceived and researched (Research Group *et al.*, 2004). More recent studies within the bidirectional model tend to develop longitudinal approaches (Archer, 2000 & 2002). Violence also occurs in homosexual couples; some studies indicate that it is as frequent as heterosexual violence (Krahé *et al.*, 2000).

Since the 1990s there have been a considerable number of prevalence studies on violence against children and young people. They focus mostly on sexual abuse, sexual harassment, parental violence and bullying in school. Despite the methodological differences between these studies, they all reveal relatively high prevalence rates of child abuse. However, prevalence is higher among girls than among boys, except for bullying. Velleman *et al.* (2008) have studied domestic violence experienced by young people living in families with alcohol problems. The study was part of the European DAPHNE project. It involved ten EU countries (Germany, Austria, England, Finland, Hungary, Ireland, Malta, Netherlands, Poland and Spain). Children affected by parental alcohol problems report having often lived under considerable stress for long periods, having to deal with family and parental environments where there was serious alcohol misuse, and serious domestic abuse, frequently moving into family violence. Findings show a complex interaction between gender, alcohol problems and child/spousal abuse.

Bussmann, Erthal and Schroth (2010) have completed a cross-national research project on the effect of banning corporal punishment in Europe. Several international studies have revealed that the prohibition of corporal punishment has contributed to the reduction of parental violence, greatly influencing the attitudes and behaviours of parents. Parents who have themselves been more exposed to parental violence used more violence on their own children. This exposure is higher in countries where parental violence is not prohibited, such as France. This shows that existing conjugal violence is a risk factor for the adoption of violent educational styles. In countries where corporal punishment has been accompanied by formative and informational campaigns and long-term measures, the violence levels tend to be lower.

1.3.5 Conclusions

This chapter is based on a life-course perspective, which enables linking of individual biographies with social and historical change. The approach adopted here also involves a particular concern for the importance of gender differences. Within this general framework, some major trends can be identified:

- Transition processes of young people have changed with the prolonged presence of young people in their family of origin and couple formation taking place later.
- New representations of partnership and parenthood emerge among young people, with changes in gender roles and in male and female identities – but the change is slow and many traditions are still strong.
- The family of today is a negotiation and affection-based family.
- The role of grandparents is important as providers of support to children and grandchildren.
- The most marked change in the field of division of paid work is the increasing level of female employment.
- The division of paid work and especially of unpaid domestic work continues to be highly gendered.
- Fathers' involvement in childrearing is increasing slightly.
- The male breadwinner model is being increasingly replaced by alternative models, with the "dual earner-female carer" model becoming the most widespread in Europe. The "one full-time and one part-time earner-female carer" model has also gained in importance.
- Family violence is still largely gender-based violence, but also a generational issue.

The growing differentiation and pluralisation of social structures mean that societies move from a single model of the family to a plurality of models. Consequently, the modalities of passing through the various developmental processes that constitute the course of family life have been transformed. The final picture that emerges of these changes is far more intricate, colourful, and multi-faceted than the one that prevailed twenty or thirty years ago – and may have more contradictions and ambivalences.

Age regulates entry into and exit from various life worlds – for example, when it is appropriate to marry and to have (or not to have) children (Elder, 1975; Elder/O'Rand, 1995). Nonetheless, today it is necessary to look at these issues with a critical eye. In contemporary society biological age and social age tend to be separated: the former is no longer an obligatory reference point for the definition of the latter. It is possible, for example, to be a child yet already have the status of an autonomous consumer or, alternatively, to be categorised in the so-called 'third age' - biological age that would qualify one as elderly - yet play a socially important role.

On a formal level, becoming an adult implies a series of changes in status and assumptions of roles that lead to a progressive independence with growing social responsibilities. However, today this construction appears to be somewhat artificial. This means that the sequences that mark the passages from one social age to another are weaker than some decades ago. Thus, for example, today the adult is no longer defined through a substantial existential stability in the family and work, or in personal rela- tions, but rather to a considerable degree through the capacity to dominate the continuous flow of changes (Saraceno, 1983).

Socio-cultural processes that contribute to weakening the life-course approach are first, the process of individualisation, and secondly, the process of the transformation of cultural norms, in particular in the direction of their increasing subjectivisation (Bozon, 2004). Individualisation is a process where individuals take upon themselves the onus of making choices and existential decisions. The dynamics of individualisation tend to "liberate" men and women from the traditional age categories and ties of gender and from familiar role models (Beck/Beck-Gernsheim, 1995; Giddens, 1990). The "subjectivisation of norms" has a deep impact on family life, for example, on sexual morality, and thus the norms that regulate cohabitation and sexual relations can be manip- ulated by individuals. There is also a widening distinction between principles and practical situations, between principles and everyday life. Norms tend to be transformed from "social" to "private" norms. The moral codes relating to the life of the couple and the family are not dissolved but rather progressively pluralised and individualised (Leccardi, 2009b).

However, there also are still strong regularities, generalities, and tradi- tions and some basic patterns of family life remain intact in Europe. You usually have to have a partner, education, a job, a house of your own and a shared future with your partner before you are ready to have your first child and marry. Even if new representations of partnership and parenthood have emerged among young people, changes seem to be rather slow, and gender divisions become more traditional after the birth of the first child. Most importantly, the division of paid and unpaid work continues to be gendered. As shown in this review, women spend less time in the labour market, they are more likely to take part-time jobs and have more career breaks than men do. At the same time they have primary responsibility for housework as well as for child-rearing. The dual carer-dual earner society remains a theoretical concept in most of the Europe.

1.4 Social inequalities and living environments

A review of existing research shows that social inequality plays a crucial role in family life, and is related to family structure and dynamics in complex ways. Families reflect social inequalities, since the unequal distribution of various resources and differentiated opportunities affect the circumstances in which family life is built up. Families also reproduce inequalities, both in the short term and intergenerationally. Research shows that family background, life-style, and resources, including both material and socio-cultural advantages, tend to affect children's lives and life chances. Transmission of wealth from older to younger generations and support in setting up family life during the transition to adulthood is significant in all European countries. In terms of intergenerational effects, families remain perhaps the most important mechanism for the transmission of unequal life chances.

1.4.1 Social inequalities, diversity and wellbeing of families

Income inequality across and within European societies

Measurement of inequalities across and within European societies relies systematically on comparative statistical data regarding levels of income. There are significant differences in levels of income across Europe. The data on mean and equivalised disposable incomes from the year 2007 (in EUR) shows that Luxembourg (over 34,000 EUR), Ireland, Denmark and United Kingdom have the highest levels of income within the EU27, but they are closely followed by a large group of other countries. Eastern and Southern European countries have lower levels of income. The lowest levels in the EU27 are in Romania and Bulgaria (under 2,000 EUR).

Income inequality within each country is also significant. Drawing on one main indicator of income inequality - the ratio of total income received by the 20% of the population with the highest income (top quintile) to that received by the 20% of the population with the lowest income (lowest quintile) - the countries with higher GDP per capita are not necessarily more equal. Together with the Nordic countries, we find lower levels of income inequality in countries such as Slovenia, Slovakia, the Czech Republic,

Hungary, and Austria. Income inequality has increased in many countries over the last decade: in Latvia, Romania, Bulgaria, Greece, Germany, France, Luxembourg, Belgium, Finland, Denmark, and Sweden. Although they are amongst the most unequal societies, Italy, Portugal, Estonia, Lithuania and Poland have reduced their income inequalities.

It has been shown that living standards, expressed as GDP per head, are generally lower in rural than in urban areas (European Commission, 2008a: 55). The available data do not permit a systematic and complete analysis of rural-urban income poverty patterns in Europe. Nevertheless, some country-specific surveys show that the gap in poverty rates between rural and urban areas is bigger in Eastern European countries than in the Western countries. In Western countries, poverty is concentrated in remote regions and, in general, regions with accessibility problems (European Commission, 2008a: 75).

Comparative datasets using the concept of class and taking up socio-professional and educational indicators to compare social categories across countries are more difficult to find. Many authors have nevertheless argued that "class", despite being a multi-faceted concept with a variety of different meanings, makes a significant contribution to understanding structured social inequality in contemporary societies (Bottero, 2004 & 2005; Devine *et al.*, 2005; Savage *et al.*, 2005; Crompton, 1998 & 2006b). The position taken by contemporary research on class is that although there has been considerable social change in European societies, and individuals may have more choices to make than in the recent historical past, class and stratification analysis is still important and useful for understanding and explaining the complex realities of inequality.

Traditionally, research and policy makers have preferred economic and social indicators of wellbeing, deemed more appropriate for measuring the development of societies. Measures of life satisfaction, happiness and generally subjective indicators of wellbeing have not been widely used in the analysis of human welfare. So far, research on subjective wellbeing has been mostly concentrated in highly developed countries. Comparisons have been based on levels of satisfaction rather than their distribution across the population, and have focused on countries with small variance in subjective wellbeing, thus diminishing possible effects of socio-economic factors. More recently, a strong case has been made for the use of subjective indicators combined with economic variables. Synthetic indicators have been proposed (Somarriba/Pena, 2009) combining various objective aspects such as income, living conditions and employment with subjective indicators like perception of quality of life.

The relationship between income inequality and subjective and objective welfare indicators is defended by Wilkinson and Pickett (2009). According to them, unequal societies tend to perform worse on objective indicators, such as life expectancy, health, crime rates, and on subjective indicators, such as trust in fellow citizens or life satisfaction. They also point out that, in the more developed countries, measures of wellbeing are no longer associated with economic performance, suggesting that the quest for increasing material wealth needs to be replaced with increased social cohesion, improved social environment and quality of life.

It is possible to say that life satisfaction seems to be related to overall societal economic performance, as life satisfaction is higher in more affluent societies (1999-2000 EVS Data Files and UNDP Data). Denmark, Malta, Ireland, Iceland and Austria are the top five countries in terms of life satisfaction. Most countries of central and northern Europe have above-average life satisfaction scores, as do more southern countries such as Slovenia, Italy, Spain, and Portugal. Croatia and Greece scores are average when compared with the overall results. The bulk of countries with low life satisfaction scores are former socialist societies. Overall, there is a consistent relationship between life satisfaction and GDP, i.e. more affluent societies tend to have more highly satisfied citizens (also Fahey/Smyth, 2004). Therefore, despite the fact that some of the richer societies in the EU are rather unequal, the level of comfort and material wellbeing achieved seems to mitigate the effect of social inequality. In poorer societies, such as those in the Eastern or Southern Europe, there seems to be a stronger link between social inequality and life satisfaction.

Economic situation of families

Since the focus is on the economic situation of families, it is meaningful to compare the incomes of different family types. Average annual net earnings can be presented as annual net income per family member, which allows comparison of different types of families and also the effect of family type on the income level, i.e. how supportive the monetary family policy measures of the state actually are.

Average annual net earnings for a single person without children across the EU were 20,208 EUR in 2007. Average earnings were a bit higher in the old Member States and considerably lower in the new ones. The lowest annual net earnings were recorded for Bulgaria at 2,048 EUR on average. This is incredibly low even compared to the other new Member States. The average annual net earnings for a two-earner married couple with two children per family member are almost everywhere about half the income of a single person. The average net income per family member in a one-earner family with four members is highest in two of the new Member States, the Czech Republic and Slovenia. The measure is also similar in Luxembourg, one of the wealthiest Member States, and average income is quite low in Cyprus, Lithuania and the United Kingdom. Poverty of families will be discussed in the next section.

Employment has a crucial effect on the financial situation of families. Families with one earner and with children are in a relatively bad financial situation when compared with two-earner families. This explains the greater need for two incomes and presents a new problem for the families: how to combine work and family life. Steady lifelong jobs are disappearing, forcing families to deal with unexpected periods of either too much or too little work, which is accompanied by income insecurity. Welfare regimes have not yet found ways to cope with these changes (Knijn/Smit, 2009: 8). Inter-relatedness between labour market developments and changing family lives has two income-related dimensions. Family formation might be frus-trated by difficulties in accessing steady jobs, since childbearing as well as marriage are rather sensitive to financial instability. Furthermore, time to care for children is only partly compensated for by paid leaves, and only marginally included in pensions. A career break for care purposes decreases one's lifelong income substantially (Knijn/Smit, 2009: 10-11).

Employment for men and women varies according to family type. Single persons without children are predominantly employed full-time, but in varying degrees (Eurostat, 2009c: 28-29). Single parents tend to work. Working full-time is fairly widespread in Bulgaria, Estonia, Greece, Latvia, Lithuania, Hungary, Portugal, Romania, Slovenia, Slovakia and Finland (with shares above 70%). In Germany, Luxembourg, the Netherlands, Austria and the United Kingdom more than 30% of single parents work part-time. More than 30% of single parents were unemployed in Belgium, Germany, the Netherlands, Poland and the United Kingdom (Eurostat, 2009c: 28-29). In most couples without children both partners work full-time. With the exception of the Netherlands (39%), this share stood above 50% in all Member States. The highest shares (above 70%) were observed in the Czech Republic, Hungary, Portugal, Slovakia and the United Kingdom. The second most relevant pattern observed among couples without children was 'one person working full-time and the other person not working'. The third type of employment pattern, with one partner working full-time and the other working part-time, was fairly common in Belgium, Germany and Austria. However, this type of working pattern was most widespread in the Neth-erlands. The fourth and last working pattern, in which both partners are unemployed, accounts for only a minor share in the overall distribution of couples without children (Eurostat, 2009c: 30).

For couples with children, the 'both working full-time' employment pattern is also the most frequent. However, this share stood above 50% in only 14 Member States. In many Eastern European Member States, the dual full-time earner model was the norm in communist times, although it became less common in recent years. In some countries, other employment patterns are prevalent. In Belgium, Germany, the Netherlands, Austria and the United Kingdom the 'one working full-time and other working part-time' pattern was the most widespread. In Spain, Italy and Luxembourg the situation where one person is employed and the other person is not employed is the group with highest shares (Eurostat, 2009c: 30). Again the situation where both persons are not working is not widespread in the Member States. With the exception of Slovenia, in all Member States the presence of a child in the household leads to a decrease in the working pattern where both persons are working full-time (Eurostat, 2009c: 30).

There is evidence that poverty is much lower in countries with an earner-carer strategy, which emphasises policy approaches meant to balance care and employment for both men and women. At the same time, poverty rates are significantly higher in countries that employ the earner strategy, which takes a market-driven approach to care issues. Poverty rates are significantly higher for single mothers and particularly single mothers of young children not only in countries that employ the earner strategy but also in those that employ the carer strategy. Policies that support care outside the home reduce poverty more for single mothers than for partnered mothers (Misra/Moller/Budig, 2007). The findings suggest that beyond the positive impact of cash benefits paid to families with children, work-family policies such as childcare and short-term leaves have powerful effects on poverty. Yet work-family policies that encourage women to take long leaves for caretaking have effects that are more ambivalent.

Employment stability plays an important role in young people's decisions to leave home and to start a family. Over the period from the late seventies to the nineties youth unemployment increased considerably. However, the importance of employment status varies across Europe. In Poland, Slovenia and Italy regular employment seems to be more important for couples starting to cohabit (in three out of four cases one of the partners was in regular employment) than in Germany (where in over 40% of cases none of the partners was in regular employment). Employment stability for at least one of the parents seems to be a necessary condition for the decision to have the first child. Besides a favourable economic situation, the need for more flexible working arrangements and other measures aimed at reconciling work and family are also indicated as relevant for family choices (Fondazione Giacomo Brodolini, 2007: 238ff.).

1.4.2 Families and poverty

In 2007, around 17% of the households in the EU27 had an equivalised disposable income that was less than 60% of the respective national median income. People living in these households were considered to be at a risk of poverty.

Different groups in society are more or less vulnerable to poverty. The unemployed are a particularly vulnerable group: 43% of unemployed persons were 'at risk of poverty', with higher rates in the Baltic Member States. Those in employment were far less likely to be at risk of poverty (8%). Women are generally at greater risk of living in a poor household: 18% of women of all ages had an income below the threshold, against 16% of men. One in every five (20%) young adults between 16 and 24 was at risk of poverty. The level of education attained also appears to play an important role: those leaving education with no more than a lower secondary education were more than three times as likely to be at risk of poverty than persons with a tertiary education.

Children and older people tend to face a higher risk than the rest of the population, even after social transfers. In 2007 one in every five children (20%) across the EU27 was at risk of poverty, with a slightly higher proportion (22%) recorded amongst older people. Old women were more at risk of poverty than old men (22% compared with 17% in 2007). This gender inequality was widest in the Baltic Member States, Slovenia and Bulgaria, but relatively narrow in Luxembourg, France and the Netherlands.

The age profile of poverty: child poverty and youth poverty

Förster and d'Ercole (2005) argue that relative poverty is, in most countries, more common among children than among the entire population, and this increased further in the second half of the 1990s. Child Poverty and Child Wellbeing in the European Union (2010) conducted a detailed analysis of child poverty and found that in 2005 19 million children lived under the poverty threshold in the EU27. In most EU countries, children are at greater risk of poverty than the rest of the population, except in the Nordic countries where 9-10% of children live below the poverty threshold (Bradshaw, 2010).

In the EU, half of all poor children live in the two types of households: 23% live in lone-parent households (against 13% for all children together) and 27% in large families (against 21% for all children together). However, the extent to which lone-parent households and large families experience greater risks of poverty depends both on their characteristics (age, education level of parents, etc.), and on the labour market situation of the parents (joblessness, in-work poverty, etc.). Children whose parents are below the age of 30 have a significantly higher risk of poverty than those living with older parents. The educational level of parents is another key determinant. It affects both the current labour market and income situation of the parents, and children's own chances of doing well at school. Children living in a migrant household face a much higher risk of poverty than children whose parents were born in the host country.

Very little research has focused on poverty among young adults. The risk of poverty for young adults was highest in Denmark (28%) where, as in other Nordic Member States, it was about twice the rate for the whole population. This may be unexpected, because one would expect youth poverty to be much lower in these Nordic welfare countries. One important answer lies in the fact that compared to other countries, young Scandinavians tend to leave home at a much earlier age. Therefore, the poverty experience of young Scandinavians is generally short-lived. In many other countries young adults continue to live in their parents' homes and are less likely to be recorded as being "at risk of poverty", since they share in their parent's income. This does not necessarily reflect their true situation, which is often characterised by a lack of access to a decent income of their own.

Both marriage and cohabitation appear to protect young individuals from poverty and deprivation, though marriage generally has a stronger effect than cohabitation. The effects of having children are less marked than the effects of marriage and cohabitation, and in the opposite direction: having children is associated with a general higher risk of poverty. The exceptions are Finland and Denmark, where children do not have any influence on the likelihood of poverty (Aassve *et al.*, 2008).

Household types and the shaping of poverty

The types of households at greater risk of poverty than others are single person households, single parent households with dependent children and households comprising two adults with three or more dependent children. Single parent households and large households have been identified as more vulnerable to poverty over the last few decades, whereas single person households - in particular of young adults - have emerged more recently as more vulnerable.

Single adult person households have been identified as more prone to poverty than other types of family (e.g. Walker/Collins, 2004). Quintano and D'Agostino (2006) carried out analysis in four European countries (Italy, France, Germany and the UK) that have welfare systems representing Mediterranean, Continental and Anglo-Saxon regimes. The effect of gender was strong in all countries, indicating that women are at a greater disadvantage than men in each country. The effect of age showed that women over seventy had a much higher poverty rate everywhere. The effect of marital status was interesting: women who had never married had low median incomes and very high poverty rates in all countries. The same effect for divorce was observed except in France, where the poverty rate did not change. On the contrary, in Germany and Italy, divorce had a strong effect on the poverty rate. The impact of divorce was worse for women than for men, and was more evident in Germany and the UK than France or Italy (see also Callens/Croux, 2009).

There is a wide consensus among researchers that lone-parent families, which in most cases are headed by a woman, are the type of household most vulnerable to poverty (e.g. Fouarge/Layte, 2005; Walker/Collins, 2004; Kröger, 2004). Lelkes and Zólyomi (2008) analysed 2006 EU-SILC data and concluded that the poverty rate of single parents reached or surpassed 30% in the majority of the countries examined. In an international comparison, the situation of single parents was relatively favourable in Denmark, Finland, and Norway, where the poverty rate of this group is not higher than 20%. However, this figure was still higher than national average poverty rates in these countries.

Large families are also among the groups more prone to poverty (Bradshaw *et al.*, 2006). Cantillon and Van den Bosch (2002) found that the poverty rate among families with three or more children was as high as that among lone-parent families in Belgium, Spain, Finland, Italy and the UK. The poverty risk of large families generally exceeded that of childless families, except in the Nordic Countries and the Netherlands. Layte and Fouarge (2004) and Whelan *et al.* (2004) examined the impact of various socio-economic factors on cross-national differences in deprivation using the European Community Household Panel (ECHP) survey. Logistic regression showed that having a larger number of children (3+) tended to lead to higher levels of deprivation across all countries, but the effect is rather small when compared to other variables, such as long-term unemployment, being a young single person aged 17-24, or lone parenthood. The negative effect of having a large family was strongest in Italy, Portugal and the UK.

Poverty dynamics: the ins and outs of poverty

Research on poverty dynamics is still rare. Jenkins and Schluter (2001) studied child poverty dynamics in the UK and Germany. Their results point to the importance of the welfare state-related differences as the principal source of differences in child poverty rates. In particular, relative to British children, German children are better protected against the consequences of adverse labour market events, and positive labour market events are more strongly reinforced. When experiencing a trigger event, Germany provides a greater cushion against adverse events and better reinforcement of positive events.

Fouarge and Layte (2005) develop a detailed analysis of the effects of family-related factors on the probability of exiting poverty. Having a female head of household slows down exit from poverty significantly, as does having a head in the oldest age group (55-64). Less favourable employment conditions for these groups or depreciated stock of human capital are possible explanations for this finding. Interestingly, although being a single parent does not seem to impact on exit, not being married does seem to be significant and negative. Although the number of adults in the household does not have a significant influence, each additional child slows exit. The effect for the number of children is not unexpected, as much of the work in this field shows that in many countries larger numbers of children are associated with a greater poverty risk. It is also clear that singles, and especially single parents, are more likely to be persistently poor and have a lower probability of exiting poverty.

A comparison between two European countries, Germany and Great Britain, vs. Canada and USA was done by Valletta (2006). According to this study, most poverty transitions, and the prevalence of chronic poverty, are associated with employment instability and family dissolution in all four countries. However, government tax-and-transfer policies are more effective in reducing the persistence of poverty in Europe than in North America. Changes in family structure are frequently associated with poverty transitions. In each country, divorce and marriage are the most common family events associated with poverty entry and exit, although poverty entries also are commonly associated with the formation of new families that split off from existing households. Among the events that are related to poverty entries, in all countries divorce is the most significant. Valletta's findings confirm widely held beliefs about the key contributions of family stability and work attachment for staying out of poverty. This suggests important roles for individual behaviour as well as public policies that strengthen family stability and work attachment. Childcare subsidies may be one example of such policies, enabling cash-strapped and time-strapped parents effectively to balance work and home commitments.

Research focusing on poverty dynamics often links this analysis to a discussion of the role of welfare regimes (e.g. Sainsbury/Morissens, 2002; Fouarge/Layte, 2005; Callens/Croux, 2009; Förster/Tóth, 2001; Cerami, 2003). Some of the major outcomes of these studies point to the fact that welfare regimes have an impact on the likelihood of poverty entry but not on the likelihood of poverty exit. Country welfare regimes, on the other hand, strongly influence long-run poverty, with Social Democratic countries reducing the level of persistent and recurrent poverty. Liberal and Southern European regime countries have both higher rates and longer durations of poverty. Despite their dissimilar patterns of poverty duration, European welfare states display rather similar patterns of exit from poverty, once we control for duration.

1.4.3 Physical living environment and housing

Living environment and housing are crucial aspects of wellbeing, health, and family life. Most important characteristics that make people satisfied with their home and environment are nice general appearance of the neighbour-hood and satisfaction with housing. The existence of crime in a neighbour-hood as well as insecurity and the fear of crime are very strong predictors of neighbourhood dissatisfaction. General environmental indicators like clean air and availability of clean water are important components of a good living environment (Parkes *et al.*, 2002: 2427; Pa Ke Shon, 2007: 2236).

The impact of environment and health is not equally distributed throughout Europe or within cities. Inequalities in quality of living environment reflect inequalities in economic, social and living conditions. Disadvantaged groups typically inhabit the worst parts of the city, beside noisy and dirty roads and industrial pollution, and are more greatly affected by the lack of green areas and public transport services. Climate change is a new and complex challenge for cities (European Environment Agency, 2009: 14ff.).

Quality of air and water has substantial effects on health and wellbeing. Measurements of air quality show that almost 90% of the inhabitants of European cities are exposed to concentrations that exceed the WHO air quality guideline level (European Environment Agency, 2009: 14f.). The highest concentration of particulate matter was found in Bulgaria and Romania. Exposure to air pollution by ozone was highest for the urban populations of Italy and Greece. Measures of air pollution were lowest in Finland, Sweden and the UK (Eurostat, 2009a: 422ff.).

The issues related to green open spaces are especially relevant because a large proportion of Europe's population lives in urban areas, where the contact with nature is often lacking. Therefore green spaces such as parks are an essential constituent of urban quality of life. Baycant-Levant *et al.* (2009: 209f.) found, that when indicators related to the availability of urban green spaces are used to determine the green performance and ranking of European cities, the Southern European (France, Italy, Spain) cities lead. However, when planning performance indicators are taken into consideration, Northern European (Belgium, Finland and Germany) cities have higher scores.

Accessibility, access to outdoor recreation, distribution and the overall design of the urban area are important to individual satisfaction and encourage daily physical activity such as walking and cycling. Access to green areas is found to be linked with several health issues like obesity, cardiovascular disease and stress levels (European Environment Agency, 2009: 17ff.; Nielsen/Hansen, 2007: 894ff.). Green areas have been found to be beneficial for children in various ways: children with good access to green open space, fewer high-rise buildings and more outdoor sports facilities are more physically active. Schoolchildren who have access to, or even sight of, the natural environment show higher levels of attention than those without these benefits (European Environment Agency, 2009: 15ff.). Neighbourhood open space, such as local parks, also plays an important role for older people in maintaining and enhancing their quality of life (Sugiyama *et al.*, 2009: 3ff.).

It also is important that people feel secure in using these areas. On average, a quarter of national populations felt unsafe walking alone in their area after dark, the figure being higher in Bulgaria, Poland, Greece, Luxembourg and Italy (over 35%) and lower in Scandinavian countries, the Netherlands and Austria (under 20%) (Van Dijk *et al.*, 2007: 127ff.). However, fear of crime and actually falling victim to it are not strongly linked to each other. Countries with a higher share of people reporting fear of crime do not experience higher rates of actual crime, while within countries, older and richer people feel more unsafe than younger and poorer people do, despite being less likely to be a victim of crime (Stiglitz *et al.*, 2009: 53).

Forty per cent of Europeans are reported as living in dwellings that are badly situated, in areas with high levels of noise, pollution or crime (Giorgi, 2003: 31) or have poor access to transport, opportunities and services (Cameron, 2009: 8ff.) and there are tendencies towards spatial segregation of different income groups (Czasny, 2004: 9). The distribution of low-income households, older people, the unemployed and lone parents, is not even or random but involves significant concentrations in particular parts of cities and regions (Musterd/Murie, 2002: 40). Groups that are especially vulnerable to spatial segregation are migrants and ethnic minorities, though the degree of spatial segregation of immigrants across the EU varies. There are multiple causes for this variation, including immigrants' income levels, discrimination in the housing market, public housing policies and degree of ethnic closure (Spencer/Cooper, 2007: 36).

The average number of people living in a household in the EU27 was 2.5 in 2005. It tends to be lower in Northern part of Europe and higher among the Mediterranean countries and those countries which have joined the EU since 2004 (Eurostat, 2009a: 252f.). In the actual quality of dwellings there is also a clear-cut break between the former Eastern Bloc countries and the countries of the EU15 (Ministry of Infrastructure of the Italian Republic/ Federcasa Italian Housing Federation, 2006: 9f.).

Some social groups, households or individuals lack access to suitable housing, because of homelessness or because accommodation has been characterised as being in "bad condition" (for example having insufficient heating); 24% of Europeans report living in accommodation which is in bad condition (Cameron, 2009: 9; Giorgi, 2003: 30f.). Problems accessing suitable housing are most relevant to poorer people, whose housing is of far lower quality than other households (Czasny, 2004: 8; European Commission, 2007b: 104f.). This is particularly true in many of the new Member States. Housing conditions in rural areas appear to be worse than those in urban areas. The urban-rural division also reflects differences according to age, income and occupational status: young people, unemployed, low-skilled and low-income people report the worst conditions. This phenomenon appears to be almost non-existent in Northern countries, but is quite severe in Eastern and some Southern countries, namely Italy, Greece, and Portugal (European Commission, 2008a: 9f.).

Overcrowded conditions are defined as when the number of people living in their homes exceeds the number of rooms in the household. The extent of crowded housing for children varies considerably between countries; in every country, at least one in ten children lives in an overcrowded home (only members of OECD are compared). Children in Eastern Europe experience overcrowding the most, and it is quite widespread in Italy and Greece, while children in the Netherlands and Spain are least likely to suffer from overcrowding (OECD, 2009b: 37ff.).

A growing mismatch between the diversity of people's life-courses and the less diverse nature of housing stock can be detected. The life-course of individuals and households has become more complex, producing an ever-greater variety of housing needs. At the same time the nature of the housing stock tends to become less diverse, with more and more people buying detached dwellings located in suburban areas. The current model of 'everyone owning their home' is unable to satisfy wide-ranging housing demands. Contrary to housing market trends, the growth of small households resulting from the fall in the birth rate and the ageing of population, together with the increase in the number of single-parent families, childless couples and people living alone, point to the need for more rental housing (Bonvalet/Lelievre, 1997: 197ff.).

According to some opinions, however, home ownership is an important aspect of wellbeing, in that it protects owners from rent fluctuations and ensures that families have a stable and safe shelter, while the value of property represents a major source of wealth for households (OECD, 2007b: 140). Home ownership is supported in most of European countries (Giorgi, 2003: 20; Priemus/Dieleman, 2002: 192) and preferred by residents in most countries (Priemus/Dieleman, 1999: 627). Due to rising housing costs, governments are trying to provide more affordable housing (Paris, 2007: 3). However, social housing does not seem to be a popular solution any more, since this sector is in decline practically everywhere (Giorgi, 2003: 25; Priemus/Boelhouwer, 1999: 644).

The switch from subsidies for dwellings to subsidies for households has dominated policy change in Europe between 1980 and 2000 (Maclennan, 2008: 424; Paris, 2007: 3). The emphasis is on satisfying the needs of vulnerable groups, not on improving the quality of life of broad segments of the population. Housing allowances are regarded as the most suitable and significant measure for assisting with housing costs (Turner/Elsinga, 2005: 108; Priemus/Kemp, 2004: 666; Paris, 2007: 3; Priemus/Boelhouwer, 1999: 644). Countries are also seeking to reduce exclusion through housing (spatial segregation) by targeting areas of poverty concentration (Cameron, 2009: 10). Urban regeneration policies are a common phenomenon in Western European countries (Kleinhans/ Priemus/Engbersen, 2007: 1069).

1.4.4 Conclusions

This section has covered a wide range of themes influencing current conditions and everyday life of European families including their financial situation, housing and environment. The main emphasis has been on families with children and inequalities, not only differences between countries but also between social groups and different families. Major trends based on this review can be identified as follows:

- Polarisation in contemporary European families is significant, particularly between low/highly qualified couples; male breadwinner/ dual earner couples; low/high income families, EU/non-EU migrant families, and the urban-rural dichotomy.
- Middle-class families, especially two-earner families with educated parents and less than three children, will have good possibilities of making a living.
- The typical loser will be a lone-parent household, headed by an unemployed woman with a low educational level and more than two children.
- The extreme vulnerability of migrant families and their children,

particularly of non-EU immigrant families, compared to other families and EU migrant families.

- The mismatch between the diversity of life-courses and housing market developments.

One central characteristic of EU countries is the value given to social equality and solidarity. In spite of growing doubts created by ethnicity, changes in class-consciousness, and a stronger belief in the values of freedom and self-determination, public opinion in the EU considers social equality to be a major value – and one not attainable by relying on market forces. This is part of a government's responsibility, and is considered a marker of the European social model. Thus social inequalities and their development play a major role, politically and socially, not only in EU Member States' thinking and policy agendas but also in the feelings of justice and wellbeing of EU citizens and families.

Two main interrelated trends in the relationship between social inequality and families can be identified. First, families reflect social inequalities, since the unequal distribution of various resources and differential opportunities affect the circumstances in which family life is built up and access to certain types of family forms, divisions of work, services or life-styles. Research shows that the tendency of individuals and couples in late modernity to organise family life and intimacy in plural ways and with more freedom beyond the external constraints of normative context and social control, does not mean that social determinants have disappeared. The formation of couples, the organisation of family life, and socialisation of children and parent-child relationships are all influenced by wider social forces and social structure. It is therefore an open question whether we can still talk about social classes. Might it be more fruitful to consider the very concept of class as being no longer useful and to focus instead on paradigms highlighting the concepts of agency, individualisation, choice, biographical diversity, gender, family form, and age?

Secondly, families reproduce inequalities, both in the short term and intergenerationally. Research shows that family background, life-style, and resources, including both material and socio-cultural advantages, tend to affect children's lives and life chances. Transmission of wealth from older to younger generations and support in setting up family life during the transition to adulthood is significant in all European countries, with more affluent families being able to transfer more material and cultural resources.

1.5 Social conditions of migrant families

Migration is an area of lively public debate and vast policy intervention in contemporary Europe. It also has gained more and more scientific interest. There are several reasons for this. Migratory movements have become increasingly visible in most of the European countries, and they defy some of the entrenched principles underlying national cultures and identities. It can be argued that international flows are currently one of the major sources of social change in Europe, and a challenge for family policy as well.

1.5.1 Demographic impact of migration

The importance of immigration - including national, intra-EU and third-country national immigrants - for European populations is widely accepted. Recent statistical data also confirm the importance and widespread character of immigration in Europe. The settlement of populations with different national backgrounds, cultures, religions, and values goes against the notion of ethnic homogeneity on which European identities are - probably mistakenly - based.

In countries such as France, a considerable amount of inflows already existed in the first half of the twentieth century. In many other developed Western countries, large inflows occurred mainly after the Second World War, against a backdrop of solid economic expansion that lasted for about 30 years. Most of the immigrants were initially supposed to be temporary guests, but many remained. From the 1970s onwards several changes occurred, including the enactment of restrictive policies and changing geography of flows and new migration patterns. From the 1980s onwards, Southern Europe and Ireland gradually became important targets of immigration, together with some Scandinavian countries. More recently, after the end of the Cold War, Central and Eastern European countries also became objects of concern, given the importance of transit and, later, durable forms of immigration (Bonifazi *et al.*, 2008). During these decades, outflows also took place from most European countries. Many of these were intra-EU flows. At the same time, a clear policy-driven difference started to emerge between intra-EU flows and others involving third countries. The contradiction between (almost) free circulation and restrictions on third-country nationals became increasingly evident.

The observation of net migration growth in Europe since the 1950s confirms several facts: the durability of inflows to the North and Western European countries to the present day; the turnaround from emigration to immigration in several countries, for example in Southern Europe; and the gradual advent of new immigration destinations. Furthermore, comparison between net migration and natural increase is a revealing indicator of how immigration is driving demographic growth. In the context of overall demographic decline in Europe, it is mainly migration that is enabling positive growth, and the smoothing out of the structural impact of ageing.

Some European countries, such as Luxembourg, Switzerland, Ireland, and Austria had a larger share of foreign-born population in 2006 than the United States, a country in which immigration is an important part of national identity. If we adopt the criterion of foreign population, these European countries are joined by Spain, Belgium, and Germany. When observing the rate of growth during recent years (1995 to 2006), both the share of foreign-born population and foreigners are on the rise in most European countries. The speed of growth has been higher in some of the recent European hosts, such as the countries of Southern Europe and Ireland. Spain is the most impressive example, having passed from a proportion of 1.6% of foreigners in the whole population in 1997 to a huge 10.3% in 2006 (OECD Factbook, 2009).

The legal channels that prospective immigrants use are diverse. In 2006, family-related migration, including family reunification and marriage migration, accounted for the majority of inflows, approaching 44% of the total. This was followed by individuals entering in the framework of free movement provisions, particularly in the case of the EU, labour migration, and humanitarian grounds (including refugees).

The demographic impact of immigration in Europe has been the object of an increasing amount of research. The reason for this is plain: in the face of the potential decline and structural ageing of the European population, the direct and indirect impact of immigration has generally been well received. The inputs resulting from (usually) young adult immigrants and their offspring have enabled increases in total population, slowed down the pace of ageing, and smoothed some of its consequences. On the other hand, the impossibility of replacement migration, in the sense of offsetting the consequences of European low fertility, has been repeatedly stated.

Studies such as that by Haug *et al.* (2002), funded by the Council of Europe, have been amongst the first to study these issues on a comparative basis. Its conclusions pointed to the fact that immigration has contributed significantly to positive demographic growth and the slower pace of ageing in a number of European countries, mainly since the 1960s. This has to do with both its sheer numbers (direct impact) and its delayed demographic effect, given the volume of immigrants' offspring (indirect impact). However, their concentration on adult fertile ages led in every case to a high proportion of births issuing from immigration – the actual basis of the second generation. In addition, their mortality rates were low, again a consequence of the young age structure and fertility rates among immigrants' offspring - although varied - tended to come down to host country levels.

On the other hand, several studies, such as Lutz and Scherbov (2006) and Bijak *et al.* (2007) confirm that immigration may be, at the most, a small part of the solution to an unavoidable problem, i.e. low demographic growth and ageing. Simulations of net migration rates over the coming decades suggest that significant immigration would be beneficial in sustaining the current quantitative level of the workforce and the current potential support ratios in most EU countries. Therefore, migrant families are an important group when we discuss social and economic conditions of European families.

1.5.2 Families, gender, generation and migration

Available research on migrant families has addressed four main topics: the decision to migrate, forms of family migration, demographic trends, and assimilation of immigration families (Wall, 2007: 2253f.). The forms of family migration are especially interesting, if we think about family. It is not easy to draw up a typology of immigrant families, though one was suggested by Kofman (2004): i) family reunification; ii) family formation or marriage migration; and iii) whole family migration.

Family reunification is the conventional form. It occurs when an immigrant, living in a host country for a certain period of time and with an already existing family back home, brings in his/her family members. Although the typical form of reunification encompasses a male immigrant and his family, there are more and more cases of processes led by immigrant women (Wall *et al.*, 2010). Family formation or marriage migration include two main sub-groups: "the first consists of second and subsequent generations of children of migrant origin who bring in a fiancé(e)/spouse from their parents' homeland or diasporic space, a particular characteristic of Turkish and North African immigrant populations [...] The second variant of marriage migration involves permanent residents or citizens bringing in a partner they have met while abroad for work, study or holiday" (Kofman/Meetoo, 2008: 155f.). Studies have shown that the volume of family formation surpassed family reunification in recent years. This was particularly true in countries with large settled immigrant communities (Kraler/Kofman, 2009).

Recent research has highlighted the growing role of women's agency in family formation (Kofman, 2004). There are an increasing number of female immigrants bringing in male spouses and fiancés from the countries of origin. At the same time, more and more marriages resulting from international contacts are the consequence of women travelling, studying, and working abroad. The third type of family migration is whole-family migration. In this case, the entire family moves at the same time. Given current legal restrictions, this is not common in Europe. The major exception involves some highly skilled immigrants, including intra-EU ones, and refugees.

One of the most relevant points raised by the literature is the impact of international migration on women's roles and power, i.e. gender relations. Some studies suggest that immigration has beneficial effects on women and gender relations. Immigration and wage-earning in Europe may lead to the increasing independence of women, a more flexible division of labour at home, less segregation in public spaces and increasing centrality of women in transnational families and networks. This helps to explain why women may be more reluctant than men about return migration. Other analyses are more negative: gender is seen as another layer of the multiple oppression of migrant women – structurally discriminated against as migrants, as women, as members of the labouring underclass, as racially stigmatised, and, finally, as accepting these oppressive structures. These studies stress that many migrant women still suffer from some specific circumstances of their community's culture and family life, which tend to collide with values of the host country (see Wall *et al.*, 2010).

At this level, many findings indicate that there is a connection between violence and migration, namely male violent behaviour against women. Some researchers explore the links between violent male behaviour and social conditions. Furthermore, violence is a problem that goes beyond households and immigrant communities. The channels of "sex, marriage and maids", as expressed by Phizacklea (1998) define some of the main avenues of female migration to Europe. The sex industry is largely demand-driven, providing opportunities for trafficking networks and prostitution, bringing in young women from less developed countries.

Studies on the second generation are crucial for understanding the integration of immigrants. This mostly results from the time perspective, which is so important in migration studies. Only a long-term perspective can measure the success or failure of migration projects. What happens to the second generation tells us about the way a society is dealing with its new members. Under the term "second generation" are usually subsumed native-born children of immigrants and usually the children who arrived before primary school.

Taking into account the major inflows that took place after the 1950s, most immigrants' offspring are still at an early stage of their lives. The majority of immigrants' descendants attended primary school in the 1980s and secondary school in the 1990s, and are now entering the labour market. This explains why most of the studies until now observed the educational attainment and the transition from school to work, but not the occupational trajectory. In general, studies have shown that, in educational terms, immigrants' children perform worse than children with no immigrant background, although better than foreign-born children. When observing their early performance in the labour market, they have lower employment rates, a vulnerability to unemployment and lower access to skilled jobs, than compared to native youngsters, although again showing better indicators than foreign-born youngsters. These gaps are justified by the low socio-economic background from which they come, reduced access to social networks in the labour market, and discrimination (Castles/Miller, 2009: 227ff.). Since many of these people have acquired national citizenship, the fact that discrimination is at least partly based on ethnic origin explains part of the problem.

The situation of second generations in Europe is however more complex (Crul/Vermeulen, 2003; King et al., 2004 & 2006). Much recent research has highlighted many differences among EU countries and among immigrant groups. On the one hand, national contexts explain a large part of the variability in integration patterns. This is often less related to immigration policies than to national education and labour market arrangements, such as type of schooling (vocational or non-vocational) and access to higher education. On the other hand, immigrant communities display heterogeneity and polarisation, between and within EU countries. This means that one may observe a fraction of second-generation youngsters performing well in some EU countries, and at risk of becoming an underclass in others.

1.5.3 Conclusions

In contemporary Europe, we continue to experience increasing flows of people, coming both internally from other Member States, and from outside its borders. These migratory flows are of huge political relevance, but they are also relevant to families and are likely to continue.

- Family-related immigration represents a long-lasting channel for entry into the EU.
- The importance of immigration for the population of Europe is universally accepted.
- Labour participation rates of the foreign-born population are relatively high, but they often perform tasks below their educational level.
- The foreign-born population is at greater risk of poverty than the rest of the population.

1.6 Media, communication and information technologies

One of the most remarkable changes in the everyday life of families has been the rapid strengthening of the media culture. It has even been claimed that the media have opened up new kinds of experiences, new learning opportunities, new ways of communicating and using time, and feelings of togetherness for family members, both children and adults. For most families in Europe the media have shifted in status from being a merely incidental, if desirable, element of private life and leisure to becoming thoroughly embedded in families' everyday life, providing the indispensable infrastructure for the domestic space and the daily timetable and, in consequence, is taken for granted as mediators of social relations within and beyond the home.

In 2007, for the first time, a majority (54%) of households in the EU27 had internet access, and the main location for accessing the internet was the home (Eurostat, 2009a). The proliferation of communication and information technologies has placed media and digital literacies at the centre of policy priorities (cf. the EC's Digital Agenda, launched in March 2010), and high on the research agenda.

Media is here articulated both as 'object' - items in the household, whose location, access, gendered usage, use for facilitating work at home or care and support for older people and the infirm have significance for the timetable, spatial arrangements and social relations of family life, and as text - where the content and reception of media messages, the ways in which they represent advantaged and disadvantaged groups, and the symbolic (and material) risks as well as opportunities they pose, influence people's perceptions of the wider world and of their place within it.

1.6.1 The changing place of media in the European home

In addition to changes in the media environment, some important social trends shape the family context within which media are accorded a place in the household. As children remain dependent on their parents for longer, their teenage and young adult years are spent in the family home, creating a demand for multiple personalised media goods to accommodate competing leisure interests. With the rise of consumerism, commerce is targeting ever-younger children and creating new markets for many forms of interactive or mass media. As the number of children in a family declines, parents are able to spend more on each child, such expenditure typically including media goods, digital toys, heavily advertised fashion items and media-related bedroom décor, sometimes with consequences for parental .

authority and values. In some countries (such as the UK) parental fears regarding the safety of their children in public places encourage a tendency to equip the home as a place of leisure entertainment to compensate for declining public provision. As the period of education extends through the late teens, and as competitive pressures to gain workplace skills increase, parents are under social and financial pressure to provide household goods, technologies and toys to support informal learning at home (Livingstone, 2002 & 2009).

As the means of communication change, requiring updated provision and new digital skills, adults too must engage in a continual process of learning – to use the technology in its own right and to use it to compete in a more flexible labour market. For diverse reasons, from the growth of an elderly population, increased migration, limits on state welfare provision and greater diversity in family structures, family communication must extend over time and place, positioning communication technologies as increasingly valuable. Finally, the shift from top-down state provision to a consumer-led model of governance places more emphasis on informed choice and varieties of technological mediation, this requiring in turn the accessible provision of information, choice and networking opportunities for connecting within and beyond communities.

Research shows a range of functions performed by media in households and families, including provision of a common focus for leisure and conversation, provision of symbolic resources for family myths and narratives, the regulation of family time and space and a means of separating or connecting family subsystems within and beyond the home. On the one hand, media experience still tends to be shared with other family members. On the other hand, media are becoming more personalised, used in private spaces, and supportive of individualised taste cultures and lifestyles within the family.

Ever since it was first introduced into the family home from the 1950s onwards, television rapidly became children's main leisure activity. The idea of media budgets (Roberts et al., 1999 & 2005) stresses that time for other activities decreases when that spent on media-related activities increases. However, they also found that heavy users of one medium are also heavy users of others. Indeed, young people seem to be multitasking and using a variety of media simultaneously. Reporting from the comparative project, Children and their Changing Media Environments, Johnsson-Smaragdi (2002: 193) found that simple media displacement is rare, given specialised media use, reallocation of media time and additive media use. Television displacing reading time has been a worry, but one without conclusive findings. Johnsson-Smaragdi's (2002: 45) findings reveal that the habitual time spent in front of the television screen has increased during the past 15 years; boys from low SES families spend the most time in front of the television, and girls from high SES families the least.

As television has been increasingly complemented, if surprisingly little displaced, by the use of new interactive technologies within the home, a new body of research developed, following Silverstone's (2006) concept of domestication of new technologies in the 1990s. The argument was that even once technologies had been purchased by the household, the process of rendering them meaningful, finding them both space and time in the life of the family, is an unfolding, ongoing process of interpretation and adjustment.

Despite rapid increases in internet access over the past decade, household access and use of the internet still varies widely across Europe, ranging from 25% in Bulgaria to 86% in the Netherlands (Eurostat, 2008a). Gender differences in internet and computer use remain inconsistent, although present, across Europe. Seybert (2007: 1) reveals that "the difference between the proportion of young women (62%) and young men (67%) in the EU25 using computers daily in 2006 was relatively small [...] slightly more young men (53%) than young women (48%) used the internet daily". While parental education and income both have a part to play, their effects may be opposed, and it is certainly not simply the more affluent who have more. Family type also matters: while two-parent households (and households with working mothers) are much more likely to provide a media-rich home, reflecting their considerably higher incomes, single parents are just as likely to provide media-rich bedrooms for their children.

For children and young people, one of the most important contributions of research has been to challenge the moral panics that commonly associate youthful media use with fears regarding their vulnerability and victimisation or, on the other hand, their engagement with new forms of mediated "hooliganism". A good example of this sensibility is research on the emergence of a media-rich bedroom culture for children (Livingstone, 2009). This could be framed in terms of children's isolation from family life and their consequent vulnerability to commercial, violent, or other media messages. Although children are hardly immune to such messages, qualitative research influenced by domestication theory adds a different understanding.

The rise of a media-rich bedroom culture suggests several consequences for the family. Children spend time in highly individualised, consumerist, and usually strongly gendered spaces. Children's media use may be more extensive, continually in the background if not also the foreground, and relatively unsupervised or unmediated by parents; the family's leisure time is more compartmentalised (Van Rompaey/Roe, 2001), with families 'living together separately' (Flichy, 1995) and with time spent together 'as a family' something that requires deliberate arrangement. Even when children are in the home, not physically co-located with friends, their leisure time may be spent in a peer context, in touch with peers as much or more than with parents (Ito *et al.*, 2010; Livingstone, 2009). Age makes a difference: older children and boys generally have more media goods in their bedroom, particularly screen entertainment media. Livingstone (2002) notes that families with sons place computers in bedrooms more often; those with daughters place them in a common space (also Johnsson-Smaragdi *et al.*, 1998).

Considerable cultural differences in bedroom culture are evidenced by cross-national differences found by "Children and their Changing Media Environment", which surveyed children in 12 countries in 1997-98. This found that a screen entertainment culture is particularly strong in the UK, with Denmark close behind. Households in Nordic countries and the Netherlands are pioneers of new technologies, including those for children. In Spain, both boys and girls are particularly likely to spend time with the family and to spend comparatively less time in the bedroom, while Swedish and Finnish teenagers are overwhelmingly more likely to spend their free time with a group of friends, also spending a smaller proportion of their free time in their own room. Indeed, Swiss teenagers spend a greater than average proportion of their time in their own rooms, while Finnish teenagers spend less than average, even though Swiss children own fewer televisions or computers and spend less time on these media while for Finnish children the opposite is the case (Bovill/Livingstone, 2001: 196).

1.6.2 Media technologies and associated risks

With 75% of European children using the internet, a figure that continues to rise (though it may soon plateau), societal concerns regarding the associated risks also increase, raising new research questions with pressing policy implications.

Half of all teenagers online give out personal information, the most common risky behaviour. Findings from the *Eurobarometer* survey (Livingstone/Haddon, 2009) suggest that according to their parents, children encounter more online risk through home than school use (although this may be because parents know little of their children's use at school).

There is evidence supporting a classification of countries based on the likelihood of children experiencing online risk. This classification suggests a positive correlation between use and risk. High use, high-risk countries are, it seems, either wealthy Northern European countries or new entrants to the EU. Southern European countries tend to be relatively lower in risk, partly because they provide fewer opportunities for use (Hasebrink *et al.*, 2009).

It seems that children's internet-related skills increase with age. Such skills are likely to include children's abilities to protect themselves from online risks, though this has not been extensively examined. It also seems that there are cross-national differences in coping. Children's perceived ability to cope with online risk reveals higher ability to cope among children in Austria, Belgium, Cyprus, Denmark, France, Germany and the UK, and lower ability to cope in Bulgaria, Estonia, Greece, Portugal and Spain (intermediate countries are the Czech Republic, Ireland, Poland, Slovenia and Sweden).

1.6.3 Parenting, media, everyday life and socialisation

Traditionally, infants and toddlers have engaged little with the media, although television, radio and music are often in the background. During primary school years, children are generally not major media users, although television and electronic games are very popular. Over the teenage years, young people begin to broaden their range of media uses and tastes, often seeking to individuate themselves from their friends via media tastes while simultaneously being absorbed in the culture of their peer group. By their late teens and early twenties, young people are negotiating a wide range of information, communication and literacy demands as they manage the transition from school to further study and/or work.

Generally, much of the available literature on media and socialisation addresses questions of media exposure and effects. Overall, the research literature points to a range of modest effects, including effects on attitudes and beliefs, effects on emotions, and, more controversially, effects on behaviour (or the predisposition towards certain behaviours). However, there are many methodological qualifications and contestations accompanying these conclusions, especially the critique of cause-effect assumptions in much socialisation theory, and concern that such research neglects the child's own agency.

In terms of family reception of media content and questions of values and tastes, the context of family viewing is a crucial determining factor in what causes offence. Research suggests that audience concern most often focuses on terms that stereotype or marginalise. Buckingham (2005) suggests that children may adopt their taste judgements from adults, including finding swearing, sex or violence distasteful or embarrassing. On the other hand, they also consider that such content in reality television, game shows and soap operas has value in offering them a kind of a projected adult future. Thus, Buckingham and Bragg (2003 & 2004) found that children may value sexual material as a means of gaining information otherwise difficult to obtain or as providing a pretext for discussing difficult issues in the family.

Parental mediation strategies for children online can be classified as active or instructive mediation, rule-making or restrictive mediation, and parental modelling or co-viewing. Research on parental mediation of the internet in fact reveals that mediation is fairly widely practised, albeit with substantial cross-national variations. The effectiveness of time restriction in European countries shows that the significance of the strategy differs with the socialisation cultures of the countries. However, evidence of 'a regulation gap', impeding parental mediation especially for the internet, shows that since parents are willing and ready to mediate television more than the internet, even though they worry more about the internet than television, it is lack of skills rather than lack of concern that results in lower levels of internet mediation. Kirwil (2009: 403) concludes that "although parental mediation is associated with fewer number of children at risk from online content, the effectiveness of several strategies seems to depend on the country's socialisation culture. In Europe, both restrictive and non-restrictive mediation may be effective in one childrearing culture, but ineffective in another one".

The economic and educational resources of the family are replicated in digital environments. To create societies in which all families are equal, it is important to understand how we can break this vicious cycle for disadvantaged families so that access to services, social relationships, education and information is not limited by cultural, social or economic background. The use of information and communication technologies (ICT) in education and learning at school and at home is the site of attention and action at the policy level, because the use of ICT for creating a positive impact on learning outcomes, achieving potentials, acquiring job skills and enhancing lifelong learning is recommended. In terms of utilising the full benefits of ICT in education and learning, Livingstone (2009: 64) identifies two hurdles: "one is attitudinal, for parents must share this educational and technological vision for their child; the other is material, for parents must possess the resources (time, space, knowledge and money) to implement this vision". Recently, there has been optimism that mobile phones may help to overcome digital divides between learners with home broadband access and those without, or that it may improve feedback from teachers. However, mobile learning necessitates a good amount of technical training, preparation and planning, production of learning material and a sequence of many other time-consuming activities. It must be admitted that as with ICT and education, the advantages of this are still unclear, and as always, these are bound to vary by demographic factors.

1.6.4 Conclusions

In contemporary Europe, the mediation of institutions and processes of family and social life has become increasingly complex. Media are an important part of the domestic environment. Some important trends in European homes today are:

- New, interactive, individualised and personalised media technologies are rapidly contributing to a diverse media environment.
- Despite privatisation of media access, media experiences are still shared with other family members.
- Children's use of the internet continues to grow. Striking recent increases are evident among younger children (6-11 years) and in the new EU Member States.
- Emergence of a media-rich bedroom culture among children.
- Socio-economic inequalities continue to matter, with patterns of digital exclusion mirroring those of social exclusion.

1.7 Family policies and social care policies

This section focuses on research on family policies and social care policies within EU Member States from a comparative perspective. In many European countries, these policies have become increasingly important. Part of the explanation is the increased political awareness of the challenges discussed in previous sections: decreasing birth rates, population ageing, diversification of family forms, and the weakening of the male breadwinner/female carer model.

The diversity of European welfare systems and family policies exceed existing country categories and welfare regime typologies. There is also significant national variation within these categories, even within the Nordic countries, which are usually classified as a joint Nordic model. Furthermore, in these comparisons and classifications the most recent EU Member States are usually missing (mostly CEE countries). However, there are some indicators that European Member States are converging in terms of their social care systems and family policies.

In his famous publication "The Three Worlds of Welfare Capitalism", Gøsta Esping-Andersen (1990) clustered welfare states in order to facilitate systematic comparisons between them. He analysed social policies in 18 countries and used the indicators de-commodification, social stratification, and interplay of state, market, and family in social provision[13]. The result was the well-known threefold typology: the Liberal, the Conservative, and the Social-Democratic welfare state regime. Feminist criticism has argued that this typology neglects gender-specific problems of the welfare state as well as the family's role in welfare provision (Orloff, 1993; Daly, 1994; O'Connor, 1996). Therefore, alternative typologies have been developed, which focus on these issues (e.g. Lewis/Ostner, 1994). Esping-Andersen (1999: 51) responded to this critique by adding the distinction between familialistic and de-familialising welfare states, concepts originally developed by Lister (1994), defining de-familialisation as "the degree to which households' welfare and caring responsibilities are relaxed either via welfare state provision or via market provision" (Esping-Andersen, 1999: 61). The distinction will be used in this section to assess the directions of family policy and social care policy changes, even though these concepts should be treated with caution.

Social care policy is in many ways linked to family policy, but it includes not only childcare but also care for older people and other forms of adult care. Since the mid-1990s it has become a core issue in social policy and in social research (Anttonen/Sointu, 2006: 4). Social care has many dimensions, which makes this research field broad and complex. In this section, the concept of social care is adopted from the EC-funded SOCCARE project. Social care is defined as the assistance and surveillance provided in order to help children or adults with the activities of their daily lives. Social care can be paid or unpaid work provided by professionals or non-professionals, and it can take place within the public as well as the private sphere. Formal service provision from public, commercial and voluntary organisations, as well as informal care from family members, relatives and others, are here included within social care (Kröger, 2004: 3). In this section, the focus is on the perspective of families and family members and on families as care providers.

1.7.1 State family policies in Europe

Family policies are part of the larger social policy context (Ferrarini, 2006: 5), which is usually quite normative and highly ideological: family policies are linked to basic assumptions about the role of the family in society (Kaufmann, 2000: 424) and to definitions of what a family is or ought to be

---

[13] De-commodification is the extent to which individuals can maintain a normal standard of living regardless of their market performance (Esping-Andersen, 1990: 86), e.g. by old-age pensions. Social stratification measures how far the welfare state itself actively orders social relations, e.g. by income-related benefits.

(Lüscher, 1999: 8). Family policies can be defined as an "amalgam of policies directed at families with children and aimed at increasing their level of wellbeing" (Gauthier, 1999). This definition acknowledges the fact that family policy is a cross-cutting issue (Gerlach, 2010: 168), which may include "topics as varied as employment, transport, food, and education policies" (Gauthier, 2002: 456). It poses a challenge to international comparisons of family policy and highlights the need for clear conceptualisations.

When family-affecting state interventions are explicitly undertaken, they are usually driven by one or several of the following motives (Kaufmann, 2000: 426ff.):

- Institutional motives, to preserve the family as an institution in its own right, often linked with conservative policy and a traditional family model.
- Demographic motives, which have increased in importance in the context of demographic changes, such as measures to increase birth rates or diminish abortions.
- Economic motives 1) to stress the importance of the family for human capital building and to assess its benefits for society; 2) to emphasise economic functionality (strengthening the workforce via childcare).
- Socio-political motives, to compensate direct and opportunity costs of family responsibilities (e.g. caretaking, income losses) and to fight poverty.
- The gender equality motive, to remove economic and social disadvantages especially for women, and to reach a more gender-equal share of family and employment tasks and set special incentives for fathers.
- Children's welfare motive, to provide the framework for public provision of children's needs.

Policy makers have a range of different instruments at their disposal, which can be distinguished as 1) regulation, 2) information and 3) financing. Regulation includes family law, protected leaves and equal opportunity laws. Information might mean family support programmes, benchmarking and performance indicators. Financing includes financing of childcare, parental leave payments, child/family allowances, social insurance, family taxation, and housing allowances (Blum/Schubert, 2010: 85).

In many European countries family policies have become a major area of reform. However, institutions, norms, and regulatory frameworks substantially limit policy makers' scope for action and the shaping of future policies. European Union regulations also influence national family policies. Nonetheless, it is reasonable to speak of diversity in European welfare systems. Family policies of the Nordic countries have been heavily influenced by the Protestant church and left-wing governments. They gradually focused on gender equality, reconciliation, and female labour market integration through de-familialising policies (Ferrarini, 2003). Nordic family policies seek to promote the interests of individuals rather than of families as units, and have no institutionalised family policies with designated ministries (except in Denmark). With regard to family law, Nordic countries are advanced in putting unmarried cohabiting couples on a par with marriage (Hantrais, 2004: 113, 133). Legal rights of same sex couples and their families have also increased in recent decades, giving them in most respects the same legal status as marriage (Eydal/Kröger, 2010: 14). They also pay particular attention to the wellbeing of children (Ostner/Schmitt, 2008), and introduced early joint custody after divorce. Legal family obligations are very weak. Family allowances are tax-funded, and government-NGO co-operation is strong (Appleton/Byrne, 2003: 212).

Family policies of the continental countries have been heavily influenced by the Catholic Church and the subsidiarity principle. The role of Christian democratic parties has been strong (Borchorst, 1994). Family policies are traditionally characterised by male breadwinner and female carer norms. In countries with Bismarckian earnings-related social security schemes (especially Austria and Germany), social protection of women and children may still be dependent on marriage and family relationships (Hantrais, 2004: 117). Austria, Germany and Luxembourg have long-standing traditions of designated family ministries, and France and Belgium have perhaps the most explicit and consistent family policies across Europe (Hantrais, 2004: 138). Several continental welfare systems prescribe state protection of the family in their constitutions. Regarding family law, some countries follow the Nordic example regarding equal treatment of non-married couples, and the majority of countries introduced same sex registration schemes (Boele-Wölki, 2007). Governmental-NGO co-operation is ambivalent: while the civil society sector has a strong role in policy implementation, its role in agenda setting and policy formulation is quite weak (Appleton/Byrne, 2003).

Anglo-American countries (Ireland, Malta, and the UK) share common ground in weak state intervention, need-oriented support, and the prominent role of the market, but differ in others. Malta and Ireland have stronger familialist traditions than the UK. Ireland prescribes family protection in its constitution and has a designated ministry for family policies. With regard to family law, divorce was made legal in Ireland only in 1996 (Hantrais, 2004: 110), and is still illegal in Malta. Family relationships are not strictly regulated in the liberal systems – but neither are duties of the state, leading to a heavy reliance on the private and voluntary sectors (Hantrais, 2004: 129). Since social welfare is an individual responsibility and the benefits directed at means-tested minimum coverage, family policy is not explicit and comprehensive, but part of general welfare policies (Rüling/ Kassner, 2007: 22). Family allowances are tax-funded in Ireland and the UK, while Malta, being a hybrid case in several respects, exhibits mixed-funding. Governmental-NGO co-operation is less integrated than in the Nordic countries, but in the UK civil society organisations are integrated into family policy (Appleton/Byrne, 2003).

The Mediterranean countries - or the Southern European countries, if we include Portugal in the group - share similarities with the continental systems in male breadwinner and female carer traditions and Catholic influences. With regard to family law, these countries legally assign mutual obligations to the extended family, and the state provides support only when these sources are exhausted (Hantrais, 2004: 129). State duties to protect the family are prescribed in the national constitutions of Greece, Italy, Portugal and Spain. Authority is often delegated to the local and regional level, leading to regional discrepancies (Hantrais, 2004: 144); the overall approach is rather fragmentary. A designated ministry for family policy exists in no country, but in 2006 Italy created a co-ordinating Department for Family Policies. Spain makes a case for institutional and legal reforms: under the social-liberal coalition in 2005 it eased the divorce law and took the lead regarding civil unions (Bertelsmann, 2008). Only Portugal recognises same sex partnerships de facto without registration (Boele-Wölki, 2007). Family allowances are generally at a low level and governmental-NGO co-operation is weak (Appleton/Byrne, 2003).

The post-socialist countries make most interesting cases, since they have faced dramatic institutional shifts in family policy. Before the Second World War, most of these countries were based on the conservative Bismarckian model and their family policies showed all signs of familialism. Then with restructuring since the 1990s, following employment-centred, universal welfare provision during the socialist era, some features exhibit path-dependence from this time (Trumm/Ainsaar, 2009: 154), and there has been institutional redesign in most cases. Case studies have shown that all former communist countries quit the path of de-familialisation and tried to reintroduce the traditional familialism-based regime (Saxonberg/Sirovatka, 2006: 186; Hantrais, 2004). Some constitutions mention protection of the family (Hungary, Estonia, Latvia, and Poland). Designated ministries exist in Romania, Slovenia, Slovakia, and in Latvia (Cunska/Muravska, 2009). Legal obligations of extended families are strong, while state duties are weak, and there are no explicit or coherent family policies. Regarding same sex partnerships, only Slovenia, the Czech Republic, and Hungary have such legislation (Boele-Wölki, 2007). Family allowances are tax-funded but underfunded.

Governmental-NGO co-operation is still weak, affected by the communist times, and the Catholic Church often holds a monopoly over NGOs. However, co-operation is developing (Appleton/Byrne, 2003: 217).

1.7.2 Childcare policies

The most widely studied topic in relation to family policies and social care has been childcare arrangements and policies including parental leave schemes, cash benefits, and day care services. This theme includes research on the division of labour and responsibilities between families and the state, but also gender division within families with regard to childcare (e.g. Gerhard/ Weckwert, 2001; Gerhard et al., 2005; Ellingsæter/Leira, 2006; Crompton et al., 2007; Lister et al., 2007; Lewis et al., 2008).

There are several reasons for the popularity of this topic. First, childcare and motherhood has been one of the main issues in feminist research, and child-care has been seen as an issue of gender equality. Secondly, the increasing employment rate of women and gender equality in working life have been important political aims in many European countries, and in the EU. This has motivated and promoted research on this topic (see e.g. Giullari/Lewis, 2005: 3f.; Plantenga et al., 2008; Plantenga/Remery, 2009).

Leave policies

Already at the end of the nineteenth century some European countries developed limited maternity leave and corresponding pay schemes (Gauthier, 2000: 3), which were intended to protect the health of the mother and the newborn child. Substantial improvements in European maternity leave systems were made during the 1960s and 1970s. Nordic countries adopted a more gender egalitarian perspective and transformed maternity leave into a gender-neutral parental leave, though reforms in most other countries were less extensive.

Since the 1970s and 1980s, there has been significant growth and diversi-fication of leave policies throughout Europe. Besides maternity leave, most of the Northern and Central European countries have adopted some system of parental leave – fostered in Southern Europe one decade later in the 1990s (Gauthier, 1996: 77). Today, the period of maternity leave varies from 14 to 20 weeks, with an earnings-related payment of between 70 and 100% (Moss/ Deven, 2009: 82). Paternity leave entitles fathers to take a short leave immedi-ately after the birth of a child. Interestingly, Portugal is the only country making paternity leave obligatory. Usually, paternity leave periods vary from two to ten days, with earnings-related payment on the same basis as in maternity leave.

Parental leave is available equally for mothers and fathers. It can be structured either as a non-transferable individual right with an equal amount of leave or as a family right, which parents can divide between themselves. According to the EU Parental Leave Directive, all Member States must provide at least three months parental leave per parent. This directive does not specify further requirements regarding payment or flexibility (Moss/Deven, 2009: 84). Concerning the length of parental leave, countries can be clustered into those providing about nine to 15 months and those providing up to three years. In many countries, the payment is earnings-related, but by contrast, in several other countries parental leave benefit is paid at a flat rate, low earnings-related rate, means-tested or for only part of the leave period. Several countries additionally have developed childcare leaves between one and three years following the maternity or parental leave period. Parental and childcare leave are creating a continuous period, sometimes with different conditions affecting payment. Furthermore, nearly all European countries offer at least unpaid time off to care for a sick child.

In recent years, more academic and political attention has been paid to options that family policies offer for men in childcare. O'Brien (2009: 194) has compared fathers' patterns of leave-taking across 24 (mostly European) countries between 2003 and 2007. The main dimensions used in her analysis are leave duration and level of income replacement. She has clustered these "father-sensitive leave models" as follows: 1) Extended father care leave with high income replacement (Finland, Germany, Iceland, Norway, Portugal, Quebec, Slovenia, Spain, Sweden); 2) Short father leave with high income replacement (Belgium, Canada, Denmark, France, Greece, Hungary, the Netherlands); 3) Short/minimalist father care leave, with low/no income replacement (Australia, Austria, Czech Republic, Estonia, Ireland, Italy, Poland, and United Kingdom); 4) No statutory father care-sensitive parental leave (United States). Her results suggest that fathers' use of statutory leave is greatest when high income replacement (50% or more) is combined with extended duration (more than 14 days). Father-targeted schemes heighten usage.

Leave policies have been an area of significant growth over recent decades. The expansion of leave arrangements in almost all European countries reflects economic and demographic motives of policy makers as well as the political pressures on governments to support parents in balancing work and family life. At the end of the 1990s, most European countries had developed a range of different types of leave for working parents and others with care responsibilities. The trend towards increasing leave time was accompanied by an expansion of part-time work. With the Nordic countries being pioneers again, in several countries working parents gained rights to reduced working time (Morgan, 2008).

In research on childcare and reconciliation of work and family life there has been a heavy emphasis on the role of the welfare state. The European Foundation for the Improvement of Living and Working Conditions, however, emphasises the role of companies, employers, and organisational culture in workplaces in how parents are able to use the options available. The authors conclude that the factors shaping the take-up patterns of parental leave include the financial and legal conditions of the statutory parental leave system, the prevailing gender division of labour, and access to measures aimed at reconciling work and family life. Other important factors are whether parental leave is accepted and supported within the organisational culture, as well as the establishment's human resources practices, and labour market conditions with regard to wage levels, job security, and unemployment (European Foundation, 2006: 6).

Childcare services

Formal childcare has been the most crucial area of family policy reform in the EU in recent years, being increasingly regarded as a vital work-family reconciliation element that contributes to multiple goals, such as female employment, gender equality, birth rates, and early education. The Barcelona European Council 2002 certainly provided a strong impetus in this regard. The summit agreed on the goals of providing childcare to at least 33% of children under 3 years of age and to at least 90% of children between 3 years old and the mandatory school age in each EU Member State by 2010. Other drivers were the OECD's research, such as PISA and 'Babies and Bosses', which pointed out the importance of early childhood education and care.

Childcare provisions in the EU countries differ substantially with regard to coverage rates, affordability and quality (see Plantenga/Remery, 2009; Da Roit/Sabatinelli, 2007; OECD, 2007a; Eurostat, 2009c; Lohmann et al., 2009). In their report based on Eurostat statistical information and national reports, Plantenga and Remery (2009: 54f.) say that in the age category 0-2 years, the use of formal childcare arrangements varies from 73% in Denmark to 2% in the Czech Republic and Poland. Only seven EU Member States (Denmark, the Netherlands, Sweden, Belgium, Spain, Portugal, and the United Kingdom) have already met the Barcelona target. The use of formal care arrangements increases with the age of children. The authors also recall the need for day-care for school age children.

Childcare services for children under three are particularly well developed in the Nordic countries. Conversely, deficient formal systems with less than 10% of children under three can be found in post-socialist Czech Republic, Hungary, Poland, Lithuania, Slovakia, but also in Austria and Malta.

Care systems for children from the age of three to mandatory school age have expanded everywhere. Regarding opening hours, part-time exceeds full-time care especially in Austria, Germany, Netherlands, but also Ireland, Malta and the UK – reflecting institutional traditions and cultural norms. Overall, full-time care is more common for the older age group than the younger one.

In childcare provision there has been a heavy emphasis on quantity, but an issue that is even more important is quality. While in national studies qualitative information has been vital, quantifiable data like staff-child ratios or education of nursery school teachers has been used in international comparisons (Lohmann *et al.*, 2009: 72). Regarding staff-child ratios, there are again considerable differences throughout Europe: for the under three age group, some countries such as Denmark (1:3) and the UK (1:3) show a favourable staff-child ratio, while others such as Italy (1:7 to 1:10) or Germany (1:6) exhibit unfavourable ratios. Similar diversity exists with regard to the educational level of childminders and pre-school teachers. Particularly low educational levels seem to exist in Belgium, Germany, the Netherlands, and the UK. Furthermore, they criticise the considerable gender imbalances among childcare staff. Staff qualifications also tend to differ between different childcare institutions within countries, and private childminders usually have a significantly lower level of education (Plantenga/Remery, 2009: 45).

Both the quantity and quality of childcare relate to national policy priorities, one indicator for which is social expenditure on childcare (Lohmann *et al.*, 2009: 70). While Nordic countries and the UK spend more on 0-2 year olds than the three to school age group, allocation in many countries is the opposite (especially in Hungary, Poland, Portugal, Slovakia). Furthermore, it can be observed that the Nordic countries and France are spending more than 0.9% of their GDP on childcare, above the EU27 average (0.6%), while particularly low spenders are Austria, Ireland, Greece and Poland.

From a policy process point of view, the reforms and expansionary or retrenchment measures of recent years are important. Plantenga and Remery (2009) show a move to higher coverage in many European countries. They state that while the Nordic countries, Belgium, France, and Slovenia have high levels of availability and direct efforts at enhancements, the UK, the Netherlands, and Germany are clearly moving towards fuller coverage of childcare services. Thus, there is an overall trend of de-familialisation through formal childcare expansion, from which a number of (mainly CEE) countries have to be excluded.

However, Central and Eastern European countries do not form a unified cluster. Szeleva and Polakowski (2008) have studied the patterns of childcare in the new member countries of the EU in Central and Eastern Europe during the period 1989-2004. Instead of a unified tendency towards familial-isation of policies, many of the post-communist countries followed different paths to familialisation, while some of them strengthened the defamilial-ising components of their policies. They distinguish between four policy types: implicit familialism (Poland), explicit familialism (the Czech Republic, Slovakia, and Slovenia), comprehensive support (Lithuania and Hungary) and female mobilising (into labour market) (Estonia and Latvia).

The dominating trend of de-familialisation is also elsewhere accom-panied by another trend, which Mahon (2002) detected for France and Finland and called "new familialism". Its emphasis is on parental choice, and the primary policy instrument for achieving this is home care allow-ances. Both Finland and France introduced such flat-rate allowances in the 1980s, Norway followed in 1998 and more recently Sweden (2008) and Germany (from 2013). There is also severe criticism of home care allow-ances for preventing female employment, and minimising equality of educational opportunities (OECD, 2007a). Saraceno and Keck (2008: 61) argue that public financial support may strengthen, incentivise, or allow familialisation of care responsibilities, and remind us that the forms public support may take are not gender-neutral. Even so, care allow-ances, together with positive de-familialisation measures, might increase parental choice. The challenge is to achieve a new equilibrium between time, money and services (Plantenga/Remery, 2009: 60) so as to develop a consistent system opening up different opportunities according to indi-vidual circumstances and preferences.

Statutory entitlements to childcare are slowly but surely moving onto national policy agendas. Finland is the only EU country with a legal right for all children under school age to attend public childcare facilities (OECD, 2007a: 160). Other Nordic countries, but also Germany and the UK, have followed in the same direction (Dörfler, 2007; Plantenga/Remery, 2009: 40). Often, however, the right is only to part-time care. Some countries have not only introduced a right to childcare, but also made the last pre-school year compulsory and free of charge. This has been the case in Austria, Poland, Cyprus, Luxembourg and Hungary (Eurostat, 2009d).

1.7.3 Cash and tax benefits for families

Cash and tax benefits for families primarily follow economic and socio-polit-ical family policy orientations: families perform valuable benefits for society and have higher costs, lower earnings capacities (due to care obligations) and often higher poverty risks than households without children (Lohmann *et al.*, 2009: 78).

OECD (2009a) provides information on the family spending proportions of cash, services, or tax measures. Most countries spend more on cash benefits than on services or tax benefits, although the "growth of in-kind benefits has outpaced the growth of cash benefits in several countries and the role of tax benefits is growing" (Gabel/Kamerman, 2006: 261). Countries clearly privilege either public childcare, leave policies, or cash and tax benefits, rather than offer mixed support (De Henau *et al.*, 2006). Cash benefits are in the majority of countries the (financially) most important family policy measures. The OECD Family Database distinguishes child/family allowances, parental leave payments, support for single parents, and public childcare support through earmarked payments to parents. Families also receive social insurance or housing benefits, but these are not considered to be direct family policy measures.

Child/family allowances exist in practically every EU country (Plantenga/ Remery, 2005: 67), while their levels vary considerably. Gauthier (2005; also Gabel/Kamerman, 2006) finds that during the 1990s in most countries family cash benefits expenditures fell as a percentage of GDP: only Germany, Italy, Luxembourg, Spain, and Switzerland spent more in 1998 than in 1993. Especially in the CEE countries, there were downward trends in the years following the collapse of the socialist regimes, but a gradual restoration has taken place since then. Between 2000 and 2005, public expenditure on family cash benefits increased in eight countries (Denmark, Germany, Ireland, Luxembourg, Portugal, Slovakia, Spain, and the UK). In five countries, they remained fairly stable (Austria, Greece, Hungary, Italy, and Sweden), while in six countries they were cut back (Belgium, Czech Republic, Finland, France, the Netherlands and Poland).

In some countries, family allowances are paid as a universal benefit, in others they depend on income, age or number of children. In five EU countries, eligibility is based on employment (and not residence), and thus excludes parts of the population. Nine countries grant family allowances only after income testing; five of these are Southern European and four post-socialist countries. Bahle (2008) concludes that in the Southern countries income testing constitutes a policy principle, while in the post-socialist ones it is done because resources are scarce. Family allowance targeting to the number of children is conducted in a small majority of the 16 countries; they represent all family policy systems. This criterion may reflect pronatalist motives, but also the fact that families with many children have higher relative costs. Those countries which have benefits varying by both age and number of children, are all "family policy pioneers [...] have long histories of family allowances [and] except for the Netherlands, they are predominantly Catholic" (Bahle, 2008: 109).

Tax credits are often not incorporated in cross-national analyses. Historically, the first forms of family allowances in European countries were introduced in the post-war period as so called "housewife bonuses" reflecting and reinforcing single-earner family norms (Dingeldey, 2001: 656). With rising female labour market participation, separate taxation was introduced in some countries (e.g. Sweden), while others kept joint taxation systems (e.g. Germany). Nineteen OECD countries had separate income taxation of spouses in 2006, while joint taxation or options for it existed in eleven countries (OECD, 2008). Child-related tax allowances exist in practically every EU country (Plantenga/Remery, 2005: 67), although they are diverse.

Adema (2009) points out that in most OECD countries net payments to governments are smaller for families with children than for similar households without children, although the differences range from very high in Austria and Hungary to quite small in Poland. However, generosity is not necessarily reflected in positive outcomes (e.g. higher birth rates or lower poverty rates), for this "crucially depends on the extent to which tax/benefits systems give parents financial incentives to work and help them combine work and care commitments" (Adema, 2009: 193). Lohmann et al. (2009: 86) conclude that dual-earner couples have lower tax rates in almost all EU countries – and significantly so in Austria, Finland, Greece and Hungary. Almost all countries either support an equal division of paid labour in families or show neutrality in the taxation of single-earner and dual-earner couples (Lohmann et al., 2009: 92).

Fouarge and Layte (2005) argue that welfare regimes strongly influence long-term poverty. Overall, the countries of the social democratic type display lower rates of poverty. The next highest rates are found in the countries of the corporatist type. In countries belonging to the residual and liberal welfare regimes, poverty is not only higher but it is also recurrent and persistent. The performance of regimes in tackling income poverty turns out to be rather different from their performance in tackling resource deprivation. Looking at the difference across regime types it became clear that deprivation and poverty tends to be more prevalent in Southern and Liberal regimes and less so in Corporatist and Social-Democratic regimes. Nonetheless, most of the variance is not explained by country or regime type differences but by common structural factors such as the needs of the household, the human capital of its members, the turnover and dynamics of the labour market and the distribution of permanent income. Particularly interesting is the large contribution of socio-economic status variables to explaining deprivation, which reflects the traditional impact of class, education, and employment status (Fouarge/Muffels, 2009).

The impact of welfare policies on child poverty has merited specific attention in recent research. Chen and Corak (2005) point out that family and demographic forces play only a limited role in determining changes in child poverty rates. Instead, labour market and the government sector are the sources of the major forces determining the direction of change in child poverty. The countries with the lowest child poverty rates are clearly those that spend most on social benefits (excluding pensions). However, a number of countries with similar wealth and similar shares of GDP invested in social benefits achieve very different child poverty rates. On average in the EU, social transfers other than pensions reduce the risk of poverty for children by 44%. In the Nordic countries, Denmark, Finland and Sweden, social transfers reduce the risk of poverty for children by more than 60%. Only France and Austria show similar results. The countries in which benefits have the strongest impact in reducing child poverty are those in which expenditures are specifically identified as family benefits (TARKI, 2010).

1.7.4 Social care for older people

A large number of comparisons have been made since the 1990s in the field of social care services for older people. However, care for older people is often studied in the framework of health/medical rather than of social care or family policy, or instead of a broader and more multidisciplinary framework. In this chapter, the focus on the care for older people is on home/family based care and services, and on the role of families as providers of care and support.

One of the most extensive EC-funded research projects on family care for older people is "Services for Supporting Family Carers of Elderly People in Europe" (Triantafillou/Mestheneos, 2006). It has provided a European review of the situation of family carers of older people in relation to the existence, familiarity, availability, use, and acceptability of support services (see Triantafillou/Mestheneos, 2006; Mestheneos/Triantafillou, 2005). According to its results, the reasons for family carers to provide care are most often physical illness, disability or other dependency of the old person. Emotional bonds constitute the principle motivation for providing care, followed by a sense of duty, personal sense of obligation or having no other alternatives. The findings show that women were predominantly both the main carers and the main older person cared for. Nearly 50% of carers were adult children of the cared-for old person, although there was national variation in this. Researchers described family care as a dynamic but long-term commitment. In terms of the financial implications, family carers had less than average disposable income as a result of having to spend time on caring (Triantafillou/Mestheneos, 2006: 4ff.).

The project also studied informal and formal support for family carers and the old people cared for. Social networks were associated with lower levels of carer stress and burden. Less than one-third of family carers had used a support service but of the cared-for old persons, a vast majority (94%) used at least one care service in the previous six months. The highest percentages were found in Sweden, Italy, and Denmark, the lowest in Greece. In all countries, there were problems with the distribution of services, especially in rural areas, and in covering hours when a carer may be working. Users and non-users of care services saw the bureaucratic and complex access procedures, high monetary costs, lack of information on available support, low quality, inadequate coverage, and the refusal of the old person to accept existing services as the main barriers to service use (*ibid.*).

Several researchers have studied the ways in which formal care services and informal care are combined. For example, Motel-Klingebiel *et al.* (2005) have studied whether formal services provided by the state "crowd out" (diminish) family care, encourage it, or create mixed responsibility. They found no evidence of a substantial "crowding out" of family help. Instead, the results support the hypothesis of "mixed responsibility" and suggest that in societies with well developed service infrastructures, help from families and welfare state services act cumulatively, but that in familialistic welfare regimes, similar combinations do not occur (*ibid.*: 863). This result could be seen as rather surprising and unexpected against the trend of "care going public" identified by many researchers (e.g. Anttonen *et al.*, 2003: 171f.). However, where there are more regular and demanding care services, "care going public", its professionalisation and institutionalisation, seems to occur in care for older people.

In the search for cost effectiveness/reduction Simonazzi (2009) has observed a convergence in how the care market is organised. According to her, all countries are moving towards home care, private provision and cash transfers. She argues that the way in which care for older people is provided and financed entails considerable differences in the creation of a formal care market. Secondly, national employment models shape the features of the care labour market, affecting the quantity and quality of care labour supply, the extent of the care labour shortage, and the degree of dependence on migrant carers.

Several researchers (e.g. Behning, 2005; Pavolini/Ranci, 2008) have suggested that all over Europe, in spite of national differences, there are at least two similar and simultaneous trends in social care for older people: first, the privatisation and marketisation of formal, professional care, and secondly, (re-)familialisation of care either with or without financial compensation. Thus, the role of public provision of social care services for older people seems to be diminishing.

Based on the findings of the SOCCARE project, Sipilä and Kröger (2004: 562ff.) conclude that European social care cultures are diverse but not completely different. They underline the importance of formal care services to needy families. They also remind us that formal social care services are strongly intertwined with informal care. From the viewpoint of families, service organisations should never be isolated institutions but flexible and capable of meeting specific human needs in individual ways. Furthermore, the idea of quality care is immediately associated with the availability of sufficient time. Carers need to be able to combine working and caring, both simultaneously and sequentially. When neither working life nor services are flexible enough, the flexibility is ultimately taken from informal sources, mainly from women.

1.7.5 Local family policies

Often family policies and services for families are implemented at the local level, and thus there is local variation in the provision of services. Policies at the local level are however much more difficult to study cross-nation-ally than national policies and service provision. Furthermore, at the local level the role of NGOs is important in complementing public services and providing support for families in almost all European countries (Council of Europe, 2009: 59).

Mingione and Oberti have observed that "local systems must be evalu-ated in terms of a varied mix of institutional and individual actors where diversity and complexity play an increasingly important role within the development of active policies, based on partnership implementation and on shared responsibility between providers and recipients" (Mingione/ Oberti, 2003). Different national welfare systems are far from being harmon-ised and thus lead to many different opportunities for developing local welfare policies. In countries such as France or the UK, there are many 'local interventions' promoted by the State. In other countries, such as Germany, Italy, or Spain, local policies are directly promoted by Regions, or even Municipalities, in a more flexible context. It is particularly because of this budgetary and administrative responsibility of the local government that one may speak of a distinct "local welfare state" (Wollmann, 2004) or of "welfare municipalities" (Kröger, 1997). The "local welfare state" is realised in Germany and Sweden, while in UK and France the welfare system is highly centralised. The comparison between different local welfare systems high-lights the involvement and responsibility of non-public organisations and groups (Wollmann, 2004).

Involvement of the community in local welfare creation has been declared fundamental in a 2004 European transnational project on work-life balance and care entitled *Equal-Tempora*. According to its authors, services have to be 'near to persons, families and their living environment', also according to the principle of subsidiarity. In the provision of local welfare services different actors such as local administrative institutions, social players, professionals, community, and family, have to participate in a concerted manner in the implementation of local welfare services (Equal Tempora 2, 2004: 14). Although there are almost no comparative studies on local policies for families (see however Kröger, 1997), the local level of welfare and political intervention in building new family policies has gained more recognition, as stated in the recent ESF Paper Partnership for more Family-Friendly living and working conditions: "European Structural Fund projects are now seeking the way to meet Lisbon targets, and family policies are therefore seen in the context of enhancing employment and gender equality, starting specifically from the local level, where families and companies are" (European Commission, 2008b: 9).

The renewed interest in the local level is also linked to the interest in the so-called 'good practice models', that is an analysis of local projects from different nations and on different topics, to highlight the variety of approaches and experiences in local family policies.

1.7.6 Conclusions

The diversity of European welfare systems and family policies exceeds existing country categories and welfare regime typologies. Most researchers today agree that the main differences in family policies and social care policies can be found between Southern and Northern parts of Europe, but there is no agreement on whether these can be called as separate regimes. There is also significant national variation within these categories; even within the Nordic countries (see e.g. Mahon, 2002). Furthermore, in these comparisons and classifications most recent EU Member States are usually missing. Still, there are some indicators that European countries are becoming more alike in their social care systems and family policies, and in problems related to them. There is a need for a more up to date and reliable family policy typology.

In family policies and social care policies at the European, national and local levels, some major trends can be identified:

- The field of family policies has become increasingly important and has expanded during recent years, and there is a trend towards institutionalisation.
- In terms of re- and de-familialisation in family and care policies across Europe, a mixture of re- and de-familialising measures can be identified.
- Leave policies have in many countries been aimed at activating

    fathers and reaching a more equal share of employment and family responsibilities between both parents.

- Childcare services have been one of the most important family policy issues and reform areas and in this field in particular, the trend towards "care going public", de-familialisation, institutionalisation and profes-sionalisation of care work and services is likely to continue.
- Social care remains a combination of formal and informal care, where the role of families and especially women in families is remarkable.
- Globalisation and internationalisation of care with its various forms and consequences will be one of the future trends.
- Growing importance of local governments with more responsibili-ties and autonomy regarding many politically relevant issues for family policies and service provision.
- Increasing role of local NGOs and networks of different actors.

The expansion of childcare facilities is high on the agenda in many European countries, and so is the expansion or reduction of child/family allowances and parental leave policies, often including elements of "active fathering". Care issues seem to have left social benefits behind on the family policy agenda even if the latter are of crucial importance e.g. in reducing poverty and diminishing social inequalities. A common trend all across Europe is reconciliation of work and family life. It seems quite evident that even if the ageing of population has been recognised as one of the biggest future chal-lenges all over Europe, childcare will remain at the heart of policy-making. That is because it is so closely related to the needs of the economy, the labour market, and gender equality policy (see e.g. Mahon, 2002; Haataja, 2005; Leira/Saraceno, 2008: 14ff.; Knijn/Smit, 2009).

    All over Europe, the field of childcare can be described as "care going public" (e.g. Anttonen/Sipilä, 2005; Geissler/Pfau-Effinger, 2005). This trend is less clear in social care for older people, where the trend seems to be twofold: on the one hand privatisation and marketisation of formal, professional care, and on the other (re-)familialisation of care either with or without financial compensa-tion. In several countries, there has been a move towards 'direct payments' or 'personal budgets'. These changes represent a tendency for the user of care services to be given considerably more say on the way her/his needs are met. This raises an increasing political and academic interest in different combi-nations of formal and informal care including intergenerational care relations. Several researchers have been interested in whether formal care replaces (crowds out) informal care or whether it complements it (crowds in). There seems to be no strong evidence for the crowding out hypothesis (e.g. Brandt et al., 2009).

In addition to the state, there are other important actors, local governments, NGOs, and especially companies and employers with their occupational family policies, which are of major importance for the reconciliation of work and family life (e.g. family-friendly working hours, workplace childcare facilities). On these policies, international comparisons are rare.

Globalisation and internationalisation with its various forms and consequences is becoming a more and more important issue for the future in family and social care policies. It means that care relations cross national borders in the forms of global care chains and transnational care, and in increasing numbers of migrant workers both in formal and informal care work. Furthermore, it means that caring is increasingly becoming an international business where multinational companies are providing care services (Anttonen *et al.*, 2009). The EU is now both in its policy and research funding investing in migration issues, but the relationships between family policy, care, gender and migration are not yet clearly identified (see European Commission, 2009).

1.8 Gaps in existing research

There is a widespread notion that existing research is somewhat oriented towards the nuclear family and largely ignores the increasing diversity in family forms and family relations. It has concentrated on families with young children, often ignoring other stages of family life and the life-course approach, which is extremely important in family research.

Moreover, most of the research does not recognise that the family is a dynamic entity: it implicitly assumes that family forms are static. From this point of view the family should be seen as the result of partnership and childbearing processes, from the life-course development of all its members (including children). Thus future research should recognise the dynamic character of families, and should be able to understand these dynamics, the transitions within traditional and new types of family forms. We need also more research on the daily and biographical practices of 'doing family'. Existing research is also adult-centred: the children's perspective (and the views of old people) on their family life, and on policies and services, are largely missing. Overall, experiences of families and individual family members within families - as policy 'targets' and service users - are largely ignored in existing research.

In connection with family structures, forms and demographic processes, knowledge on fertility and demographic trends has been most extensive in terms of the availability of indicators as well as the countries and time spans covered. We also know quite a lot about the major changes and trends in family structures (e.g. changes in marriage, birth, and divorce rates, numbers of lone-parent families, etc.). The consequences of increasing life expectancy and the causes of low fertility should, nonetheless, be analysed further. Increasing life expectancy points to the problems of households of older people: to loneliness, coping with decreasing biological, physical and psychological resources, and to the need for care, a suitable environment, etc. Macro-social reasons and individual rights also point to the need for research on developments in childbearing behaviour. Low fertility puts a question-mark over the sustainability of the social system, so it is essential to understand the processes behind it. From an individual perspective, the growing gap between intentions and realisation should be studied, especially in times when low fertility has negative consequences.

So called 'new and rare' family types are understudied, and the data available is very fragmentary. For example, both the quality and quantity of data on gay and lesbian (rainbow) families differs widely between the European countries. Similarly, knowledge on multi-generation households as a specific family form and arrangement is rather scarce. Patchwork/reconstituted families are another central and increasingly common family form which has been much debated in political and public discourse, but comparative research and data are virtually non-existent at the European level.

Gender and generations in families is also a broad research area, which has mainly concentrated on the gender division of paid and unpaid work in families and reconciliation of work and family life. There should be greater linkage between family research and labour market research, with specific emphasis on the role of job markets, companies and employers in decision-making within families.

Gender is influential in partnering and parenting, and gendered practices within families seem to change very slowly. Nonetheless it is legitimate to speak of 'a new parenthood', and men are taking a more active role in families. This means that we need research from a male perspective on partnership, parenting and caring work within families, on the 'child effect' on men in paid work, and on (de-)gendering strategies of young parents, who struggle with or adapt to the re-traditionalisation of gender roles after the birth of the first child. In general, there seems to be something of a blind spot regarding transitions to parenthood, especially in comparative research, and future research should concentrate on illuminating the decision-making processes, including practices and self-concepts of young women and men in the process of becoming (or not becoming) mothers and fathers.

In terms of generations, the recent focus has been on care relations between generations, but further research is also needed in this area. However, generational relations are not only a care issue. There is a need for future research on the relations between parents and children, and between young people and their parents (especially bearing in mind the trend that young people are leaving home later than before and that this has consequences for their transition to adulthood), on the role of grandparents, and on multigenerational family as a family form. Violence in families is highly gendered but also a generational issue. It is important to implement cross-national comparative prevalence studies in order to compare particular features of violence against women, men, children, and older people.

Given the major role of the family in the reproduction and transmission of social inequality, it is essential to have more information on social inequalities and social mobility; how they are evolving in European societies and what are the best measures to be used in order to capture and describe social inequalities today. Secondly, there is very little research comparing types of family interactions and functioning, and how they relate to social and economic inequalities across the different EU countries. There is a third research gap in connection with inequality "strategies" and "processes", whereby families reproduce as well as deepen social advantage. Given the overriding importance of educational attainment, it would be important to focus in greater depth on the nexus between families' inequality strategies and their relationship to the educational context and systems. It is also important to recognise the differences between families and children living in rural or peripheral and urban areas in terms of education levels, educational opportunities, and attainment.

On families and poverty, the dominant focus in comparative research on income poverty provides a very specific outlook, one which ignores the experience of poverty and how it affects family life and individuals within families (e.g. loss of dignity, choice, and control; limited access to social capital and to assets of other kinds; poor health; poor housing; few opportunities; and an uncertain future). Social analysis of families and poverty would also benefit from a reinforcement of the household/family as a unit of analysis. Housing, neighbourhood, and physical environment are closely linked with questions of social inequality. It is important to study those components of living environment as a whole, as they are strongly interrelated, and influential in the everyday life of families.

Research on immigration in Europe is extensive and plural. There are still many gaps in current research, particularly on family and integration issues and migrant families. It may be argued that family migration has been much less studied than other related issues. New forms of mobility are becoming common (including work-related migration within the EU, student migration, retirement migration, and female-led migration), surpassing the more traditional settlement migration. Areas also deserving better scrutiny are the impact of family reunification on immigrants' strategies; the impact of specific legislation on family life; the consequences of recent policy restrictions for family reunification and formation; and the use of irregular channels by family members for immigration and integration.

Research in the area of family wellbeing and media environments needs to merge the family studies literature within sociology with the media and communications literature. Much of the research done so far comes from the field of media and communications studies, although all questions at the heart of family priorities (for example parenting, child-raising practices, and relationships within the family) are intensely mediated questions. Not much is known about different age groups, especially the media consumption of older people. While media use, especially by young people, is being heavily researched, little research distinguishes or compares "youth" or "children" by age and other sociological variables. Furthermore, findings across Europe on social class, ethnicity, and cultural differences remain scarce in terms of media literacy, education, and civic participation. More research is also required on the opportunities, skills, and risks related to media and new technology, and how these are divided between social classes and educational levels in society.

Family policies and social care policies in Europe have been studied more than other family-related issues at the European comparative level, probably on account of the crucial political and economic importance of these policies for the labour market, work-family reconciliation, and gender equality. The diversity of family policy measures is a challenge for comparative research while systems and situations are often very country-specific. Furthermore, because of the cross-cutting nature of family policy and the impact of other policies on families, the family aspect of the whole policy-making process (e.g. employment, health, urban development) should be further studied.

In political terms, there are three important fields and questions: countries learning from each other, reactions to the current economic crisis, and responding to future challenges. Learning from each other depends on understanding why a certain measure was introduced in another place, why it worked there or why it did not; negative lessons may also be helpful. With regard to the current economic crisis, it is important to research the (long-term) consequences for national family policies. An interesting area for policy makers is international benchmarking: often, only international comparison can show how family-friendly a country is in a specific area, e.g. childcare or cash benefits. Another field of interest is of course the situation of families, their current and future challenges. For this, family policy reporting in the Member States is important. In addition, the evaluation of policies has become increasingly important.

Family policy actors other than the state are extremely important in comparative research. The role of NGOs, local family policies, and occupational family policies, are of major importance for the reconciliation of work and family life. With regard to local policies, a new methodological Good Practice Policy model would be useful; an evaluation system is needed to define and share good practices.

Informal care and the role of families are crucial and should be studied more, as the overwhelming majority of care is (and probably will be) provided in families. It is important to note that reconciliation of work and care is not only a childcare issue, but should include adult care as well. For example, spousal care and men as caregivers are widely ignored in existing research. Abuse in care relations is still an understudied issue.

Within the field of cross-national, comparative research, there are different methodological orientations. The main division is between macro-level multi-national comparisons using quantitative data and micro-level, small-scale studies using qualitative or mixed methods. Most of the large multi-national projects reviewed in this report have used either national statistical information, statistics provided by Eurostat, and/or large multinational surveys and databases, such as The European Community Household Panel (ECHP), Gender and Generations Surveys (GGP), Survey of Health and Retirement of Europe (SHARE), and the European Social Survey (ESS). Even if the situation has improved over the last decade, the need for comparative, harmonised, and often longitudinal data has been identified, as well as the need for interdisciplinary approaches and more in-depth qualitative research that would give us better insight into family life and its changes and decision-making processes. Small-scale qualitative comparisons can also advance theory building, while large-scale comparisons mainly test existing theories.

Even if indicators are available, cross-national comparability of data on families frequently proves to be a problem due to the frequent need to employ data from different statistical sources. Administrative data comes initially from various nation-specific statistical offices that frequently differ in the definition of central terms and concepts due to the application of different statistical traditions. In data-gathering and collection even the seemingly straightforward categories, e.g. women's involvement in paid work, might be measured in various different ways. The definition and measurement of unpaid work varies even more.

One of the problems in existing comparative research is country coverage. Some countries have been more popular than others in existing European comparative research, which means that some countries are understudied. Today, large survey-based studies already cover a large number of Member

States (often twenty or more), but in more detailed and focused comparisons there are still some "favourite" countries like Germany, the UK, Sweden, France, with the Netherlands, Italy and Spain in a second tier of favourites. Often the countries studied have been chosen to represent a particular welfare (or care) regime typology (usually Esping-Andersen's), which has been shown in this review to be a rather misleading starting-point for country selection because of the diversity and exceptions within the different country clusters. Regional comparisons are mostly based on geographical closeness: Northern Europe, Southern Europe/Mediterranean countries, Western Europe/old Member States, and Eastern Europe/CEE countries/new Member States. The last group is the one least covered. These kinds of groupings seem to be of varying usefulness, depending on the issues studied. For example, for economic issues and housing the groups based on geography are useful, since differences between regions are evident and countries in each group have quite a lot in common. Such groups are however not very helpful when comparing family policy systems.

Unpaid work and use of time in families are examples of potential problems in data availability. The main methods of collecting information on housework, for example, have been survey questions and time diaries, but other methods such as qualitative in-depth interviews, direct observation, and discourse analyses have also been used. Low cost and high response rates are the main advantages of survey instruments. However, in different time use surveys direct questions vary considerably in their wording. Time diaries provide an alternative instrument. Despite some drawbacks, such as low response rates, and the problems of accounting for parallel activities, time diary methodology provides reliable estimates of time use patterns in households. Because time use survey methods have not been harmonised for a long time, nationally produced data are not comparable. Eurostat started to support projects aimed at harmonising time use statistics in the EU in the early 1990s and was mandated to develop guidelines for Harmonised European Time Use Surveys (HETUS) in order to ensure that surveys of the Member States are comparable. The most recent guidelines were published in 2009 (Eurostat, 2009b), and most European national statistical institutes have taken them into account since the late-1990s. Some countries, however, differ from the recommendations to a varying degree. National time use surveys are therefore only comparable up to a certain level. Currently the online HETUS database contains comparable data from 15 countries.

The variety of ways domestic work is measured limits the comparability of the results of different studies. Research in the field of both paid work and unpaid work could benefit from applying longitudinal survey data. Such information however is rare. The only comparative research programme that strives towards this end is the Generations and Gender Programme, with several EU Member States producing panel survey data on demographic and social developments – including information on the division of work. The great variability of methods and approaches applied places serious restrictions on establishing trends over time. Findings from two or more distinct studies, carried out on data from different points in time, are rarely comparable. There are only a few comparative studies which attempt to explore changes over time.

There are slightly different problems in studying specific groups of individuals or families e.g. immigrant families. Again, there is a need for more complete and comparative data. Concepts and sources vary, and several areas are poorly captured. Beyond the description of some major variables, an in-depth comparative study of immigrants' characteristics, families and the second generation is barely possible using current datasets. Longitudinal studies, which enable us to understand the mobility experience, are also generally not available.

One of the problems with the existing databases is that the data available is not necessarily suited to the specific research interests in question, and such data goes out of date rather quickly, especially when dealing with family policies or formal service systems in individual countries. Quite often comparative studies rely on national expert reports from individual countries, expertise of the research team and/or previous research available. The methodological problem with data sources of this type is their reliability and coverage. Expert reports might vary depending on who the expert is. There might be also problems in finding previous research and other written documents, especially from smaller language areas. Therefore, there is a need for databases which are comparable across Europe.

There is also need for more qualitative comparative research that would make it possible to analyse decision-making processes within families, for example, and the perspectives of individual family members, so as to gain greater insight into family life. Many research questions concerning families and family policies cannot be answered by using quantitative methodology. In the field of family policy research, qualitative designs make it possible to study the specificities of individual cases while keeping the comparative advantages. Meso (or even micro) level analyses are much less frequent, though they have many insights to offer. In policy analysis, more in-depth, qualitative comparisons are needed to understand and explain policy reforms and processes. In terms of the policy cycle, for example, this would help us to understand how problems are defined, how the political agenda is set and policies formulated, how political decisions are made, implemented, possibly evaluated and finally either terminated or re-formulated.

For the time being, there are 27 Member States in the EU. There are differences and variations between countries and within countries. Europe is rich in differences. On the other hand, based on the Existential Field reports, there are certain general trends discussed above, some weaker and some stronger, running through all European Union countries. We are probably heading into more similarities in many areas of family life, family relations, and family policy.

One question which is often asked in connection with existing empirical findings pointing at diversity is whether it is still possible to identify country clusters or welfare regimes. To some extent, it is possible to recognise the division into the three regimes identified by Esping-Andersen (liberal, conservative, and social-democratic), or a more nuanced division into five regions: Nordic (or Northern) countries, Southern (sometimes Mediterranean) countries, Continental countries, the UK, and Central and Eastern European (CEE) countries. Although these groupings work in some respects, there are always exceptions and qualifications. More information is needed on regional, national and cultural differences in order to resolve the question of what kind of country cluster and welfare regimes we actually have in Europe.

1.9 References

- Aassve, A., Betti, G., Mazzuco S. & Mencarini L. (2006). *Marital Disruption and Economic Well-Being: A Comparative Analysis. ISER Working Paper 2006-07*. Colchester: University of Essex. Available from: *http://www. iser.essex.ac.uk/publications/working-papers/iser/2006-07.pdf* [accessed 1.4.2011].
- Aassve, A., Busetta, A., & Mendola, D. (2008). *Poverty permanence among European youth. ISER Working Paper 2008-04*. Colchester/ Palermo: University of Essex, University of Palermo. Available from: *http://www.iser.essex.ac.uk/publications/working-papers/iser/2008-04.pdf* [accessed 1.4.2011].
- Aboim, S. (2010). *Plural Masculinities: The remaking of self in private life*. Farnham: Ashgate.
- Adema, W. (2009). *Family support: Lessons from different tax/benefit systems*. In: Von der Leyen, U. & Spidla, V. (eds.) *Voneinander lernen – miteinander handeln. Aufgaben und Perspektiven der Europäischen Allianz für Familien*. Baden-Baden: Nomos Verlag, 189-202.
- Ahn, N. & Mira, P. (2002). *A note on the relationship between fertility and female employment rates in developed countries*. Journal of Population Economic 15 (4), 667-682.
- Albertini, M., Kohli, M. & Vogel, C. (2007). *Intergenerational transfers of*

time and money in European families: common patterns different regimes. Journal of European Social Policy 17 (4), 319-334.

- Aliaga, C. (2006). *How is the time of women and men distributed in Europe? Eurostat, Statistics in focus, Population and Social Conditions 4/2006.* European Communities. Available from: *http://epp.eurostat.ec.europa. eu/cache/ITY_OFFPUB/KS-NK-06-004/EN/KS-NK-06-004-EN.PDF* [accessed 17.1.2010].

- Aliaga, C. & Winqvist, K. (2003). *How women and men spend their time – Results from 13 European countries. Eurostat, Statistics in focus, Population and Social Conditions 12/2003.* European Communities. Available from: *http://epp.eurostat.ec.europa.eu/cache/ITY_OFFPUB/KS-NK-03-012/EN/KS-NK-03-012-EN.PDF* [accessed 17.1.2010].

- Allen, K. & Demo, D. (1995). *The families of Lesbians and Gay men: A new frontier in Family research.* Journal of Marriage and the Family, 57 (1), 111-127.

- Anttonen, A., Sipilä, J. & Baldock, J. (2003). *Patterns of social care in five industrial societies: Explaining diversity.* In: Anttonen, A., Baldock, J. & Sipilä, J. (eds.) *The young, the old, and the state: Social care systems in five industrial nations.* Cheltenham: Edward Elgar, 167-198.

- Anttonen, A. & Sipilä, J. (2005). *Comparative approaches to social care: diversity in care production modes.* In: Pfau-Effinger, B. & Geissler, B. (eds.) *Care and Social Integration in European Societies.* Bristol: Policy Press, 115-134.

- Anttonen, A. & Sointu, L. (2006). *Hoivapolitiikka muutoksessa. Julkinen vastuu pienten lasten ja ikääntyneiden hoivasta 12 Euroopan maassa.* [Care politics in transition. Public responsibility on the social care of young children and old people] Hyvinvointivaltion rajat -hanke. Helsinki: Stakes.

- Anttonen, A., Sointu, L., Valokivi, H. & Zechner, M. (2009). *Lopuksi.* In: Anttonen A., Valokivi, H.& Zechner, M. (eds.) *Hoiva. Tutkimus, politiikka ja arki* [Care: Research, policy and everyday life] Tampere: Vastapaino, 238-254.

- Appleton, L. & Byrne, P. (2003). *Mapping Relations between Family Policy Actors.* In: Social Policy & Society, 2 (3), 211-219.

- Arai, L. (2009). *Teenage pregnancy: the making and unmaking of a problem.* Bristol: Policy Press.

- Archer, C. (2000). *Sex differences in aggression between heterosexual partners: A meta-analytic review.* Psychological Bulletin 126, 651-680.

- Archer, C. (2002). *Sex differences in physically aggressive acts between heterosexual partners. A meta-analytic review.* Aggression and Violent Behaviour, 7 (4), 313-351.

- Arnett, J. J. (2004). *Emerging Adulthood: The Winding Road from the Late Teens through the Twenties*. New York: Oxford University Press.
- Arnett, J. J. (2006). *Emerging Adulthood in Europe: A Response to Bynner*. Journal of Youth Studies, 9 (1), 111-123.
- Bahle, T. (2008). *Family policy patterns in the enlarged EU*. In: Alber, J., Fahey, T. & Saraceno, C. (eds.) *Handbook of Quality of Life in the Enlarged European Union*. London/USA/Canada: Routledge, 47-73.
- Banens, M. (2010). *Mariage et partenariat de même sexe en Europe. Vingt ans d'expérience. Politiques sociales et familiales*. Comparaisons internationales, n° 99.
- Batalova, J. & Cohen, P. (2002). *Premarital Cohabitation and Housework: Couples in Cross-National Perspective*. Journal of Marriage and Family 64, 743-755.
- Baycant-Levant, T., Vreeker, R. & Nijkamp, P. (2009). *A Multi-Criteria Evaluation of Green Spaces in European Cities*. European Urban and Regional Studies 16 (2), 193-213.
- Beck, U. & Beck-Gernsheim, E. (1994). *Individualisierung in modernen Gesellschaften – Perspektiven und Kontroversen einer subjektorientierten Soziologie*. In: Beck, U. & Beck-Gernsheim, E. (eds.) *Riskante Freiheiten*. Frankfurt am Main: Suhrkamp.
- Beck, U. & Beck-Gernsheim, E. (1995). *The Normal Chaos of Love*. London: Polity Press.
- Becker, G. (1981). *A treatise on the family*. Cambridge: Harvard University Press.
- Behning, U. (2005). *Changing long-term care regimes: a six-country comparison of directions and effects*. In: Pfau-Effinger, B. & Geissler, B. (eds.) *Care and Social Integration in European Societies*. Bristol: Policy Press, 73-91.
- Berk, S. F. (1985). *The gender factory*. New York: Plenum.
- Bertelsmann Stiftung (2008). *Internationaler Reformmonitor, Sozialpolitik, Arbeitsmarktpolitik, Tarifpolitik, Ausgabe 10/11*. Gütersloh: Bertelsmann Stiftung.
- Bianchi, S. M., Milkie, M. A., Sayer, L. C., & Robinson, J. P. (2000). *Is anyone doing the housework? Trends in the gender division of household labor*. Social Forces, 79, 191-228.
- Bickel J. F. & Lalive d'Epiney C. (2001). *Les styles de vie des personnes âgées et leur evolution récente: une étude de cohorts*. In: Legrand, M. (ed.) *La retraite: une revolution silencieuse*. Ramonville-Saint-Agne: l'Harmattan.
- Biggart, A. & Walther, A. (2006). *Coping with Yo-yo Transitions: Young Adults Struggle for Support, between Family and State in Comparative Perspective*. In: Leccardi, C. & Ruspini, E. (eds.) *A New Youth? Young People, Generations*

and Familiy Life. Aldershot: Ashgate, 41-62.

- Bijak, J., Kupiszewska, D., Kupiszewski, M., Saczuk, K., Kicinger, A. (2007). *Population and labour force projections for 27 European countries, 2002-2052: impact of international migration on population ageing*. European Journal of Population 23, 1-31.
- Blossfeld, H-P., Klijzing, E., Mills, M. & Kurz, K. (eds.) (2005). *Globalization, uncertainty and youth in society*. London/New York: Routledge.
- Blum, S. & Schubert, K. (2010). *Politikfeldanalyse*. Wiesbaden: VS Verlag.
- Boele-Wölki, K. (2007). *The Legal Recognition of Same-Sex Relationships Within the European Union*. Tulane Law Review 82, 1949-1981.
- Boele-Woelki, K. & Fuchs, A. (2002). *Legal Recognition of Same-Sex Couples in Europe. European Family Law Series*. Antwerp/Oxford/New York: Intersentia.
- Bonifazi, C., Okólski, M. Schoorl, J. & Simon, P. (eds.) (2008). *International Migration in Europe – New Trends and New Methods of Analysis*. Amsterdam: Amsterdam University Press.
- Bonvalet, C. & Lelievre, E. (1997). *The Transformation of Housing and Household Structures in France and Great Britain*. International Journal of Population Geography 3 (3), 183-201.
- Borchorst, A. (1994). *Welfare State Regimes, Women's Interests and the EC*. In: Sainsbury, D. (ed.) *Gendering Welfare States*. London: Sage.
- Bottero, W. (2004). *Class identities and the identity of class*. Sociology 38 (5), 985-1003.
- Bottero, W. (2005). *Stratification: Social Division and Inequality*. London: Routledge.
- Bovill, M. & Livingstone, S. (2001). *Bedroom culture and the privatization of media use*. In: Livingstone, S. & Bovill, M. (eds.) *Children and their changing media environment: A European comparative study*. Mahwah, NJ: Lawrence Erlbaum Associates, 179-200.
- Bozon, M. (2004). *Famille, couple, sexualité*. In: Marquet, J., Bastien, D., Bozon, M. & Chaumont, J.M. (eds.) *La nouvelle normativité des conduites sexuelles ou la difficulté de mettre en cohérence les expériences intimes*. Louvain-la-Neuve: Academie Bruylant.
- Bradshaw, J., Finch, N., Mayhew, E., Ritakallio, V. & Skinner, C. (2006). *Child Poverty in Large Families*. Bristol: Policy Press.
- Bradshaw, J. (2010). *Child poverty and child wellbeing in the European Union. Policy overview and policy impact analysis. A case study: UK*. In: TARKI (ed.) *Child Poverty and Child Well-being in the European Union*. Report prepared for the DG Employment, Social Affairs and Equal Opportunities (Unit E.2) of the European Commission. Budapest: TARKI.
- Brandt, M., Haberkern, K. & Szydlik, M. (2009). *Intergenerational Help and*

Care in Europe. European Sociological Review 25 (5), 585-601.

- Brines, J. (1993). *The exchange value of housework*. Rational Sociology 5, 302-340.
- Buber, I. & Neuwirth, N. (eds.) (2009). *Familienentwicklung in Österreich. Erste Ergebnisse des Generation and Gender survey 2008/2009*. Wien ÖIF and Vienna Institute of Demography. Available from: *http://www.ggp-austria.at/fileadmin/ggp-austria/familienentwicklung.pdf* [accessed 28.3.2011].
- Buckingham, D. (2005). *The media literacy of children and young people. A review of the research literature*. London: Ofcom.
- Buckingham, D. & Bragg, S. (2003). *Young people, media and personal relationships*. London: BBC and Broadcasting Standards Commission.
- Buckingham, D. & Bragg, S. (2004). *Young people, sex and the media: The facts of life?* Basingstoke: Palgrave Macmillan.
- Bundeskanzleramt (2009). *17.11.2009. Bundeskanzler Faymann: Kompromiss für Eingetragene Partnerschaft ist wichtiger Schritt zur Gleichstellung*. Available from: *http://www.bka.gv.at/site/cob__37209/currentpage__0/6856/default.aspx* [accessed 18.3.2010].
- Burchell, B., Fagan, C., O'Brien, C. & Smith, M. (2007). *Working conditions in the European Union: the gender perspective*. EUROFOUND. Available from: *http://www.eurofound.europa.eu/pubdocs/2007/108/en/1/ef07108en.pdf* [accessed 17.1.2010].
- Bussmann, K., Erthal, C. & Schroth, A. (2010). *The Effect of Banning Corporal Punishment in Europe: A Five-Nation Comparison*. In: Smith, A. B. & Durrant, J. (eds.) *Global pathways to abolishing physical punishment: Realizing children's rights*. New York: Routledge.
- Callens, M. & Croux, C. (2009). *Poverty Dynamics in Europe. A Multilevel Recurrent Discrete-Time Hazard Analysis*. International Sociology, 24 (3), 368-396.
- Cameron, S. (2009). *KATARSIS Growing Inequality and Social Innovation: Alternative Knowledge and Practice in Overcoming Social Exclusion in Europe*. WP 1.3 Housing and Neighbourhood, Available from: *http://katarsis.ncl.ac.uk/wp/wp1/D13papers/D13.pdf* [accessed 17.1.2010].
- Cantillon, B. & Van den Bosch, K. (2002). *Social Policy Strategies to combat income poverty of children and families in Europe*. Working paper no. 336. Syracuse/New York: Maxwell School of Citizenship and Public Affairs, Syracuse University.
- Castles, S. & Miller, M. (2009). *The Age of Migration – International Population Movements in the Modern World*, 4th ed. New York: The Guilford Press.
- Cerami, A. (2003). *The Impact of Social Transfers in Central and Eastern*

Europe. Luxembourg Income Study Working Paper No. 356. Available from: *http://ssrn.com/abstract=1123574* [accessed 17.1.2010].

- Chen, W-H. & Corak, M. (2005). *Child Poverty and Changes in Child Poverty in Rich Countries since 1990*. Innocenti Working Paper No. 2005-02. Florence: UNICEF Innocenti Research Centre.
- Coleman, L. & Cater, S. (2006). *'Planned' Teenage Pregnancy: Perspectives of Young Women from Disadvantaged Backgrounds in England*. Journal of Youth Studies 9 (5), 593-614.
- Coltrane, S. (2000). *Research on household labor: Modeling and measuring the social embeddedness of routine family work*. Journal of Marriage and the Family 62, 1208-1233.
- Coté J. (2000). *Arrested Adulthood. The Changing Nature of Maturity and Identity*. New York/London: University Press.
- Council of Europe (2009). *Family Policy in Council of Europe Member States. Two expert reports commissioned by the Committee of Experts on Social Policy for Families and Children*. Strasbourg: Council of Europe.
- Crompton, R. (1998). *Class and Stratification*. Cambridge: Polity.
- Crompton, R. (2006a). *Employment and the family: the reconfiguration of work and family life in contemporary societies*. Cambridge: Cambridge University Press.
- Crompton, R. (2006b). *Class and Family*. Sociological Review (November 2006), 658-679.
- Crompton, R., Lewis, S. & Lyonette, C. (eds.) (2007). *Women, Men, Work and Family in Europe*. Basingstoke: Palgrave Macmillan.
- Crul, M. & Vermeulen, H. (2003). *The Second Generation in Europe*. International Migration Review 37 (4), 765-986.
- Cunska, Z. & Muravska, T. (2009). *Social Policy Implementation in Latvia Post EU Accession*. Riga: University of Latvia. Available from: *http://www.politika.lv/index.php?f=1442* [accessed 27.11.2009].
- Czasny, K. (ed.) (2004). *SOCOHO The importance of Housing Systems in Safeguarding Social Cohesion in Europe – Final Report*. Available from: *http://www.srz-gmbh.com/socoho/report/downloads/socoho-final-report_2005.pdf* [accessed 17.1.2010].
- D'Addio, A. C. & D'Ercole, M. M. (2005). *Trends and Determinants of Fertility Rates: The Role of Policies. OECD Social, Employment and Migration Working Papers 27, OECD*. Directorate for Employment, Labour and Social Affairs. Available from: *http://ideas.repec.org/p/oec/elsaab/27-en.html, http://ideas.repec.org/s/oec/elsaab.html* [accessed 28.3.2011].
- Daly, M. (1994). *Comparing welfare states: towards a gender friendly approach*. In: Sainsbury, D. (ed.) *Gendering Welfare States*. London: Sage Publications.

- Da Roit, B. & Sabatinelli, S. (2007). *The Cost of Childcare in EU Countries. Study Requested by the European Parliament's Employment and Social Affairs Committee*. Brussels: European Parliament.
- Davis, S. & Greenstein, T. (2004). *Cross National Variations in the Division of Household Labor*. Journal Marriage and Family 66, 1260-1271.
- Davis, S. N., Greenstein, T. N. & Marks, J. P. G. (2007). *Effects of union type on division of household labor: Do cohabiting men really perform more housework?* Journal of Family Issues 28, 1246-1272.
- De Henau, J., Meulders, D. & O'Dorchai, S. (2006). *The Childcare Triad? Indicators assessing three Fields of Child Policies for Working Mothers in the EU-15*. Journal of Comparative Policy Analysis 8 (2), 129-148.
- Devine, F., Savage, M., Scott, J. & Crompton, R. (2005). *Rethinking Class*. Basingstoke: Palgrave Macmillan.
- Dingeldey, I. (2001). *European Tax Systems and Their Impact on Family Employment Patterns*. Journal of Social Policy 30 (4), 653-672.
- Dürnberger, A., Rupp, M. & Bergold, P. (2009). *Zielsetzung, Studienaufbau und Mengengerüst*. In: Rupp, M. (ed.) *Die Lebenssituation von Kindern in gleichgeschlechtlichen Lebenspartnerschaften*. Köln: Bundesanzeiger-Verlags-Gesellschaft, 11-50.
- Dykstra, P. (1997). *Employment and caring*. Working paper No. 7. Haag: NIDI.
- Dörfler, S. (2007). *Kinderbetreuungskulturen in Europa. Ein Vergleich vorschulischer Kinderbetreuung in Österreich, Deutschland, Frankreich und Schweden*. ÖIF Working Paper Nr. 57. Vienna.
- Eggen, B. & Rupp, M. (2010). *Gleichgeschlechtliche Paare und ihre Kinder: Hintergrundinforationen zur Entwicklung gleichgeschlechtlicher Lebensformen in Deutschland* [Same sex couples and their children – Background information on the development of same sex unions in Germany]. In: Rupp, M. (ed.) *Gleichgeschlechtliche Lebensgemeinschaften mit und ohne Kinder*. Sonderheft der Zeitschrift für Familienforschung 2010, Opladen: Barbara Budrich.
- EGRIS (European Group for Integrated Social Research) (2001). *Misleading Trajectories: Transition Dilemmas of Young Adults in Europe*. Journal of Youth Studies 4 (1), 101-118.
- Elder, G.H. Jr. (1975). *Age Differentiation and the Life Course*. Annual Review of Sociology 1, 165-190.
- Elder, G. H. & O'Rand A. M. (1995). *Adult Lives in a Changing Society*. In: Cook, K. S., Fine, G. A. & House, J.S. (eds.) *Sociological Perspectives on Social Psychology*. Needham Heights, MA.: Allyn and Bacon.
- Ellingsæter, A. L. & Leira, A. (2006). *Politicising parenthood in Scandinavia: gender relations in welfare states*. Bristol: Policy Press.
- Engelhardt, H. & Prskawetz, A. (2004). *On the Changing Correlation*

between Fertility and Female Employment over Space and Time. European Journal of Population 20 (1), 35-62.

- Equal Tempora – Groupe de travail transanational 2 (2004). *Nouveaux services locaux pour la qualité de la vie.* Available from: *http://www. mtin.es/uafse_2000-2006/equal/ProductosEqual/archivos/AD_437_ producto_13.pdf* [accessed 11.4.2011].
- Esping-Andersen, G. (1990). *The Three Worlds of Welfare Capitalism.* Cambridge: Polity Press.
- Esping-Andersen, G. (1999). *Social Foundations of Postindustrial Economies.* Oxford: Oxford.
- European Commission (2007a). *Study on Poverty and Social Exclusion among Lone Parent Households.* European Communities. Available from: *http://www.apb.hu/download.php?ctag=download&docID=14778* [accessed 24.3.2010].
- European Commission (2007b). *Social Inclusion and Income Distribution in the European Union 2007.* Available from: *http://www.tarki.hu/adatbank-h/ kutjel/pdf/b251.pdf* [accessed 20.4.2011].
- European Commission (2008a). *Poverty and Social Exclusion in Rural Areas.* Available from: *http://ec.europa.eu/employment_social/spsi/docs/social_ inclusion/2008/rural_poverty_en.pdf* [accessed 24.3.2010].
- European Commission (2008b). *Partnership for more family-friendly living and working conditions. How to obtain support from the European Structural Funds.* Luxembourg: Office for Official Publications of the European Communities. Available from: *http://bookshop.europa. eu/eubookshop/publicationDetails.action?pubuid=608957&offset=0* [accessed 24.3.2010].
- European Commission (2009). *Moving Europe: EU research on migration and policy needs.* Directorate-General for Research. Socio-economic Sciences and Humanities. EUR 23859 EN. Brussels. Available from: *ftp://ftp.cordis.europa.eu/pub/fp7/ssh/docs/ssh_research_ migration_20090403_en.pdf* [accessed 31.3.2011].
- European Environment Agency (2009). *Ensuring Quality of Life in Europe's cities and Towns.* Available from: *http://www.eea.europa.eu/publications/ quality-of-life-in-Europes-cities-and-towns?&utm_campaign=quality- of-life-in-Europes-cities-and-towns&utm_medium=email&utm_ source=EEASubscriptions* [accessed 24.3.2010].
- European Foundation for the Improvement of Living and Working Conditions (2006). *Working Time and Work-Life-Balance in European Companies. Establishment Survey on Working Time 2004/5.* Dublin. Available from: *http://www.eurofound.europa.eu/pubdocs/2006/27/en/1/ ef0627en.pdf* [accessed 28.3.2011].

- Eurostat (2003). *Time use at different stages of life – Results from 13 European countries. Working Papers.* Luxembourg: Office for Official Publications of the European Communities. Available from: *http://epp. eurostat.ec.europa.eu/cache/ITY_OFFPUB/KS-CC-03-001/EN/KS-CC-03-001-EN.PDF* [accessed 17.1.2010].
- Eurostat (2004). *How Europeans spend their time – Everyday life of women and men. Data 1998-2002. Pocketbooks.* Luxembourg: Office for Official Publications of the European Communities. Available from *http:// www.unece.org/stats/gender/publications/MultiCountry/EUROSTAT/ HowEuropeansSpendTheirTime.pdf* [accessed 17.1.2010].
- Eurostat (2008a). *Internet access and use in the EU27 in 2008.* Available from: *http://europa.eu/rapid/pressReleasesAction.do?reference=STAT/08/ 169&format=HTML&aged=0&language=EN&guiLanguage=nl* [accessed 28.3.2011].
- Eurostat (2008b). *The Life of Women and Men in Europe – A Statistical Portrait.* Luxembourg: Office for Official Publications of the European Communities. Available from: *http://epp.eurostat.ec.europa.eu/cache/ ITY_OFFPUB/KS-80-07-135/EN/KS-80-07-135-EN.PDF* [accessed 28.3.2011].
- Eurostat (2009a). *Europe in Figures – Eurostat Yearbook 2009.* Available from: *http://epp.eurostat.ec.europa.eu/portal/page/portal/publications/ eurostat_yearbook* [accessed 28.3.2010].
- Eurostat (2009b). *European time use surveys – 2008 guidelines, Methodologies and working paper. Theme: Population and social conditions.* Luxembourg: Office for Official Publications of the European Communities. Available from: *http://epp.eurostat.ec.europa.eu/ cache/ITY_OFFPUB/KS-RA-08-014/EN/KS-RA-08-014-EN.PDF* [accessed 17.1.2010].
- Eurostat (2009c). *Reconciliation between Work, Private and Family Life in the European Union.* Luxembourg: Office for Official Publications of the European Communities. Available from: *http://epp.eurostat.ec.europa. eu/cache/ITY_OFFPUB/KS-78-09-908/EN/KS-78-09-908-EN.PDF* [accessed 17.1.2010].
- Eurostat (2009d). *Key Data on Education in Europe 2009.* Brussels: European Commission. Available from:*http://epp.eurostat.ec.europa. eu/cache/ITY_OFFPUB/KS-80-07-135/EN/KS-80-07-135-EN.PDF* [accessed17.1.2010].
- Eurostat (2010). *Datenübersicht.* Available from: *http://epp.eurostat. ec.europa.eu/portal/page/portal/statistics/search_database* [accessed 24.3.2010].
- Eydal, G. B. & Kröger, T. (2010). *Nordic family policies: constructing contexts for social work with families.* In: Forsberg, H. & Kröger, T. (eds.) *Social work*

and child welfare politics. Through Nordic lenses. Bristol: Policy Press, 11-27.
- Facchini C. & Rampazi M. (2009). *No Longer Young, Not Yet Old. Biographical Uncertainty in Late-Adult Temporality.* Time and Society 18 (2/3), 351-372.
- Fahey, T. & Smyth, E. (2004). *Do subjective indicators measure welfare?* European Societies 6 (1), 5-27.
- Federal Ministry for Family Affairs, Senior Citizens, Women and Youth. (2004). *Violence against men. Men's experiences of interpersonal violence in Germany. Results of the pilot study.* Berlin: Federal Ministry for Family Affairs, Senior Citizens, Women and Youth. Available from: *http://www.cahrv.uni-osnabrueck.de/conference/SummaryGermanMenstudy.pdf* [accessed 31.3.2011].
- Fenstermaker, S. & West, S. C. (2002). *Doing Gender, Doing Difference, Inequality, Power, and Institutional Change.* New York: Routledge.
- Ferrarini, T. (2003). *Parental Leave Institutions in Eighteen Post-war Welfare States.* Swedish Institute for Social Research, dissertation series No. 58. Stockholm: Swedish Institute for Social Research.
- Ferrarini, T. (2006). *Families, States and Labour Markets. Institutions, Causes and Consequences of Family Policy in Post-War Welfare States.* Cheltenham/Northampton: Edward Elgar.
- Ferree, M. M. (1990). *Beyond separate spheres: Feminism and family research.* Journal of Marriage and the Family 52, 866-884.
- Festy, P. (2006). *Legal Recognition of Same-Sex Couples in Europe.* Populations 61 (4), 417-453.
- Fisher, K. & Robinson, J. P. (2009). *Average Weekly Time Spent in 30 Basic Activities Across 17 Countries.* Social Indicators Research 93 (1), 249-54.
- Fisher, K., McCulloch, A. & Gershuny, J. (1999). *British Fathers and Children: A Report for Channel 4 Dispatches.* Institute for Social and Economic Research. Colchester: University of Essex. Available from: *http://www.iser.essex.ac.uk/press/doc/2000-12-15.pdf* [accessed 17.1.2010].
- Flichy, P. (1995). *Private communication. Dynamics of Modern Communication.* London: Sage, 152-171.
- Fondazione Giacomo Brodolini (2007). *Job Instability and Family Trends.* Available from: *http://ec.europa.eu/social/keyDocuments.jsp?type=1&p olicyArea=0&subCategory=0&country=0&year=2007&advSearchKey=Jo b+instability+and+family+trends&mode=advancedSubmit&langId=en* [accessed 17.1.2010].
- Förster, M. F. & Tóth, I. G. (2001). *Child Poverty and Family Transfers in the Czech Republic, Hungary and Poland.* Journal of European Social Policy 11 (4), 324-341.
- Förster, M. & d'Ercole M. M. (2005). *Income distribution and poverty in OECD*

countries in the second half of the 1990s. Directorate for Employment, Labour and Social Affairs OECD Social, Employment and Migration Working Papers. Paris: OECD.

- Fouarge, D. & Layte, R. (2005). *Welfare Regimes and Poverty Dynamics: The Duration and Recurrence of Poverty Spells in Europe*. Journal of Social Policy 34 (3), 407-426.
- Fouarge, D. & Muffels, R. (2009). *The Role of European Welfare States in Explaining Resources Deprivation (February, 09 2009)*. OSA Working Paper No. WP2003-22. Available from: *http://ssrn.com/abstract=1340192* [accessed 17.1.2010].
- Frejka, T., Sobotka, T., Hoem, J. M. & Toulemon, L. (2008). *Summary and general conclusions: Childbearing Trends and Policies in Europe*. Demographic Research 19 (2), 5-14.
- Furlong A. & Cartmel F. (1997). *Young People and Social Change: Individualization and Risk in Late Modernity*. Buckingham: Open University Press.
- Fuwa, M. (2004). *Macro-Level Gender Inequality and the Division of Household Labor in 22 Countries*. American Sociological Review 69 (6), 751-767.
- Fuwa, M. & Cohen, P. (2007). *Housework and social policy*. Social Science Research 36, 512-530.
- Gabel, S. G. & Kamerman, S. B. (2006). *Investing in Children: Public Commitment in Twenty-one Industrialized Countries*. Social Service Review 80 (2), 239-263.
- Gaspar, S. & Klinke, M. (2009). *Household division of labour among European mixed partnerships. e-WORKING PAPER*. Lisbon: CIES. Available from: *http://www.cies.iscte.pt/destaques/documents/CIES-WP78GaspareKlinke.pdf* [accessed 17.1.2010].
- Gauthier, A. H. (1996). *The State and the Family. A Comparative Analysis of Family Policies in Industrialized Countries*. Oxford: Clarendon Press.
- Gauthier, A. H. (1999). *The sources and methods of comparative family policy research*. Comparative Social Research 18, 31-56.
- Gauthier, A. H. (2000). *Public Policies Affecting Fertility and Families in Europe: A Survey of the 15 Member States*. Unpublished paper for the European Observatory on Family Matters. Available from: *http://www.iesf.es/fot/Policies-affecting-fertility-2000.pdf* [accessed 28.3.2011].
- Gauthier, A. H. (2002). *Family Policies in Industrialized Countries: Is There Convergence?* Population 3 (57), 447-474.
- Gauthier, A. H. (2005). *Trends in policies for family-friendly societies*. In: Macura, M., MacDonald, A. & Haug, W. (eds.) *The New Demographic Regime: Population Challenges and Policy Responses*. New York/Geneva:

United Nations, 95-100.

- Gauthier, A. H., Smeedeng, T. M. & Furstenberg Jr., F. (2004). *Are Parents Investing Less Time in Children? Trends in Selected Industrialized Countries.* Population and Development Review 30 (4), 647-71.
- Geissler, B. & Pfau-Effinger, B. (2005). *Change in European care arrangements.* In: Pfau-Effinger, B. & Geissler, B. (eds.) *Care and Social Integration in European Societies.* Bristol: Policy Press, 3-19.
- Geist, C. (2005). *The Welfare State and the Home: Regime Differences in the Domestic Division of Labour.* European Sociological Review 21(1), 23-41.
- Gerhard, U. & Weckwert, A. (2001). *Thematic Network: "Working and Mothering: Social Practices and Social Policies" Final Report.* TSER Programme of the European Commission Area III: Research into Social Integration and Social Exclusion in Europe. Available from: *http://ec.europa.eu/research/social-sciences/projects/117_en.html* [accessed 22.1.2010].
- Gerhard, U., Knijn, T. & Weckwert, A. (eds.) (2005). *Working Mothers in Europe. A Comparison of Policies and Practices.* Cheltenham/Northampton: Edward Elgar.
- Gerlach, I. (2010). *Familienpolitik.* Wiesbaden: VS Verlag.
- Giddens, A. (1990). *The Consequences of Modernity.* Cambridge: Polity Press.
- Giorgi, L. (ed.) (2003). *The housing dimension of welfare reform – final report.* Available from: *http://www.iccr-international.org/impact/docs/final-report.pdf* [accessed 17.1.2010].
- Giullari, S. & Lewis, J. (2005). *The adult worker model family, gender equality and care: the search for new policy principles, and the possibilities and problems of a capabilities approach.* Social policy and development programme paper No. 19. Geneva: United Nations Research Institute for Social Development.
- González, M. J., Jurado-Guerrero, T. & Naldini, M. (2009). *What Made Him change: An Individual and National Anlysis of Men's Participation in 26 Countries.* DemoSoc Working Paper, Paper Number 2009/30, Barcelona: Department of Political Social Studies, Universitat Pompeu Fabra. Available from: *http://www.recercat.net/bitstream/2072/41841/1/DEMOSOC30[1].pdf* [accessed 17.1.2010].
- Gornick, J. C., Meyers, M. K., & Ross, K. E. (1997). *Supporting the Employment of Mothers: Policy Variation Across Fourteen Welfare States.* Journal of European Social Policy 7 (1), 45-70.
- Gupta, N., Smith, N., & Verner, M. (2008). *Perspective Article: The impact of Nordic countries' family friendly policies on employment, wages, and children.* Rev Econ Household 6, 65-89.

- Haas, B. (2005). *The Work-Care Balance: Is it Possible to Identify Typologies for Cross-National Comparisons?* Current Sociology 53 (3), 487-508.
- Haas, B., Steiber, N., Hartel, M. & Wallace, C. (2006). *Household employment patterns in an enlarged European Union*. Work, Employment & Society 20 (4), 751-771.
- Haataja, A. (2005). *Family leave and employment in the EU: transition of working mothers in and out of employment*. In: Pfau-Effinger, B. & Geissler, B. (eds.) *Care and Social Integration in European Societies*. Bristol: Policy Press, 255-278.
- Hagestad, G. & Herlofson, K (2007). *Micro and macro perspectives on intergenerational relations and transfers in Europe*. Report from United Nations Expert Group Meeting on Social and Economic Implications of changing Population Age Structures. New York: United Nations/ Department of Economic and Social Affairs, 339-357.
- Hakim, C. (2003). *A New Approach to Explaining Fertility Patterns: Preference Theory*. Population and Development Review 29 (3), 349-374.
- Hank, K. & Buber, I. (2009). *Grandparents caring for their grandchildren. Findings from the 2004 Survey of Health, Ageing, and Retirement in Europe*. Journal of Family Issues 30, 53-73.
- Hantrais, L. (2004). *Family Policy Matters. Responding to family change in Europe*. Bristol: Policy Press.
- Hasebrink, U., Livingstone, S., Haddon, L. & Ólafsson, K. (2009). *Comparing children's online opportunities and risks across Europe: Cross-national comparisons for EU Kids Online, 2nd ed.* London: LSE.
- Haug, W., Compton, P. & Courbage, Y. (eds.) (2002). *The Demographic Characteristics of Immigrant Populations*. Strasbourg: Council of Europe.
- Hiller, D. (1984). *Power dependence and division of family work*. Sex Roles 10, 1003-1019.
- Hobcraft J. & Kiernan K. (1995). *Becoming a Parent in Europe*. In: *Revolution in European Population Vol. 1*. European Population Conference Milano 1995. Milano: Franco Angelipp, 27-65.
- Hook, J. L. (2006). *Care in Context: Men's Unpaid Work in 20 Countries, 1965-2003*. American Sociological Review 71, 639-660.
- Ministry of Infrastructure of the Italian Republic & Federcasa Italian Housing Federation (2006). *Housing Statistics in the European Union 2005/2006*. Available from: *http://www.federcasa.it/news/housing_statistics/Report_housing_statistics_2005_2006.pdf* [accessed 18.3.2010].
- ILGA (2010). *Fourth ILGA-Asia Conference in Indonesia in March 2010*.
- Inglehart, R. (1990). *Culture Shift in Advanced Industrial Society*. Princeton: Princeton University Press.
- Ito, M., Baumer, S., Bittanti, M., Boyd, D., Cody, R., Herr, B., *et al.* (2010).

Hanging Out, Messing Around, Geeking Out: Living and Learning with New Media. Cambridge: MIT Press.

- Jansen, E. (2010). *Regenbogenfamilien. Alltäglich und doch anders. Beratungsführer für lesbische Mütter, schwule Väter und familienbezogenes Fachpersonal*. Available from: *http://www.family.lsvd. de/beratungsfuehrer/* [accessed 18.03.2010].
- Jenkins, S. & Schluter, C. (2001). *Why are Child Poverty Rates Higher in Britain than in Germany? A Longitudinal Perspective*. ISER Working Paper 2001-16.
- Johnsson-Smaragdi, U. (2002). *A Swedish perspective on media access and use*. Paper presented at the New Generations – New Media Conference.
- Johnsson-Smaragdi, U., D'Haenens, L., Frotz, F. & Hasebrink, U. (1998). *Patterns of old and new media use among young people in Flanders, Germany and Sweden*. European Journal of Communication 13 (4), 479-501.
- Jönsson, I. & Letablier, M.-T. (2005). *Caring for children: the logics of public action*. In: Gerhart, U., Knijn, T. & Weckwert, A. (eds.) *Working Mothers in Europe. A comparison of Policies and Practices*. Cheltenham/Northampton: Edward Elgar Publishing.
- Kalmijn, M. & Rigt-Poortman, A. (2006). *His or Her Divorce? The Gendered Nature of Divorce and its Determinants*. European Sociological Review 22 (2), 201-214.
- Kan, M.Y. (2008). *Does gender trump money? Housework hours of husbands and wives in Britain*. Work Employment and Society 22, 45-66.
- Kangas, O. & Rostgaard, T. (2007). *Preferences or institutions? Work-family life opportunities in seven European countries*. Journal of European Social Policy 17 (3), 240-256.
- Kapella, O., Rille-Pfeiffer, C., Rupp, M. & Schneider, N. F. (eds.) (2009). *Die Vielfalt der Familie: Tagungsband zum 3*. Europäischen Fachkongress Familienforhschung. Opladen: Barbara Budrich.
- Kaufmann, F. (2000). *Politics and Policies towards the Family in Europe: A Framework and an Inquiry into their Differences and Convergences*. In: Kaufmann, F-X, Kuijsten, A., Schulze, H-J & Strohmeier, K. P. (eds.) *Family Life and Family Policies in Europe 2: Problems and Issues in Comparative Perspective*. Oxford: Clarendon Press, 419-490.
- Kiernan, K. (2003). *Changing European families: Trends and issues*. In: Scott, J., J. Treas & M. Richards (eds.) *Blackwell Companion to Sociology of the Family*. Oxford: Blackwell Publishing, 17-33.
- Kiernan, K. (2004). *Unmarried Cohabitation and Parenthood in Britain and Europe*. Law and Policy 26 (1), Oxford: Blackwell Publishing: 33-55.
- King, R., Thomson, M., Fielding, T. & Warnes, T. (2006). *Time, Generations*

and Gender in Migration and Settlement. In: Penninx, R., Berger, M. & Kraal, K. (eds.) *The Dynamics of International Migration and Settlement in Europe – a State of Art.* Amsterdam: AUP (Imiscoe Joint Studies), 233-267.

- King, R. *et al.* (2004). *Gender, Age and Generations – State of art report.* Working Paper nº5, IMISCOE.
- Kirwil, L. (2009). *Parental mediation of children's internet use in different European countries.* Journal of Children and Media 3 (4), 394-409.
- Kleinhans, R., Priemus, H. & Engbersen, G. (2007). *Understanding Social Capital in Recently Restructured Urban Neighbourhoods: Two Case Studies in Rotterdam.* Urban Studies 44 (5/6), 1069-1091.
- Knickerbocker, L., Heyman, R. E., Smith Slep, A. M., Jouriles, E. N., McDonald, R. (2007). *Co-occurrence of child and partner maltreatment.* European Psychologist 12 (1), 36-44.
- Knijn, T. & Smit, A. (2009). *The Relationship Between Family and Work: Tensions, Paradigms and Directives. Working paper on the reconciliation of work and welfare in Europe.* Edinburgh: RECWOWE Publication, Dissemination and Dialogue Centre. Available from: *http://www. socialpolicy.ed.ac.uk/__data/assets/pdf_file/0018/32319/REC-WP_1109_ Knijn_Smit.pdf* [accessed 28.3.2010].
- Kofman, E. (2004). *Family-related migration: a critical review of European studies.* Journal of Ethnic and Migration Studies 30 (2), 243-62.
- Kofman, E. & Meetoo, V. (2008). *Family Migration.* In: International Organization for Migration (ed.) *World Migration Report 2008,* Geneva: IOM, 151-172.
- Kohler, H. -P., Billari, F. C., & Ortega, J. A. (2006). *Low fertility in Europe: Causes, implications and policy options.* In: Harris, J. A. (ed.) *The Baby Bust: Who will do the Work? Who Will Pay the Taxes?* Lanham, MD: Rowman & Littlefield Publishers.
- Kovacheva, S., Kabaivanov, S. & Andreev, T. (2007). *Comparative Report on the Institutional Context of Work and Quality of Life.* Available from: *http://www.projectquality.org/files/D3.2_comparative%20report_ FINAL_0.pdf* [accessed 28.3.2010].
- Krahé, B., Schütze, S., Fritsche, I. & Waizenhöfer, E. (2000). *The prevalence of sexual aggression and victimization among homosexual men.* The Journal of Sex Research 37 (2), 142-150.
- Krahé, B., Scheinberger-Olwig, R. & Bieneck, S. (2003). *Men's reports of non-consensual sexual interactions with women: prevalence and impact.* Archives of Sexual Behavior 32 (2), 165-175.
- Kraler, A. & Kofman, E. (2009). *Family Migration in Europe: Policies vs. Reality.* Imiscoe Policy Brief, 16.
- Kröger, T. (1997). *The Dilemma of Municipalities: Scandinavian Approaches*

to Child Daycare Provision. Journal of Social Policy 26 (4), 485-507.
- Kröger, T. (ed.) (2004). *New kinds of families, new kinds of social care. Families, Work and Social Care in Europe. A qualitative study of care arrangements in Finland, France, Italy, Portugal and the UK.* SOCCARE Project. Final report. European Commission 5th Framework Programme.
- Kröger, T. & Sipilä, J. (eds.) (2005). *Overstretched. European Families up Against the Demands of Work and Care.* Oxford: Blackwell.
- Lavee, Y. & Katz, R. (2002). *Division of labor, perceived fairness, and marital quality: The effect of gender ideology.* Journal of Marriage and Family 64, 27-39.
- Layte, R. & Fouarge, D. (2004). *The dynamics of income poverty.* In: Berthoud, R. & Iacovou, M. (eds.) *Social Europe. Living Standards and Welfare States.* Cheltenham: Edward Elgar, 202-224.
- Leccardi C. (2009a). *Sociologie del tempo. Soggetti e tempo nella società dell'accelerazione.* [Sociologies of Time. Subjectivities and Time in the 'Acceleration Society'], Roma/Bari: Laterza.
- Leccardi C. (2009b). *Le trasformazioni della morale sessuale e dei rapporti fra i generi* [Trasformations of Sexual Morality and Gender Relations]. In: Sciolla L. (ed.) *Processi e trasformazioni sociali. La società europea dagli anni Sessanta a oggi* [Social Processes and Transformations. European Society from the Sixties to Today]. Roma/Bari: Laterza.
- Leccardi, C. & Ruspini, E. (eds.) (2006). *A New Youth? Young People, Generations and Family Life.* Aldershot: Ashgate.
- Leira, A. & Saraceno, C. (2008). *Childhood: Changing Contexts.* In: Leira, A. & Saraceno, C. (eds.) *Childhood: Changing Contexts.* Comparative Social Research 25, Bingley: Emerald JAI, 1-24.
- Lelkes, O. & Zólyomi, E. (2008). *Poverty Across Europe: The Latest Evidence Using the EU-SILC Survey.* European Centre, Policy Brief, October 2008. Available from: *http://www.euro.centre.org/data/1226583242_93408.pdf* [accessed 31.3.2011].
- Lesthaeghe, R. & van de Kaa, D. (1986). *Twee demografische transities?* In: Lesthaeghe, R. & van de Kaa, D. (eds.) *Bevolking: groei en krimp, Mens en Maatschappij, 1986 book supplement.* Deventer: Van Loghum Slaterus, 9-24.
- Levin, I. (2004). *Living Apart Together: A New Family Form.* Current Sociology 52, 223-240.
- Lewis, J., Campbell, M. & Huerta, C. (2008). *Patterns of paid and unpaid work in Western Europe: gender, commodification, preferences and the implications for policy.* Journal of European Social Policy 18 (1), 21-37.
- Lewis, J. & Ostner, I. (1994). *Gender and the Evolution of European Social*

Policies. ZeS-Arbeitspapier Nr. 4. Bremen: Zentrum für Sozialpolitik.

- Lister, R. (1994). *'She Has Other Duties': Women, Citizenship and Social Security*. In: Baldwin, S. & Falkingham, J. (eds.) *Social Security and Social Change: New Challenges*. Hemel Hempstead: Harvester Wheatsheaf.
- Lister, R., Williams, F., Anttonen, A., Bussemaker, J., Gerhard, U., Heinen, J., Johansson, S., Leira, A., Siim, B., Tobio, C. & Gavanas, A. (2007). *Gendering Citizenship in Western Europe: New Challenges for Citizenship Research in a Cross-national Context*. Bristol: Policy Press.
- Livingstone, S. (2002).*Young people and new media: Childhood and the changing media environment*. London: Sage Publications.
- Livingstone, S. (2009). *On the mediation of everything, ICA Presidential address*. Journal of Communication 59 (1), 1-18.
- Livingstone, S. & Haddon, L. (2009). *Introduction*. In: Livingstone, S. & Haddon, L. (eds.) *Kids online*. Bristol: The Policy Press, 1-18.
- Lohmann, H., Peter, F. H., Rostgaard, T. & Spiess, K. C. (2009). *Towards a Framework for assessing family policies in the EU*. Final Report. OECD Social, Employment and Migration Working Papers 88. Paris: OECD.
- Lutz, W. & Scherbov, S. (2006). *Future demographic change in Europe: the contribution of migration*. In: Papademetriou, D. G. (ed.) *Europe and Its Immigrants in the 21st Century – A New Deal or a Continuing Dialogue of the Deaf?* Washington/Lisbon, MPI/FLAD, 207-222.
- Lüscher, K. (1999). *Familienberichte: Aufgabe, Probleme und Lösungsversuche der Sozialberichterstattung über die Familie. Forschungsschwerpunkt "Gesellschaft und Familie", Arbeitspapier Nr. 32*. Konstanz: Universität Konstanz. Available from: *http://www.ub.uni-konstanz.de/kops/volltexte/2002/769/pdf/AP32.pdf* [accessed 31.3.2011].
- Maclennan, D. (2008). *Trunks, Tails, and Elephants: Modernizing Housing Policies*. European Journal of Housing Policy 8 (4), 423-440.
- MacRae S. (2003). *Choice and Constraints in Mothers Employment Careers: MacRae Replies to Hakim*. British Journal of Sociology 54 (4), 585-592.
- Mahon, R. (2002). *Child Care: Toward What Kind of "Social Europe"?* Social Politics 9 (3). 343-379.
- Margherita, A., O'Dorchai, S. & Bosch, J. (2009). *Reconciliation between work, private and family life in the European Union*. Eurostat, Luxembourg: Office for Official Publications of the European Communities. Available from *http://epp.eurostat.ec.europa.eu/portal/page/portal/product_details/publication?p_product_code=KS-78-09-908* [accessed 17.1.2010].
- Martin, C. & Théry, I. (2001). *The PACS and marriage and cohabitation in France*. International Journal of Law, Policy and the Family 15, 135-159.
- Martinez, M. & Schröttle, M. (2006). *State of European research on the*

prevalence of interpersonal violence and its impact on health and human rights. Valencia/Bielefeld: CAHRV.
- McKeever M. & Wolfinger N. H. (2001). *Reexamining the Economic Costs of Marital Disruption for Women.* Social Science Quarterly 82 (1) (March), 202-217.
- Mestheneous, E. & Triantafillou, J. on behalf of the EUROFAMCARE group (2005). *Supporting Family Carers of Older People in Europe – the Pan-European Background.* Available from: *http://www.uke.de/extern/ eurofamcare/publikationen.php* [accessed 10.2.2010].
- Mingione, E., Oberti, M. (2003). *The Struggle Against Social Exclusion at the Local Level. Diversity and Convergence in European Cities.* European Journal of Spatial Development. Available from: *http://www.nordregio.se/ EJSD/refereed1.pdf* [accessed 17.1.2010].
- Misra, J., Moller, S. & Budig, M. J. (2007). *Work Family Policies and Poverty for Partnered and Single Women in Europe and North America.* Gender & Society 21, 804.
- Morgan, K. J. (2008). *Caring time policies in Western Europe: trends and implications.* Prepared for delivery at the 2008 Annual Meeting of the American Political Science Association, Boston MA., August 28-31, 2008: Copyright by the American Political Science Association.
- Moss, P. & Deven, F. (2009). *Country Notes 2009: introduction and main findings.* In: Moss, P. (ed.) *International Review of Leave Policies and Related Research 2009, Employment Relations Research Series No. 102.* London: Department for Business, Enterprise & Regulatory Reform, 77-99.
- Motel-Klingebiel, A., Tesch-Roemer, C. & Von Nondratowitz, H-J. (2005). *Welfare states do not crowd out the family: evidence for mixed responsibility from comparative analyses.* Ageing and Society 25 (6), 863-882.
- Musterd, S. & Murie, A. (eds.) (2002). *The Spatial Dimensions of Urban Social Exclusion and Integration.* Amsterdam: URBEX. Available from: *http:// ec.europa.eu/research/social-sciences/pdf/finalreport/soe2ct983072-final-report.pdf* [accessed 17.1.2010].
- Nave-Herz, R. (2004). *Ehe- und Familiensoziologie. Eine Einführung in Geschichte, theoretische Ansätze und empirische Befunde.* Weinheim/ München: Juventa.
- Nielsen, T. S. & Hansen, K. B. (2007). *Do Green Areas Affect Health? Results from a Danish Survey on the Use of Green Areas and Health Indicators.* Health & Place 13. 839-850.
- Nordenmark, M. & Nyman, C. (2003). *Fair or unfair? Perceived fairness of household division of labour and gender equality among women and men – the Swedish case.* European Journal of Women's Studies 10, 181-209.
- Norris P. & Inglehart R. (2004). *Secular and Sacred: Religion and Politics*

Worldwide. Cambridge: Cambridge University Press.
- O'Brien, M. (2009). *Fathers, Parental Leave Policies, and Infant Quality of Life: International Perspectives and Policy Impact.* The Annals of the American Academy of Political and Social Science 624 (1), 190-213.
- O'Connor, J. S. (1996). *Gendering welfare state regimes.* Current Sociology 44 (2), 1-130.
- OECD (2006). *OECD Family database.* Available from: *http://www.oecd. org/els/social/family/database* [accessed 17.1.2010].
- OECD (2007a). *Babies and Bosses. Reconciling Work and Family Life. A Synthesis of Findings for OECD Countries.* Paris: OECD. Available from: *http:// puck.sourceoecd.org/vl=2865013/cl=19/nw=1/rpsv/~6682/v2007n20/s5/ p82* [accessed 17.1.2010].
- OECD (2007b). *OECD Regions at a Glance.* Paris: OECD. Available from: *http://puck.sourceoecd.org/vl=2865013/cl=19/nw=1/rpsv/~6685/ v2007n19/s28/p254.*
- OECD (2008). *Taxing Wages, 2006-2007.* Paris: OECD.
- OECD (2009a). *OECD Family Data Base.* Available from: *http://www.oecd. org/els/social/family/database* [accessed 15.01.2010].
- OECD (2009b). *Doing Better for Children.* Available from: *http://puck. sourceoecd.org/vl=2865013/cl=19/nw=1/rpsv/~6670/v2009n14/s1/p1l* [accessed 15.1.2010].
- OECD Factbook (2009). Available from: *http://dx.doi.org/10.1787/factbook- 2010-en.*
- Orloff, A. S. (1993). *Gender and the social rights of citizenship: the comparative analysis of gender relations and welfare states.* American Sociological Review, 58, 303-328.
- Ostner, I. & Schmitt, C. (eds.) (2008). *Family Policies in the Context of Family Change. The Nordic Countries in Comparative Perspective.* Wiesbaden: VS Verlag.
- Pa Ke Shon, J.-L. (2007). *Residents' Perceptions of their Neighbourhood: Disentangling Dissatisfaction, a French Survey.* Urban Studies 44 (11), 2231-2268.
- Paris, C. (2007). *International Perspectives on Planning and Affordable Housing.* Housing Studies 22(1), 1-9.
- Parkes, A., Kearns, A. & Atkinson, R. (2002). *What Makes People Dissatisfied with their Neighbourhoods?* Urban Studies 39 (13), 2413-2438.
- Pavolini, E. & Ranci, C. (2008). *Restructuring the welfare state: reforms in long-term care in Western European countries.* Journal of European Social Policy 18 (3), 246-259.
- Pettit, B. & Hook, J. L. (2005). *The Structure of Women's Employment in Comparative Perspective.* Social Forces 84 (2), 779-801.

- Peuckert, R. (2008). *Familienformen im sozialen Wandel.* Wiesbaden: VS Verlag.
- Philipov, D., Thevenon, O., Klobas, J., Bernardi, L. & Liefbroer, A. C. (2009). *Reproductive Decision-Making in a Macro-Micro Perspective (REPRO) State-of-the-art review.* European Demographic Research Papers. Available from: *http://www.oeaw.ac.at/vid/repro/assets/docs/ed-researc hpaper2009-1.pdf* [accessed 17.1.2010].
- Phizacklea, A. (1998). *Migration and globalisation: a feminist perspective.* In: Koser, K. & Lutz, H. (ed.) *The New Migration in Europe.* London: Macmillan, 21-38.
- Phoenix, A. (1991). *Young mothers.* Cambridge: Polity.
- Pittman, J. F. & Blanchard, D. (1996). *The effects of work history and timing of marriage on the division of household labour: a life-course perspective.* Journal of Marriage and the Family 58, 78-90.
- Plantenga, J. & Remery, C. (2005). *Reconciliation of work and private life: A comparative review of thirty European countries.* Brussels: European Communities.
- Plantenga, J., Remery, C., Siegel, M. & Sementini, L. (2008). *Childcare services in 25 European Union Member States: The Barcelona targets revisited.* In: Leira, A. & Saraceno, C. (eds.) *Childhood: Changing Contexts.* Comparative Social Research 25. Bingley: Emerald JAI Press, 27-53.
- Plantenga, J. & Remery, C. (2009). *The provision of childcare services. A comparative review of 30 European countries.* European Commission's Expert Group on Gender and Employment Issues (EGGE). European Commission Directorate-General for Employment, Social Affairs and Equal opportunities. Luxembourg: Office for Official Publications of the European Communities.
- Priemus, H. & Boelhouwer, P. (1999). *Social Housing Finance in Europe: Trends and Opportunities.* Urban Studies 36 (4), 633-645.
- Priemus, H. & Dieleman, F. (1999). *Social Housing Finance in the European Union: Developments and Prospects.* Urban Studies 36 (4), 623-631.
- Priemus, H. & Dieleman, F. (2002). *Social Housing Policy in the European Union: Past, Present and Perspectives.* Urban Studies 39 (2),191-200.
- Priemus, H. & Kemp, P. A. (2004). *The Present and Future of Income-Related Housing Support: Debates in Britain and the Netherlands.* Housing Studies, 19 (4), 653-668.
- Pronovost G. (1992). *Générations, cycles de vie et univers culturels.* Loisir et societé 15 (2), 437-460.
- Prskawetz, A., Vikat, A., Philipov, D. & Engelhardt, H. (2003). *Pathway to Stepfamiliy Formation in Europe: Results from the FFS.* Demographic Research 8, 107-150. Available from: *http://www.demographic-research. org/volumes/vol8/5/8-5.pdf* [accessed 23.03.2010].

- Quintano, C. & D'Agostino, A. (2006). *Studying inequality in income distribution of single-person households in four developed countries.* Review of Income and Wealth 52 (4), 525-546.
- Ramos X. (2005). *Domestic work time and gender differentials in Great Britain 1992-1998: what do 'new' men look like?* International Journal of Manpower 26 (3), 265-295.
- Rindfuss, R. R., Guzzo, K. B & Morgan, S. P. (2003). *The Changing Institutional Context of Low Fertility.* Population Research and Policy Review 22 (4), 411-438.
- Roberts, D., Foehr, F., Rideout, V., & Brodie, M. (eds.) (1999). *Kids and Media @t the New Millenium.* Menlo Park: Henry J. Kaiser Family Foundation.
- Roberts, D., Foehr, U. & Rideout, V. (2005). *Generation M: Media in the lives of 8-18 year olds.* Menlo Park, CA: Kaiser Family Foundation.
- Rost, H. (2009). *Familienhaushalte im europäischen Vergleich.* In: Mühling, T. & Rost, H. (eds.) *Ifb-Familienreport Bayern 2009.* Schwerpunkt: Familie in Europa. Bamberg: Bayrisches Staatsministerium für Arbeit und Sozialordnung, Familie und Frauen, 6-32.
- Rupp, M. (ed.) (2009). *Die Lebenssituation von Kinder in gleichgeschlechtlichen Lebenspartnerschaften.* Köln: Bundesanzeiger-Verlags-Gesellschaft.
- Rüling, A. & Kassner, K. (2007). *Familienpolitik aus der Gleichstellungsperspektive. Ein europäischer Vergleich.* Berlin: Friedrich-Ebert-Stiftung.
- Sainsbury, D. & Morissens, A. (2002). *European Anti-Poverty Policies in the 1990s: Toward a Common Safety Net?* Luxembourg Income Study Working Paper No. 307. Available from: *http://ssrn.com/abstract=324881* [accessed 17.1.2010].
- Saraceno C. (1983). *Il tempo nella costruzione dei ruoli e delle identità* [Time and the Construction of Roles and Identities]. Rassegna Italiana di Sociologia 24 (1), 105-130.
- Saraceno, C., Olagnero, M. & Torrioni, P. (2005). *First European Quality of Life Survey. Families, Work and Social Networks.* European Foundation for Improving Working and Living Conditions. Luxembourg: Luxembourg Office for Official Publications.
- Saraceno C. & Naldini, M. (2007). *Sociologia della famiglia* [Sociology of the Family]. Bologna: il Mulino.
- Saraceno, C. & Keck, W. (2008). *The institutional framework of intergenerational family obligations in Europe: A conceptual and methodological overview.* The first deliverable of WP1 of the Multilinks project funded by the European Commission under the 7th framework programme. Available from: *http://www.multilinks-project.eu* [accessed

15.12.2009].

- Savage, M., Warde, A. & Devine, F. (2005). *Capitals, assets and resources: some critical issues.* British Journal of Sociology 56 (1), 31-47.
- Saxonberg, S. & Sirovátka, T. (2006). *Failing Family Policy in Post-Communist Central Europe.* Journal of Comparative Policy Analysis 8 (2), 185-202.
- Scharle, Á. (2007). *The effect of welfare provisions on female labour supply in Central and Eastern Europe.* Journal of Comparative Policy Analysis: Research and Practice 9 (2), 157-174. Available from: *http://www. informaworld.com/smpp /title~db=all~content=t713672306~tab=issuesli st~branches=9 - v9* [accessed 17.1.2010].
- Seybert, H. (2007). *Gender differences in the use of computers and the internet.* Luxembourg: Eurostat.
- Shelton, B. A. & John, D. (1996). *The division of household labour.* Annual Review of Sociology 22, 299-322.
- Silverstone, R. (2006). *Domesticating domestication: reflections on the life of a concept.* In: Berker, T., Hartmann, M., Punie, Y. & Ward, K. J. (eds.) *The domestication of media and technology.* Maidenhead: Open University Press, 229-48.
- Simonazzi, A. (2009). *Care regimes and national employment models.* Cambridge Journal of Economics 33 (2), 211-232.
- Sipilä, J. & Kröger, T. (2004). *Editorial Introduction. European Families Stretched between the Demands of Work and Care.* Social Policy & Administration 38 (6), 557-564.
- Somarriba, N. & Pena, B. (2009). *Synthetic Indicators of Quality of Life in Europe.* Social Indicators Research 94, 115-133.
- Spéder, Z. (2005). *Diversity of Family Structure in Europe – Selected characteristics of partnerships, childhood, parenting, and economic wellbeing across Europe around the millennium.* Presentation paper. Available from: *http://www.iussp.org/France2005/jpe/fichiers/speder.pdf* [accessed 25/03/2010].
- Spencer, S. & Cooper, B. (2007). *Social Integration of Migrants in Europe: A Review of the European Literature 2000-2006.* OECD Literature Review. Oxford: COMPAS.
- Stauber, B. & du Bois-Reymond, M. (2006). *Familienbeziehungen im Kontext verlängerter Übergänge. Eine intergenerative Studie aus neun europäischen Ländern* [Family relationships within transitions. An intergenerational study in nine European countries]. ZSE - Zeitschrift für Soziologie der Erziehung und Sozialisation 26 (2), 206-221.
- Steinbach, A. (2008). *Stieffamilien in Deutschland. Ergebnisse des "Generation and Gender Surveys 2005".* Zeitschrift für

Bevölkerungswissenschaft 33 (2), 153-180.

- Stier, H., Lewin-Epstein N., & Braun, M. (2001). *Welfare Regimes, Family-Supportive Policies, and Women's Employment along the Life-Course.* The American Journal of Sociology 106 (6),1731-60.
- Stier, H. & Lewin-Epstein, N. (2007). *Policy effects on the division of housework.* Journal of Comparative Policy Analysis 9 (3), 235-259.
- Stiglitz, E., Sen, A. & Fitoussi, J.-P. (2009). *Report by the Commission on the Measurement of Economic Performance and Social Progress.* Available from:*http://media.ft.com/cms/f3b4c24a-a141-11de-a88d-00144feabdc0.pdf* [accessed 17.1.2010].
- Sugiyama, T., Ward Thompson, C. & Alves, S. (2009). *Associations Between Neighborhood Open Space Attributes and Quality of Life for Older People in Britain.* Environment and Behavior 41(1), 3-21.
- Szeleva, D., & Polakowski, M. P. (2008). *Who cares? Changing patterns of childcare in Central and Eastern Europe.* Journal of European Social Policy 18, 115-131.
- TÁRKI Social Research Institute Applica (2010): Child poverty and child well-being in the European Union. Available from: http://www.tarki.hu/en/research/childpoverty/report/child_poverty_final%20report_jan2010.pdf [accessed 21.6.2011].
- Théry I. (1993). *Le démariage.* Paris: Jacob.
- Thomson, E. (2004). *Step-families and Childbearing Desires in Europe.* Demographic Research, Special Collection 3, 117-134. Available from *http://www.demographic-research.org/special/3/5/s3-5.pdf* [accessed 25.3.2010].
- Triantafillou, J. & Mestheneos, E. (2006). *Services for Supporting Family Carers of Elderly People in Europe: Characteristics, Coverage and Usage. Summary of main findings from EUROFAMCARE.* Hamburg: EUROFAMCARE. Available from: *http://www.uke.de/extern/eurofamcare/documents/overview_teusure.pdf* [accessed 28.3.2010].
- Trumm, A. & Ainsaar, M. (2009). *The welfare system of Estonia: past, present and future.* In: Schubert, K., Hegelich, S. & Bazant, U. (eds.) *The Handbook of European Welfare Systems.* London: Routledge 153-170.
- Turner, B. & Elsinga, M. (2005). *Housing Allowances: Finding a Balance Between Social Justice and Market Incentives.* European Journal of Housing Policy 5 (2), 103-109.
- United Nations (2005). *Generations & Gender Programme: Survey Instruments.* New York/Geneva: UN.
- Uunk, W. (2004). *The Economic Consequences of Divorce for Women in the European Union: The Impact of Welfare State Arrangements.* European Journal of Population 20, 251-285.

- Uunk, W., Kalmijn, M. & Muffels, R. (2005). *The Impact of Young Children on Women's Labour Supply. A Reassessment of Institutional Effects in Europe.* Acta Sociologica 48 (1), 41-62.
- Valletta, R. G. (2006). *The ins and outs of poverty in advanced economies: government policy and poverty dynamics in Canada, Germany, Great Britain, and the United States.* Review of Income and Wealth 52 (2), 261-284.
- Van der Lippe, T. (2001). *The Effect of Individual and Institutional Constraints on Hours of Paid Work of Women: An International Comparison.* In: Van der Lippe, T. & Van Dijk, L. (eds.) *Women's Employment in A Comparative Perspective.* New York: Aldine, 221-243.
- Van der Lippe, T., & Van Dijk, L. (2002). *Comparative Research on Women's Employment.* Annual Review of Sociology 28, 221-241.
- Van Dijk, J., Van Kesteren, J. & Smit, P. (2007). *Criminal Victimization in International Perspective.* Available from: *http://rechten.uvt.nl/icvs/pdffiles/ICVS2004_05.pdf* [accessed 1.4.2011].
- Van Rompaey, V. & Roe, K. (2001). *The Home as a Multimedia Environment.* Communications 26 (4), 351-370.
- Vannoy, D., Rimashevskaya, N., Cubbins, L., Malysheva, M., Meshterkina, E. & Pisklakova M. (1999). *Marriages in Russia: Couples during the economic transition.* Westport, CT: Praeger.
- Velleman, R., Templeton, L., Reuber, D., Klein, M. & Moesgen, D. (2008). *Domestic abuse experienced by young people living in families with alcohol problems: results from a cross-European study.* Child Abuse Review 17, 387-409.
- Verbrugge, L.M., Gruber-Baldini, A.L. & Fozard, J.L. (1996). *Age Differences and Age Changes in Activity.* Journal of Gerontology, Social Sciences, 51(1), 30-41.
- Verschraegen, B. (2009). *Rechtliche Absicherung der Lebens- und Familienformen – Ein europäischer Überblick.* In: Kapella, O., Rille-Pfeiffer, C., Rupp, M. & Schneider, N. F. (eds.) *Die Vielfalt der Familie. Tagungsband zum 3.* Europäischen Fachkongress Familienforschung. Opladen/Farmington Hills: Barbara Budrich, 431-443.
- Vlasblom, J.D. & Schippers, J. (2006). *Changing dynamics in female employment around childbirth. Evidence from Germany, the Netherlands and the UK.* Work, Employment & Society 20 (2), 329-347.
- Voicu, M., Voicu, B. & Strapcova, K. (2009). *Housework and Gender Inequality in European Countries.* European Sociological Review 25 (3), 365-377.
- Walker, R. & Collins, C. (2004). *Families of the Poor.* In: Scott, J., Treas, J. & Richards, M. *The Blackwell Companion to the Sociology of Families.* Malden: Blackwell publishing, 193-217.
- Wall, K. (2007). *Immigrant Families.* In: Ritzer, G. (eds.) *Blackwell*

Encyclopaedia of Sociology, 5. Oxford: Blackwell Publishing, 2252-2255.

- Wall, K., Aboim, S. & Cunha, V. (2010). (eds.) *A Vida Familiar no Masculino Negociando Velhas e Novas Masculinidades*. Lisboa: Comissão para a Igualdade no Trabalho e no Emprego.
- Walther A. & Stauber B. (eds.) (2002) *Misleading Trajectories. Integration Policies for Young Adults in Europe?* Opladen: Leske + Budrich.
- Walther, A., du Bois-Reymond, M. & Biggart, A. (eds.) (2006). *Participation in Transition. Motivation of Young Adults in Europe for Learning and Working*. Frankfurt a. M./Berlin/Bern/Bruxelles/New York/Oxford/Wien: Peter Lang.
- Walther, A., Stauber, B. & Pohl, A. (2009) *UP2YOUTH – Insights into Youth as Actor of Social Change by an Agency-Perspective*. Final Report for the UP2YOUTH-Project. Available from: *http://87.97.212.72/ne/images/stories/Up2YOUTHFinalreportwithoutAnnex.pdf* [accessed 17.01.2010].
- Wegener, A. (2005) *Regenbogenfamilien. Lesbische und schwule Elternschaft zwischen Heteronormativität und Anerkennung als Familienform*. Feministische Studien 23 (1), 53-67.
- Weigel, D. J. (2003). *A communication approach to the construction of commitment in the early years of marriage*. The Journal of Family Communications, 3, 1-19.
- West, C. & Zimmermann, D. H. (1987). *Doing gender*. Gender & Society 1, 125-151.
- Whelan, C. T., Layte R. & Maître, B. (2004). *Understanding the Mismatch Between Income Poverty and Deprivation: A Dynamic Comparative Analysis*. European Sociological Review (2004) 20 (4): 287-302.
- Wilkinson, R. & Pickett, K. (2009). *The Spirit Level. Why More Equal Societies Almost Always Do Better*. London: Allen Lane.
- Wollmann, H. (2004). *The "local welfare state" in European Countries – in comparative perspective. Concepts, patterns and trends*. Paper for the international WRAMSOC Conference. Berlin. Available from: *http://www.kent.ac.uk/wramsoc/conferencesandworkshops/conferenceinformation/berlinconferenc/thelocal%20welfarestateineuropeancountries.pdf* [accessed 1.4.2011].
- Work Changes Gender (2007). *Towards a new organization of men's lives – emerging forms of work and opportunities for gender equality. Final Project Report*. Luxembourg: Office for Official Publications of the European Communities. Available from *http://ec.europa.eu/research/social-sciences/pdf/workchangesgender-final-report_en.pdf* [accessed 17.1.2010].

Chapter 2: Critical Review of Research on Families and Family Policies in Europe

Karin Wall, Mafalda Leitão & Vasco Ramos

Introduction

The aim of this chapter is to describe and report on the results of the international Conference *"Families and Family Policies in Europe – A Critical Review"*, which took place in Lisbon, at the Institute for Social Sciences (University of Lisbon), in May 2010[1]. This three day Conference was organised by FAMILYPLATFORM with the purpose of carrying out a critical review of existing research on families and family policies by setting up a dialogue between scientific experts, representatives of family associations, social partners and policy makers. Drawing on expert reviews of the state of the art of research, critical statements by stakeholders and policy makers, and debate on the major challenges for research and policies, the Conference was organised with a view to providing a major forum for discussing and identifying the design of future family policies and research.

Presentation of the critical review process carried out in this chapter is based on qualitative analysis of written documents (texts/suggestions/critical statements sent in by chairs, rapporteurs, stakeholders and experts before and after the Conference) as well as audio-tapes of the debates in all the working groups[2]. Analysis of the critical review process was built up through various levels: firstly the main topics were discussed, then there were the contributions of participants, and then finally discussion of gaps in research and methodology.

Conference structure

The Conference was organised around the following types of sessions: plenary sessions with keynote speeches; focus groups on the topics of the Existential Fields (eight in all, with 15 to 20 participants each for which

---

[1] The full version of the Conference Report based on the overall discussions (including on-line contributions to the process) and contributions from all participants comprising the different views, points of critique and perspectives for future research on families in Europe, including key policy issues, was published in September 2010 and is available at: *http://hdl.handle.net/2003/27687*.

[2] All plenary sessions (keynote speeches followed by open debate; feedback and reporting from rapporteurs, followed by open debate) were videotaped and all focus groups and workshops (16 working groups) were audiotaped.

FAMILYPLATFORM produced state-of-the-art reports); workshops on *key issues for policy and family wellbeing* (eight in all, with 20 to 40 participants each); plenary sessions where rapporteurs summarised the debate/conclusions of the working groups; and a final plenary session with closing speeches and a presentation of the on-going foresight exercise. These 16 working groups were structured so they could carry out three main tasks: discuss the major trends in family change and developments in research and policies for each *Existential Field/key policy issue*; understand if these trends/issues represent important challenges for the wellbeing of families in the future; identify major gaps in research and to discuss possible new developments and future tasks for research.

Conference participation

The Conference brought together a total of 140 participants, more women (90) than men (50). Among the total of participants, experts on family from university/research institutions (60 experts plus 11 junior researchers) and stakeholders (58 from about 50 family-related organisations) were almost equally represented. In the selection of the participants the aim was to be as inclusive as possible of the plurality of perspectives and agendas regarding families in Europe. The main criterion for the selection was diversity, meaning different approaches to the family and to family policies, different types of organisation and different countries of origin. However it was not easy to establish a balance. Some groups and organisations were, in fact, less well represented, in particular policy makers/social partners (11), especially unions and employers associations, as well as some types of family associations (e.g. lone parent families and ageing families). On the other hand, there was a general agreement that some fields of research/disciplines were also missing, such as psychology, economics, medicine, neurobiology, urban planning.

Dialogue between experts and civil society – bringing together different relevant actors

Drawing on the dialogue between these relevant actors - experts and civil society - the Conference was considered by the participants as a stimulating and innovative forum of discussion, thus representing a new experience and point of departure, by bringing together specialists from different work communities who do not normally engage intensively with each other's thoughts, understandings, agendas and work.

The review set out in this chapter therefore seeks to provide information on this forum on the basis of two perspectives: firstly, to allow for a detailed description of the structure and main contributions which took place; secondly, to bear witness to some of the interactions and processes of the Conference, consisting of questions, arguments, and discussions, which were overall lively and mutually enriching, but also imparted diverse and sometimes contrasting perspectives on the wellbeing of families in European societies and the issues that ought to be included in the Research Agenda.

While stakeholders were more goal and policy oriented, clearly stating the objectives and claims of their organisations, experts were more focussed on mapping the state of the art and the gaps in research. The dialogue between experts and civil society reveals that stakeholders have an important role in drawing attention to the problems of specific and vulnerable families or members of families; they are more sensitive to local contexts and to the risks and problems affecting many families with children; and reminded researchers about their difficulties in communicating and exchanging with civil society. On the other hand, experts also had an important role in drawing attention to the results of research, while revealing greater sensitivity to the diversity of families and the need to confront family and gender changes in contemporary societies in order to design a viable research and policy strategy for the future; they also reminded stakeholders about the need for a balanced approach between the knowledge deriving from their field experience with families and knowledge deriving from research. All in all the Conference represented a unique opportunity to increase mutual understanding and to discuss on-going and future research.

The description which follows will give readers an idea of the current cross-roads and patchwork of thoughts, doubts and agendas concerning families and family policies which exist across Europe.

The structure of this chapter follows closely the structure of the Conference, though focusing attention mainly on the working groups. In Section 2 we present the working groups of the eight Existential Fields. For each one we present an overall summary of the *organisation of the workshop and keynote speeches*, the *general discussion and contributions by stakeholders*, and finally examine the *major gaps and challenges for research* identified and debated within the working group. Section 3 of the chapter looks at the working groups of the eight workshops on key policy issues along the above-mentioned lines. Section 4 presents a summary of all suggestions regarding methodological gaps and challenges which were common to all working groups. In Section 5, drawing on the discussions, statements, keynote speeches, and other documents examined in Sections 2 and 3, we have made a preliminary selection of the main concerns and research issues, as well as some suggestions for *project topics* for the future European Research Agenda.

2.1 Focus group sessions

2.1.1 Existential Field 1 - Family structures and family forms

Organisation of the focus group and keynote speeches

The session began with a presentation from Elisa Marchese (University of Bamberg) on the major trends in *Family structures and family forms* in Europe. She highlighted the main results by giving an empirical overview of the following topics: fertility and childbearing; the institution of the family; new and unusual family forms. Following this presentation there was a brief discussion which was also enriched by presentations from the three keynote speakers:

1. Andreas Motel-Klingebiel (German Centre of Gerontology) emphasised some basic aspects of the report: the increasing diversity of families; the postponement of births and the decrease in fertility; the delay in marriage and the decrease in partnership stability; the lower rate of increase in births out of wedlock. He also considered the importance of adding other relevant aspects related to family life-course and dynamics: the household perspective; the parent-child unit; spatial aspects and demographic trends. He concluded that information on current trends in more or less complex family and partnership patterns is important, but stressed that what is really needed is a discussion on the goals of family policies as well as agreement on such goals, a task for society as a whole and particularly for policy makers.

2. Anália Torres (ISCTE, University Institute of Lisbon) made a presentation on "*Family structure and family forms in Europe - Trends and policy issues*". She gave a general overview of the main results of the European Social Survey according to a cluster analysis using indicators and data on main trends in family. She concluded that:

   • The transformations of the family in Europe follow the same patterns but but differ culturally and temporally. Each region has particular configurations and combinations between old and new patterns. It still makes sense, analytically, to differentiate between the northern and the southern European countries (although there are also internal differences within the groups of countries).
   • The participation of women in the labour market is not a constraint on achieving a higher fertility rate. On the contrary, it seems that it enhances it. If both partners of a couple are in paid employment then the chances

of them making a decision to have children are increased.

- Women want to invest in both family and work. However in the majority of the countries they have to pay a price for maintaining both investments (overload, not giving up a career, feelings of guilt, unfulfilled identities); gender equality is continually at stake.
- Family is still the main sphere of personal investment for both men and women. What is changing are the models of family life, the meanings and forms of investment in the family. Although there is an increasing diversity, the (heterosexual) nuclear family is still the predominant model. The importance of feelings and emotional life is universally stressed – family, friends, leisure.
- Private matters are a subject for public and political agendas. Employment, care and gender equality should be linked together.

3.   Maks Banens (MODYS, Université de Lyon) focussed on a comparative analysis of same sex unions in Western Europe. He addressed the following questions: how same sex union registration laws were adopted in Europe; what may be hidden behind the different legal status of registered unions; and how to understand the huge differences in same sex union registration.

Main topics discussed and contributions from stakeholders

The experts' presentations and the contributions from stakeholders underlined several key questions which were considered and discussed by the participants. The following topics summarise the main points of discussion:

Comparatively pronounced changes in family forms and structures throughout recent decades

> *"General trend of a decline in institutionalised relationships";*
> *"move away from the previously dominant 'nuclear family model'*
> *towards a variety of different family forms"; "simultaneous growth*
> *in other family forms where research is still scarce, particularly*
> *on new and rare family forms (foster families, multi-generational*
> *households, rainbow families, commuter families, families living*
> *apart together, patchwork families)".*

Overall postponement of family formation and childbirth for both men and women (individualisation? insecurity? wealth?). Generally downward trend in fertility rates though future developments remain in doubt. Some important issues to be examined:

"To what extent do young people today consider that they have the prerequisite conditions for having children? Do governments and local authorities make sufficient efforts to enable young people to have children, taking working and living conditions into account?"

"Fertility intentions still exceed fertility behaviour: possible diffusion of 'low-fertility ideal'; 'one-child trend'?"

"Union dissolution is much less investigated than union formation (comparative research is scarce). Since partnership remains an important prerequisite for childbearing, dynamics of family formation (for example, increasing popularity of cohabitation, moving in and out of unions among young adults) and its consequences for fertility should receive more attention in the future".

The transformations of the family in Europe follow the same patterns in spite of calendar differences and cultural variants. Each region has its particular configurations and combinations between the old and the new. Even so:

"It still makes sense, analytically, to differentiate the northern and the southern European countries; there is no uniform European trend but significant cross-national variations... De-standardisation of the family is more pronounced in Scandinavia compared to the high standardisation which characterises southern Europe".

"Scandinavian countries have high fertility rates compared to many other European countries; at the same time Scandinavian countries also have high proportions of cohabitation, divorces and remarriage: is there a direct connection between the family formation patterns described above and fertility rates?"

Discussion on same sex families

"Same sex union registration laws were not just the outcome of local political circumstances inside each country: transforming family values and practices seems to be the main social force behind same sex union legislation. They seem to be the necessary conditions, maybe even sufficient conditions, for obtaining same sex union recognition".

Discussion on European fertility rates. Questions raised:

"Is it good or not for Europe to have a high fertility rate?"

"Is low fertility in Europe a real problem or does low fertility also imply positive aspects and opportunities for societies (especially in a global context, where population growth will probably lead to more resource distribution conflicts in forthcoming decades); increasing fertility might create more problems; many countries cannot afford to have care facilities…there are economic consequences of high fertility rates such as unemployment (there is a high unemployment rate among young people today, a problem created by high fertility in the 1970s)".

"Fertility is high where female activity is also high (e.g. services and childcare facilities combined with being active = higher fertility rates)".

"Difference between aspirations and number of children… the research has to go deeper into the reasons why people did not have the number of children they wanted. What do families expect from governments and policies? What do they wish for? We only look at these issues form the point of view of the job market and the economy".

"What is the principle according to which society should decide that having children, having families is good for society or not? This is a very fundamental question. Answers will probably differ from one country to another".

"Policies and political changes have an effect on family changes e.g. fertility rates in Scandinavian countries, where from the sixties onwards their fertility went down and later on, from the eighties onwards, it began to rise again; France combines different policies which also have effects on the rise in fertility; for example, the southern countries had a fall in fertility from the eighties onwards, but in eastern countries the fall started in the nineties, which means that political changes had an impact on fertility and employment".

"With respect to the arguments and ideas put forward above regarding fertility and marriage aspirations we would call for continued research and exchanges on measures that aim to support

marriage and thereby the family as the basic unit of society and on the means (financial, services and time) that help families to reconcile work and family life. Investing in these areas can be seen as an investment in our future by considering that families are the future of Europe".

Discussion on the inflexibility of the labour market

"Family structures are being impacted upon by the inflexibility of labour market, which is still based on older forms; there is a lot of debate on how it is necessary to change labour market policy in order to suit the new circumstances of globalisation, but there is very little debate on how it needs to change to suit changes in the structure of the family structures and of society".

Discussion on methodology and data availability

"Cross-sectional demographic indicators tend to be well covered and easily available for most of the European countries. However, data on families and family forms are more difficult to obtain via existing statistical data sources. There are several problems. For example, the definition of families/families with children varies across European countries, and data collection at the national level is not carried out systematically. Some forms of families (cohabiting unions sometimes even with children, same sex couples, and multi-generational families) may not exist in statistical data sources, as is acknowledged in the expert report".

"In order to study fertility and family formation dynamics more thoroughly, we need longitudinal data sets and different types of indicators. Cross-sectional indicators or survey data (if not retrospective) cannot cover many important aspects of family formation. Life trajectories of the young today are more fragmentary, in terms of educational histories, working life, and family formation, than they were a few decades ago. Many of the indicators are not designed to capture the multitude of transitions during individual life".

"Researchers and policy makers should try to identify trends but at the same time detect diversity. There is a need for more qualitative research in order to capture the diversity in terms of small groups, because these groups raise new issues".

Major gaps and challenges for research

- Micro-level research on fertility development and its determinants is needed to understand not so much whether the couple will have children, but when they do so. Only a small group of European countries (15 in 2009) have participated in Generations and Gender Survey (GGS) so far; there is a need to understand the different reasons for fertility postponement, consequences of postponement and the magnitude of recovery.
- More research on fertility trends behind the EU average: on the differences between family forms, qualifications, social classes, regional data, housing and its costs, etc.
- A recent and probably increasingly important theme is the possibility and acceptance of medical aid for fertility and its impact on family life e.g., postponement of childbearing or single-parenthood.
- There are significant research gaps in the study of new family forms. In particular, there is a need to re-examine the relationship between homosexuality and the family of origin in the different European areas. What is the relationship between same sex unions and a) welfare systems and b) community solidarity? New registration logistics are to be studied in more detail. There is a need for differentiated and comparable data on new family and conjugal living arrangements throughout Europe.
- Another very interesting research field could be the general relationship between values and behaviour. Furthermore, very little is known about the reciprocal influence of institutions and attitudes. A better knowledge of the relationships between these areas is crucial for the future implementation of political measures, as it is still quite unclear how decision-making processes of couples and families are influenced and affected by other components (e.g. value or political systems).
- Furthermore, demographic research should concentrate on the differential effects of rising migration and mobility as well as rising life expectancy in Europe. The Existential Field report pointed out that the standard nuclear family model is on the decline in Europe and is increasingly complemented by a large variety of other family forms. Further research should consider and include these developments and also pay more attention to the resources and networks of families in Europe. In this context, change in intergenerational support is a very important topic for analysis.
- There is relatively little information on union formation and dissolution among the older population. So far, much of the research on family formation and family forms has focussed on young adults. Although cohabitation is still more common among the young, its role is also

increasing among middle-aged or older people (particularly in countries which have been forerunners in cohabitation).

2.1.2 Existential Field 2 - Development processes in the family

Organisation of the focus group and keynote speeches

Two presentations opened the debate in this focus group. The first presentation came from Carmen Leccardi (University of Milano-Bicocca) who was
responsible, together with Miriam Perego, for the report on the general topic
of Existential Field 2: "*Family developmental processes*". The second one came
from keynote speaker Karin Jurczyk (German Youth Institute) and focussed
mainly on the theme of "doing family" today.

Carmen Leccardi started her presentation by explaining that in their report
the meaning of development, a concept she recognises to be ambiguous, is
connected with two types of transformation within the family over the last few
decades: changes in family forms (growing plurality in the ways of making a
family) and changes in the identities of the several family members (young,
adults, elderly), both being important to trace out developments in the trajectories of families in the new century. She also referred to the role of time and its
impact on social changes and life-course changes. Accordingly she highlighted
four processes involved in these changes affecting European families:

1. individualisation;
2. transformation in gender relations;
3. the pluralisation of role models;
4. the 'subjectivisation' of norms associated with the family and
   the couple.

Carmen Leccardi continued her presentation by identifying four main trends
emerging from developmental processes in the family:

1. The prolonged presence of young people within their family of
   origin (the role played within it by: the negotiation and affection-
   based family; the de-standardisation of the life-course; yo-yo transitions; labour market instability; the parents' home as a shelter for
   fragmented transitions to adulthood).
2. Young people and parenthood – the new representations of parenthood among young people (new models of parenting, changing
   roles and obligations as regards gender).
3. Conjugal instability, preconditions, modalities and their social and

cultural consequences for family life, gender identities, and divorced fathers and mothers.

4. The new role of grandparents (new active biographical trajectories of grandparents involving care support of younger family members).

Following Carmen Leccardi's presentation, Karin Jurczyk started her presentation on "*Doing family – a new approach to understanding family and its developments*". Her proposal focussed on a discussion of "What does doing family mean?" The starting point for Karin Jurczyk is that there is a need for new approaches and theoretical discussions:

*   The state of the art concerning the report on developmental processes in the family (Existential Field 2) "*presents a lot of empirical details but lacks concepts related to social changes and what is going on in contemporary families*".
*   There is a "*need to frame contemporary families within the trends of late modernity and eroding traditions of so-called normal biographies and biographical regimes, as well as the erosion of structural contexts and the trend towards individualisation*".

The main point Karin Jurczyk brought to the discussion was that "*Family is more than ever a practice which has to be done permanently over the whole life-course; family has no nature, no given resource and no fixed institutional framework of private life and individual biography*".

Commenting on the four major trends identified by Carmen Leccardi, Karin Jurczyk once again stressed her suggestion that a better understanding of those trends within contemporary society requires a radical turn to theories of 'praxeology'. In her view, there is a need to know how people do their families; how they live their concrete daily lives; not so much know what their values and attitudes are, but find out what their practices are. According to Karin Jurczyk, there is a lack of knowledge on what the dimensions of daily life are – what is really going on with the practices: "*we have to understand how people do family*", by differentiating unreflected practice/ routine and focusing on intentional action. "*The challenge is that we have to understand the daily and biographical shaping of a common life as a family, as a whole, as a group, not only the daily life of a woman, of a man and of a child, but the integration of these different perspectives into family life [...] the integration of the individual is not only the addition of different actors; there are conflicts between solidarities, intentions, demands, there are tensions between the individual 'me' and the 'us'. They are not at the same level, there is gender and generations [...] this biography is interlinked with shared life context, into family as a group* [...] *family is an actor in itself".* For Karin Jurczyk, *"another aspect which is neglected, for example, is a bodily dimension of family, family is also physical".*

Summarising Jurczyk's perspective, 'doing family' today must be understood essentially not so much in the light of theoretical approaches but on the basis of a new 'praxeology' which sees as protagonists the individual components of the family (children, mothers and fathers, siblings, grandparents, neighbours and so on) and the relations of solidarity/conflict that they construct on a daily basis through reciprocal interaction. From the point of view of this analytical approach there are numerous phenomena involved: bodily, emotional, cognitive, media-related, social, temporal and spatial aspects. Family policies, in order to be effective, must in turn come to grips with this multiplicity.

While commenting on the four major trends identified by Carmen Leccardi, Karin Jurczyk also identified some research gaps and challenges for further research (see research gaps). She suggested a better understanding is needed of the "huge gap" that still exists between attitudes and practices regarding gender roles (not only concerning men's roles, but also with regard to mothers' ambivalence when they demand more participation from men in the daily life of the family yet simultaneously restrict that very participation). She also found ambivalences and contradictions in the developments of welfare regimes and stated, as an example, the case of Germany where some laws push families towards modern forms of family while, at same time, relying on traditional forms. On the other hand, she considered that the generational perspective of the family has been underestimated and that researchers speak a lot about couples but neglect the role of children as active actors. There is also the dual role of the elderly, who are both care receivers and caregivers. She also pointed out the importance of studying the family as a network, an extended perspective of the family, taking into consideration new developments of the life-course, especially those emerging from divorce (patchwork families), as well as the spatial dimension of the family in respect to 'multilocality' (as a cause of divorce but also of professional mobility). For Karin Jurczyk it is important to study the impact of multilocality in families due to the fact that multilocality can create virtual families and *"there is a limit to possible virtuality in family life".* Finally, she emphasised the lack of knowledge on the concrete procedures for negotiating and practicing partnership and parenthood as well as on what she considers to be a big research gap: the interaction process of becoming a parent (*"we have studies about men, women's wishes, child's wishes, but no research on how to become a parent"*).

These two experts raised several key questions, which the group discussed. The following paragraphs summarise this discussion.

Main topics discussed and contributions from stakeholders

Among the comments and critical ideas put forward by the participants, it became clear that Karin Jurczyk's proposal to conceptualise the family as an actor itself, not limited to the experience of couples and committed to the idea of an interactional network with emotional and bodily (physical) aspects, was very much appreciated by all the participants in this focus group.

There was agreement that these different facets of interaction, reinforced by Karin Jurczyk's presentation, tend to be completely forgotten in European studies. The proposal to turn to 'praxeology' was also very well received and considered a pertinent approach to capturing diversity, as well as an important basis for grounded policies (and as an alternative to structural policies constructed on the basis of generality).

The following topics summarise the key issues discussed in this focus group:

What does "doing family mean"? What does it imply from the point of view of methodology and theoretical approaches?

"First of all, this means focusing on relations between people within the family and on their practices in everyday life. In this respect, we have to be aware that family is not a given, but is a living thing, in constant change, which is done and re-done constantly [...] if we are able to raise adequate theoretical questions in investigating families, then we are also able to put forward a 'praxeology' in this respect. This means being able to understand the daily practices of 'doing family' and constructing the interactions between family members".

"Pick up the bodily and emotional aspects in the life of the family and how the practices in everyday life can shape the family, do the family".

"Families are constructed through interactions and also through bodily and emotional interaction, and this means that interactions have to deal with the family as a whole, and that inside the family we have different generations, different genders, we have children, adults and the elderly, and all these subjects intervene through their interactions and also through their bodily and emotional interactions".

"The importance of studying not only the family as a whole, but the family as a series of interactions, requires us to go further with the theoretical tools that we use to obtain good empirical data. That also implies going beyond the fact that some fields of sociology are sectored: sociology of the family, of youth, of education. Then there is the need to look at reality from an interaction perspective, covering the process as a whole from the beginning: sociology usually studies the fixed time or a fixed moment and neglects how things interact and how people in the family negotiate over the long term".

The centrality of the phenomenon of negotiation within the family: between partners, between parents and children, between grandparents and grandchildren, and so on. Questions raised:

"Is it possible in this regard to affirm that the family today is characterised by fully-fledged models of negotiation (matching the various components of the family)? If so, what is the character of these and in what way are they constructed? What effects do they produce on the life of the family and on the wellbeing of families? From this point of view what role is played by the processes of individualisation brought to light in the introductory presentations?"

"How can negotiations harmonise the level of individual and the couple's goals, and how do macro social circumstances affect these decisions, how to match these three levels? How can we carry out empirical research on this topic?"

"We must take into account the role of education in the developmental processes of the family, e.g. the importance of education for negotiating, how to solve problems, how to communicate with children… negotiation of conflicts between the older generations who stay longer in the labour market and the needs of family to have them at home caring for children".

Families with a large number of children, and the importance of their educational role in the current panorama of transformations in the family

"What messages do these traditional forms of family offer to the social world today, at a time when the general tendency is to limit the

number of children per family? And in what way can welfare policies take this type of family into consideration in a concrete way?"

The role of welfare policies

"A big challenge for family policy is to construct family policy that covers diversity; all follow one particular model of family either implicitly or explicitly. Is it necessary that family policies should always focus on one particular model, or is diversity and plurality a model that can be supported by politics? How might that work?"

"Unpaid work of women and family carers (e.g. handicapped children) is an important value for society and Gross Domestic Product (GDP) (states economise a great deal) but does not appear in statistics, even though it is fundamental for families' lives; family carers of disabled children are often not paid. Also to be researched are their conditions of life and care [...] family care should be included in GDP".

Possible relationship between scientific experts engaged in the study of the family and stakeholders (as well as policy makers)

"What is the nature of this relationship today and how could it be improved? How might it contribute to helping more vulnerable families or to the rights of children within the family? More generally, how can this relationship throw light on questions that are central to the wellbeing of families and also facilitate the development of appropriate public policies?"

"The main points regarding this relationship between researchers and stakeholders are awareness by researchers and NGOs that we deal with representations of reality. If we find there are common grounds for these representations then a common voice can also be found".

Children's rights and wellbeing

The theme of the rights of children proved central to the discussion and was analysed from various points of view. In particular, attention was drawn to the issue of the consequences of divorce on the life of young children. The importance of the regularity of contact over time, and the interactions between separated fathers and mothers and their children, were highlighted as essential elements in the wellbeing both of children and their parents. More generally, emphasis was placed on the importance of maintaining a high level of awareness of the effects of developmental processes in the family on the wellbeing of children: *"we should start by focusing on children as subjects and equals in the family and not just as minors or subject to a hierarchical condition".*

The need to look at divorce as a start of a new form of family

"Since the late sixties divorce has increased significantly, so a lot of people are sons or daughters of divorced parents. We do not yet know what kind of family life-course histories they have experienced. Maybe these changes and the instability of family and marriage will take us to new forms of family in the present. Too much research focuses on women's problems and less on fathers; we also have to look at the family from the perspective of men, in particular the role of fathers after divorce".

Same sex families can be a starting point for a new way of looking at the family

"Due to the fact that there seem to be no gender differences it is interesting to understand the way they manage their individual perspectives of family life". "What does it mean to be a gay father or a lesbian mother in relation to the children's future and adulthood?"

"We are facing different ways of conceiving families".

The importance of conducting empirical research of a comparative kind in relation to European Member States

First of all, empirical research of this type is important in order to understand in a detailed way what lies behind the differences between European countries as far as the family is concerned (the timing of family life, the ways of 'doing family', the relationship between parents and children, the balance between family, work and personal life – in connection with the reality of the labour market and welfare policies), and the ways in which family choices (and individual choices within the family) are made and negotiated.

Major gaps and challenges for research

* The lack of knowledge on the concrete procedures for negotiating the practices of partnership and parenthood; on the interactions of becoming a parent – we have studies on men, women, and children's

wishes, but no research on the interactions between parents and the process of how to deal with becoming a parent.

- There is a lack of studies on same sex families.
- More research on the huge gap between attitudes and practices relative to gender roles, not only from the point of view of men but also of women/mothers who tend to restrict men's participation in family life; it is important to look at these ambivalences.
- Research must also look into the bodily and emotional aspects of the family. More specifically, attention should be drawn to the physical dimension of motherhood.
- An effort must be made to promote greater understanding of new forms of families today, including the relationship between children and parents. We refer here, for example, to 'patchwork' families (after divorce), to migrant couples and parenthood, or to same sex couples.
- It is important to focus on children who are experiencing or have already experienced critical events both in new and traditional families (for example, in same sex couples) and to understand what the risks are for children in high-conflict family situations, in order to develop sensitive policies to support them.
- Need for research on how education for family life, marriage, conflict handling, etc., can contribute to changes in the attitudes of young adults towards family values.
- The generational perspective has been underestimated. "We speak a lot about couples but neglect the role of children as active actors, as equals within families and not just minors". There is a need for further research on the importance of mutual care between generational and gender groups, and cross-national comparisons on the role of grandparents in childcare.
- In summary, there is a major research gap in the linkages between daily life and development of the family, the daily and biographical processes of doing family as an interaction.

2.1.3 Existential Field 3 - State family policies

Organisation of the focus group and keynote speeches

Sonja Blum (University of Münster) opened the session with a presentation on the results of the Existential Field report on *State family policies*, which she authored together with Christiane Rille-Pfeiffer (University of Vienna). According to Sonja Blum, state family policies in Europe have gained tremendous importance in recent years due to major challenges European societies are facing today such as ageing, growing diversity of families and the reconciliation of work and family life. These challenges have made family policy one of the few expanding areas of welfare. However, in comparison with other social policy fields, family policy is characterised by a low degree of institutionalisation, even though it cuts across other policies related to employment, education, housing and urban development that impact on families.

Family policies across EU Member States are characterised by great diversity. Sonja Blum focussed on major trends in family policy in Europe in relation to regulatory frameworks, leave policies, care services, and cash and tax benefits. Her approach was based on a geographical typology of family policies, which she adopted as a temporary solution (Nordic, Continental, Anglo-American, Mediterranean and Post-Socialist countries); on the family policy database of the Council of Europe; on the information made available by the annual review of the International Network on Leave Policy and Research; and on the data emerging from a small questionnaire her work team sent to welfare state researchers in all EU27 countries.

In her conclusion, Sonja Blum identified some European trends in terms of either re-familialisation or defamilialisation. While family policy in Nordic countries seems to keep the sense of re-familialisation and Mediterranean countries keep their orientation toward defamilialisation, conservative and Anglo-American countries both show moves towards defamilialisation, while Post-Socialist countries are very heterogeneous, between defamilialisation and re-familialisation.

A second presentation came from Kathrin Linz (Institute of Social Work and Social Education, Frankfurt) entitled "Hurdles to overcome in comparative research on family policy", focusing on what she experienced as obstacles while doing comparative research on family policy in Europe. In her presentation she mentioned two comparative studies, both conducted in 2009: one on the "reporting on policies for families in the EU Member States", the other on "policies for families in times of the economic crisis – reactions of the EU Member States".

Family support systems are being shaped and developed independently in each country. National policy measures are developed in a context of differences in cultural conceptions, socio-political targets, welfare state configurations and financing possibilities. Therefore Kathrin Linz considers that when doing cross-national research on family policies in European countries "we need to take into account the different traditions of dealing with policies for the wellbeing of families as well as the development of institutions in this field", namely the political structures dealing with families in different states. Is the overall support system for families stronger in countries where family policy is explicit, and where there is a designated ministry for family affairs? *"In some Member States it was not easy to find the right person to talk to because by the time we conducted the study there were only nine ministries in the Member States which had 'family' in their title".*

When conducting comparative studies on family policy *"we need to know more about how family policy is culturally and institutionally embedded in each country".* Why is the word *"family"* part of the ministry name in some countries? How can we explain differences in the development of explicit family policies in some Member States?

Kathrin Linz mentioned a study (by Franz Rothenbacher, University of Mannheim) which shows that there is a strong connection between state expenditure on families, the standing of family policy in society and the development of institutions. The study also concluded that the development of explicit family policies is to be expected to a higher degree in countries where Catholic values correlate with high socio-economic development. In countries where Catholic values correlate with weak socio-economic performance there is often an effect on the development of institutional structures in family policy.

According to Kathrin Linz, when comparing major trends in current debates on family policies in European countries it seems that changes in policies are increasingly focussed on facing up to demographic challenges, for example the expansion of childcare facilities in Germany. Policy makers and stakeholders are also worried about the impact of financial cuts on families. What impact has the economic crisis had on national family policy measures in Europe? Research results showed that the impact of the economic crisis on public finances was particularly significant in relation to measures and programmes for families, which means that funding strongly shapes policies for families. However, responses to the economic crisis varied widely. There are also contradictions in the changes introduced by countries, with some changes resulting in a higher level of support for families, and others going in the opposite direction. Kathrin Linz considered that changes in family policy as a response to the economic crisis are a fertile ground for further research, and stated that *"we should also look at changes in family policy resources over time".* As national family policies are subject to constant change, it is important to understand which changes have a positive and which changes have a negative impact on families. More longitudinal comparative research would be useful to increase knowledge of developments and outcomes of family-oriented policies.

A third presentation came from Jorma Sipilä (University of Tampere), and focussed on cash for care: *"Cash for childcare: an exquisitely debated subject",* an issue which he recognised to be a small but crucial detail in the whole process of family policy. He regarded it as an interesting detail because it is related to emotions as well as being a controversial subject: "*most social researchers as well as policy makers, especially those involved in the economic field, have either not been interested in this topic or are against it*".

The major question is: "*should the state pay parents for taking care of their children at home?*" According to Jorma Sipilä, there are two alternative ways of doing this: one is the American method, where parents are given cash to purchase care as they wish (parents may pay for care or provide care themselves). The alternative is the Nordic method, which is "*about giving cash instead of day care*", meaning that if parents do not use subsidised day-care they are eligible to receive cash for care. Jorma Sipilä focussed on this latter measure, which is more common among Nordic countries, where there has been a broader use of day-care and high expenditure on families. However, there is no mainstream model in Scandinavia, but a variety of different principles, particularly at the local level.

Jorma Sipilä put forward a set of arguments for and against these extensively debated home care allowances. Among the arguments against are the following: the risks to career development for women; poverty and female unemployment (the state is spending money to reduce female unemployment while every political program demands the opposite); the extra costs it represents for the state; increasing marginalisation among mothers as well as greater risks for children. It is also seen as an advantage for mothers but not for children; it is particularly poor and under-educated parents who prefer this cash care solution; it might be problematic for immigrant children, who will be raised separately from other children; there is no guarantee of quality because the state cannot intervene and examine what is going on; and it creates problems in terms of gender equality with respect to formal and informal work.

Among the arguments in favour, Jorma Sipilä named the following: the benefit allows people to protest against the lack of reconciliation between work and family life (very popular among young people), and increases family time (also a very popular argument reflecting the growing value of maternal care and children). In fact, about 90 per cent of people in Finland have used cash for care for some length of time. From the perspective of research, explanations for this are related to the following: insecurity in the transition to the labour market; family care capital at times when unemployment is on the rise; parents are less and less able to afford to stay at home without benefits; privatisation and non-formalisation of care as a cause of economic crisis; neoliberal emphasis on individual choice; motherhood and gender differences are somehow glorified by the entertainment industry; the existence of this kind of benefit legitimises its use (social policy benefits function as normative recommendations, intensifying the social obligation to make time for care); advantages of day-care and family care according to medicine and child psychiatry.

In conclusion, considering the importance this benefit still has for families and taking into account the current controversial debate over home care allowances, Jorma Sipilä made some recommendations for improving the benefit. One is that parents should not stay exclusively at home i.e. the benefit should not promote exclusion from the labour market, but should encourage a combination of part-time work and part-time childcare; parents should not be encouraged to reject the labour market; the benefit should be shared between parents, as care leaves also tend to be shared; the introduction of father's quotas in cash for care would also improve gender equality. A point stressed by Jorma Sipilä was that the right to the cash benefit should never exclude the right to day-care, as happens today in Finland (though not in Norway); children's participation in group activities should be a condition for the cash benefit, in order to prevent children growing up in closed families.

A final presentation came from Jonas Himmelstrand (HARO, Sweden) who focussed on Swedish family policies with a presentation entitled "*Are the Swedish state family policies delivering?*" His main point was to challenge Sweden's perfect image regarding family policy, i.e. as having the best state welfare model in international benchmarking. According to Jonas Himmelstrand, Sweden today has a culture and a form of political commitment which considers state-provided professional childcare as the most suitable form of care for the child's development, while family care is regarded as a lesser choice. Gender equality is a core issue in the debates on childcare.

Overall Sweden is known for having great statistics in respect of low infant mortality, very high life expectancy, relatively high birth rate, low child poverty, high spending on education, equality and gender equality, and the best parental leave. However, Jonas Himmelstrand argues that quality must be also balanced with quantity: "*are we actually producing a next generation which has the psychological maturity and ability to handle stress, and manage the challenges of future life?*"

Sweden is known as having one of the best parental leave schemes. However, one of the main ideas Jonas Himmelstrand wanted to stress is that after the 16 months of well-paid parental leave (13 months at 80 per cent of salary plus another three months at a lower level) the 'door closes'. He also pointed out that cash for care depends on municipalities, and only one-third of them are providing it. On the other hand, the high Swedish tax rates are designed for dual-earner households; family policy emphasises a work policy saying that "*everybody should work after parental leave*"; parental leave is expected to be split in equal shares between men and women. Therefore the overall family policy model is becoming "*children in day-care and parents working*".

In relation to this family policy model of childcare, Jonas Himmelstrand brought up some "*uncomfortable statistics*", namely: the severe decrease in psychological health among youth; the very high rates of sick leave among women; day-care staff at the top of the sick leave statistics; rapidly decreasing quality in Swedish schools; plummeting educational results in Swedish schools; severe discipline problems in Swedish classrooms; deteriorating parental abilities, even in the middle classes; a highly segregated labour market. Among the main possible causes, based on current knowledge, Jonas Himmelstrand reinforced the negative impact of early separation of children and parents as well as of early exposure of children to large groups of peers.

Jonas Himmelstrand concluded that "*Swedish state family policies are not emotionally sustainable and thus not sustainable in terms of health, psychological maturation or learning [...] Swedish State family policies may not even be democratically sustainable, as there are definitive difficulties in even discussing these policies*".

Main topics discussed and contributions from stakeholders

Childcare and cash for care. Questions raised:

> "*Should cash for care also be introduced for elderly persons and regardless of income? Home care allowances have been introduced as a trend in childcare expansion. Is it feasible to have this for older persons? (It is surprising that faced with an ageing population care services for older persons have gained less attention)*".

> "*There are many indicators on the quantity aspects of childcare but there is a lack of indicators on quality. There are often two indicators, which are also covered by OECD family database: child staff ratio and educational levels of childcare employees, but these are very poor indicators for comparative research, and even for these indicators we still do not have reliable and comparative data (e.g. childcare expenditures and outcomes)*".

> "*Family care/maternal care or childcare/non-maternal care? We need research for the long term. There is a lot of political talk but very little research on what is best for children, particularly smaller children under the age of one. It is important to carry out early childcare longitudinal studies in different countries*".

"It is also important to look at the effects of early childcare on parenthood. How does it affect being a parent, their health, their psychological maturation?"

"There is a lack of information on tax systems. In Austria there is a proposal to guarantee a minimum income for families no matter how many children they have. It is important to give families financial security and to compare tax systems in Europe".

"We have to support families' freedom of choice in relation to care arrangements: there is significant investment in day-care, but what do we give to the families that look after the children themselves? Families do not have equal opportunities to fulfil their wishes as long as family policies support certain forms of family and neglect others. We have to focus on the wishes of families, and they are very different".

"Lone parent families and blended families seem to be more highly valued, they are regarded as modern families, and married young couples are looked at as traditional families".

The inevitable and increasingly important link between family policies and employment policies

"One of the most important aspects of family policy in the future is flexible working conditions in the labour market. It is utopian to expect that all children will be cared for at home by families. Families are also needed in the employment market. So the question is: can we be there and also take care of our children and of our parents when they grow old? It is crucial to focus on the simultaneous combination of employment and care [...] nowadays family policies go along with the situation of the labour market".

"Flexibility might also mean less job security [...] flexible working conditions should take account of employee points of view as well as employers"'.

Employers' points of view

"Small and medium-sized employer firms find it harder to replace people taking leave. When talking about leaves we often underestimate the employer side, we tend to emphasise the state's point of

view or the child's point of view or the parents' and families' point of view, but these situations also affect employers. Employers must be involved in the discussion of these policies".

Wider focus when looking at state family policy

"The focus on state family policies and on central level or even federal state structures might lead us to miss some substantial developments and aspects of family policies. One of the major developments is that there is an increase in actors and stakeholders who are discussing and debating family policies. The implementation of family policy measures is increasingly carried out at the regional and local levels, so it is a huge challenge to try to capture any comparison between all Member States of the Union, for example through case studies".

"Are there differences between national and local levels of policies (and also between countries) regarding which type of families they are addressing? There is a need to address all types of families in terms of an approach to social justice, and sometimes that can be more evident at the local level".

The crucial role of time management in family policies

"Family policies are usually looked at as a tool kit of three policies – benefits in cash, benefits in kind (different types of childcare services), and time and time management. Increasingly public authorities try to convince employers to do more about time management (flexible working hours), because they have their whole agenda of employment levels, getting people into employment, and keeping people in employment (specially women and mothers), so time management is also a trend and should be of greater interest for the immediate future".

"Love is the main reason for founding a family. The main reason we have so many divorces and separations is that there is no time to cultivate this love, so the love disappears and then the partnership is dissolved. So love, time, money and resources are the main reasons for families to work or not to work".

Mainstreaming family

> *"Gender mainstreaming is on the agenda everywhere, so perhaps it would be interesting to introduce family mainstreaming as a new attitude for family policy makers and family science experts. A family impact report should be a standard starting point for the policy decision-making process".*

> *"The lack of consensus on a definition of family is one of the reasons why there is no platform for action for family as there is for youth, for old people or for people with disabilities, for example. From an international perspective, if it is not possible to reach a common definition on family, at least there should be an agreement on what family functions are, because the definitions of those functions could help to design good policies for families [...] and establishing a regional framework of family rights".*

> *"It is important not to define the family and look at all sorts of family models because children do not choose the type of family they are born into".*

Framing family policies over the life-course and from an intergenerational perspective

> *"In most countries, family polices relate to pregnancy, birth and early childhood and then when school starts family policies seem to be out of sight [...] they could be important again in connection with parents' supporting adolescents [...] it is crucial to try to see policies over the life-course and according to relevant family transitions in order to support and try to contribute to the wellbeing of families. Time management should also take into account the 'sandwich generation', those who have to take care of both children and old parents".*

Policy evaluation and its consequences in the long term

> *"It is important to evaluate policies that are not explicit family policies but nevertheless impact and influence family outcomes. It is also important to understand the impact of evidence-based policy-making [...] What are the consequences for families of policy measures?"*

"Consultation on family policies: it would be interesting to see how family policies are made and how they are being implemented. Is there a consultation process on what families actually expect from government and on their different needs for reconciling family and work life?"

How to evaluate family policies?

"Without knowing the aims of the policies how can we evaluate them? For evaluation we would need information on what the goals of special family policies are. Politicians do not usually state the aims of these policies explicitly. We would also need to know how much money was invested as well as with what results. This kind of effectiveness is difficult to evaluate".

"The aim of any family policy is social justice; standards of living between families in democratic societies require more equality; family policies are prevention policies against poverty and social exclusion, so maybe this can be a form of benchmarking and policy evaluation: social justice between different forms of families".

Typologies/classifications

"We need to improve typologies and we need specific typologies for post-socialist countries". "Is it possible to consider the diversity of family-oriented policies in a single 'pot'? Typologies do not always help, at least from the perspective of family organisations".

Family policies - should we all have family policies?

"Family must be nurtured from the inside. There might be too much family policy and too much control from the State. How do we foster that inner motivation for family? Are family policies a good thing in that context?"

Major gaps and challenges for research

- There is a need to broaden the focus of analysis, going behind state family policies (e.g. more research on government-NGO relations, occupational family policies, regional and local family policies, in order to capture differences between all Member States of the Union, for example through case studies).

- We need to know more about the effects of policies which are not explicit family policies, but which impact on families (for example employment policies).
- We need to know more about the belief systems and the policy ideas of family policy makers (and the differences across countries).
- More data is needed on the total expenditure on family policy in order to assess the impact of policy changes on overall support for families.
- There is insufficient data on intergenerational transfers within families, and the overall contribution of older persons to the wellbeing of their families.
- Effects of early childcare on parenthood and adult maturation: research is needed on the impact of maternal and non-maternal care (e.g. effects on child-parent relations).
- More research is needed on family policies and men's role in the family (e.g. why they take less parental leave).
- It is important to examine the impact of certain leave schemes on employers and according to company size. Small and medium-sized employers find it harder to replace people taking leave.

2.1.4 Existential Field 4 - Family, living environments and local policies

Organisation of the focus group and keynote speeches

The focus group began with an introductory presentation from Leeni Hansson (University of Tallinn) on the subject: "Family life and living environment: different ways of development". She started with a theoretical approach to the concept of living environment according to the Urie Bronfenbrenner's perspective, which defines four social environmental systems affecting family life:

1. Micro-system: immediate environments and settings (e.g. home, school, informal networks, etc.).
2. Meso-system: a system comprising connections between immediate environments (i.e. home and a child's school).
3. Exo-system: external environmental settings which indirectly affect family life (e.g. parents' workplace settings).
4. Macro-system: large cultural and social contexts (economy and labour market, legislation, educational system, etc.).

Focusing on the macro-system, and drawing attention to poverty rates, Leeni Hansson stated that today's EU is characterised by similarities and contrasts between countries. For example, according to the Eurobarometer, two adults without children face a lower risk of poverty than the average of the total population. Families with three children are at greater risk of poverty. Single parents with dependent children, single elderly people, and especially single elderly female are the household types with the highest risk of poverty. These general poverty trends are similar in the majority of the Member States, and differences between countries are not so significant. However, when comparing poverty rates of the total population with poverty rates of two adults with three or more children, there are huge differences between and within countries. Families with three or more children are far more exposed to poverty than the total population in some countries, while in other countries poverty rates between the total population and large families are more even. According to Leeni Hansson, the countries where there are no significant differences in poverty rates between total population and large families are those where social security benefits are well organised to support families. Ending her presentation, she focussed on research gaps as well as on what is needed in order to measure living environments and to carry out cross-country comparisons of family life and living environments.

After Leeni Hansson's presentation, there were some brief comments, and a further two presentations followed. Epp Reiska and Ellu Saar (both from University of Tallinn and authors of the Existential Field report on *Family and living environments...*) summarised major trends and research gaps according to six sub-topics of the general topic of this focus group: economic situation, employment, education, environmental conditions, housing and local politics. The last presentation, before the debate, came from Francesco Belletti (Forum delle Associazioni Familiari) who focussed on *Local politics - programmes and best practice models*, a sub-topic of Existential Field 4.

All presentations were commented on by participants. Several key questions and issues were raised and discussed as follows.

Main topics discussed and contributions from stakeholders

The role of family policy; family policy mainstreaming

Leeni Hansson's presentation generated a discussion on the role of social and family policy and the need to monitor its effects on families' wellbeing according to more subjective and comparable indicators.

It was also pointed out by participants that many countries do not have a specific department for family policy, while in others family policy is integrated or diluted in social policy. There is accordingly a need for creating specific departments for family policy and to bring family policy into the mainstream: *"in many European documents we find the concept of cohesion, social, economic, political; the first model of cohesion is the family; we cannot speak about cohesion if there is no cohesion in families; this approach stressing the integration of policies might be interesting for the future of mainstream family policy [...] family at the heart of several policies. It is precisely in cohesion policy that we find local development, regional development, sustainability [...] we should try to mainstream family in many European policies"*.

The need to monitor policies and their effects on family wellbeing was also a major point in the discussion: *"One of the statements of the Lisbon Strategy 2000 was that poverty should be diminished by 10 per cent in 2010, but actually poverty increased by 15 per cent. So what advances have been made? Poverty and social exclusion still exists [...] in order to achieve some success we should be monitoring policy advancement and how the policy of Member States reflects on family welfare and family wellbeing. Where is family in European strategies?"*

How to measure living environment? How to define the 'friendly-family environment' concept?

Another key issue in the debate was the question of how to measure a *"living environment"*. Living environment was presented by Leeni Hansson as a multi-dimensional concept with different key elements. She suggested some indicators for measuring what a good living environment consists of: functional housing environments, adequate work places in appropriate locations, adequate educational and child care facilities, adequate services in appropriate locations, a wide range of parks and recreational areas, functional transport networks, a functional municipal infrastructure, and an unpolluted, noise-free environment.

It was pointed out that there is a lot of data on the performance of economies and that GDP is the most widely-used measure of economic activity. However, it was recognised that it measures only market production and not economic wellbeing. Material living standards are more closely associated with measures of real income and consumption but do not tell us anything significant in terms of families' wellbeing. There was agreement that what is missing is a perspective that goes beyond GDP: the household perspective as well as a family perspective, e.g.

consumer patterns and their unequal distribution according to households and family types.

The importance of parents and families as agents influencing and designing their living environment was also emphasised (their role in defining indicators to measure their wellbeing): *"family wellbeing cannot only be measured by economic indicators such as GDP; the pertinence of the capability approach (access to basic rights, education, being able to care for the people we like... in this context). Research should go to the community and to the local specificities, and involve people and their own definitions of their wellbeing [...] we need to construct more subjective indicators related to household and family perspectives".*

Do we have a family-friendly environment? Work versus family, or balance?

> *"One important impact of the FAMILYPLATFORM project should be to change policies and create more friendly environments for families... flexible working time does not exist in reality, employers prefer to not employ women with children under five or women who plan to have children [...] For European citizens it is not so important to have strategies, road maps, white papers, or green papers, but rather to understand how the policy of Member States reflects on their own life and on more friendly environments for the family".*

The issue of family-friendly enterprises was also raised in the discussion. The need for a unified definition was stressed: *"what does it actually mean for a company to be family-friendly?"* Examples such as childcare facilities and the role in caring for retired employees were mentioned as characteristics that could be included in the definition.

The crucial importance of a dual approach when studying family and environment

This Existential Field was considered essential for capturing macro changes and carrying out macro-level analysis of how family changes affect the environment and how environment influences families. There was general agreement on the lack of data on the impact of families on the environment:

> *"In the report there is some information on the environment, but it does not go further on sustainability indexes and on the linkages between household behaviour and family behaviour and sustain-*

ability. What do internal family changes mean for sustainability and related polices? [...] It is very important to try to link internal family changes and decision-making within families and to establish what their impact is on the sustainability of the environment".

Sustainability was considered to be a key challenge. It was recognised that it is important to identify what the changing patterns are in families in Europe (fewer marriages, more single parents), and what they mean from the point of view of sustainability. In this respect families are very important and *"specially mothers because they make the daily decisions on, for example, purchasing, using energy, etc.; so they have a huge impact on daily activities which impact environmental management [...] it is important to try to link internal family changes and decision-making processes with their impact on the sustainability of the environment".*

"There are changes within the family – how are these changes impacting outside? There are changes in the relationships inside the family: what are they bringing to the overall changes in society and the environment?"

The gender perspective was also discussed: there is a need to consider new forms of fatherhood and the increasing movement towards gender equality within families, and to understand how the improvement of work and family life balance connects with environments. A question was raised concerning the gendered configuration of public spaces: the most frequent example given by fathers who are interviewed is the difficulty of caring in public spaces, which are often designed just for women (e.g. nappy-changing facilities in women's toilets).

How to do family policy at local level? How to harmonise the different responsibilities of managing the municipality as a public actor?

Francesco Belletti's presentation stressed the importance of family policy at a local level, as well as the need to spread best practices models. According to him, the local level is acquiring more and more relevance because at local level actions can be targeted to specific needs and problems can be tackled in a more *"rounded and responsive way".*

Focusing on the local level also means stressing the family as an important actor and therefore the importance of family associations, Volunteers and NGOs were also mentioned: *"If we want to know families we have to know their local representatives; it is a way to get into national families; that is why NGOs are so important in the family platform".*

The main point highlighted with respect to local family policies was that there is a need for more research on local welfare in order to carry out comparative approaches on the good or best practices models. It was recognised, however, that the collection of good practices needs to be more systematic, to allow for comparisons and an evaluation of the results achieved.

Emphasis was put on the importance of defining a research agenda and developing a monitoring system for local family policies. There was general agreement on the crucial importance of qualitative approaches, which seem to be more productive in detecting the complex and interactive mechanisms of local networks and finding which actors determine a local policy's effectiveness: stakeholders, institutions, etc. There was also general agreement on the need to conduct further research into the portability and reproducibility of good practices.

Discussion on methodological approaches; the limitations of existing statistical data

Specific statistical data is mostly available at the macro or country level. There are, however, no specific family-focussed data (for example, by different family types). There are also difficulties in interpreting data, as different concepts are not always well defined (e.g. *"if we have to compare families with children: in some countries this refers to families under 16, in some countries 16 is included, in others it is children under 8, in Italy children under the age of 25, when they live with their parents… What are we comparing?"*)

"There are surveys which have questions on satisfaction with life, family life, housing, leisure time… but who is satisfied? Is the answer only from the person who is answering the questionnaire or is it shared by the partner or other family members as well? We do not know, we do not have the family perspective".

"How is poverty measured? Is it really a poverty line? What is the meaning of poverty? Income is a good measure, but it is not enough to measure poverty in rural areas – it does not reflect reality. There is a need for qualitative designs; averages do not help us in telling who needs what. What kind of families are having difficulties? What kind of classification do we need in order to disaggregate data? There are different types of classifications which need to be developed further".

The importance of longitudinal data was also stressed, and the example of the life cycle was quoted, to reflect the fact that responsibilities change over time.

Major gaps and challenges for research

Economic situation:

- There is a need to include a household and family perspective. More information is required on how families obtain and use resources such as money, material resources, available services, time, etc. There is a lack of information on incomes and consumer patterns and their distribution across family types. Overall there is a lack of subjective measures of economic wellbeing; and even when available, data are not comparable across countries.

Education:

- The absence of data on access to education (e.g. access to primary education in rural areas and access to lifelong learning opportunities), percentage of children attending crèches and pre-school (rural and urban areas, according to age), school drop-out and comparable surveys exploring the connections between education and other outcomes related to family wellbeing.
- The need for more research on the role of family in primary socialisation as a component of education, on parents' involvement in children's schools, and on parents' skills and parenting support. There is a lack of studies on education and schools for minority groups.
- Difficulties finding data on rural families and differences in relation to urban areas.
- A lack of data on flexible working time arrangements by household and family type; lack of data on cross-border employment.
- The need to monitor family-friendly policies at local, regional, national and cross-country levels (e.g. companies with childcare and elderly care facilities).

Environmental conditions:

- There is a need for an agreement between researchers on what elements constitute a 'family-friendly environment'.
- There is a lack of environmental indicators when considering the families' point of view: for example, there is no data on the amount of people or special groups of people exposed to different contaminants in the environment. There is also a lack of comparable data on the existence and quality of green areas in European cities.
- There is also the need to carry out research on the gendering of public spaces.

- There is a need for double-sided research on how changes in families affect environmental conditions and how environmental conditions affect families.

Housing:

- Existing data needs to be updated and made relevant and comparable, as there are some subjectivities in conceptual definitions (for example, the number of rooms is used as an indicator for living conditions of families in countries with different stages of development, and affordable decent housing is still an ambiguous concept without a common definition).

2.1.5 Existential Field 5 - Family management

Organisation of the focus group and keynote speeches

The focus group started with a presentation from Marietta Pongrácz (Hungarian Demographic Research Institute and leader of this Existential Field) who highlighted the main results of the working report on the general topic *"Patterns and trends of family management in the European Union"*. The presentation covered three aspects of family management: allocation of tasks and gender roles; parenting and childrearing; family and work. After this presentation, a brief discussion took place between all participants.

A second presentation was made by Michael Meuser (Technical University of Dortmund), focusing on the changing culture of fatherhood and on how fathers put fatherhood into practice. During his presentation on *"Fathers and family management - expectations, pretensions and social practice"*, Michael Meuser identified several research challenges in the field and also contributed to an interesting discussion on the *"new cultural idea of the new father"* and its connections with labour market structure and the role fathers play in family management.

A third and final presentation came from Gordon Neufeld (University of British Columbia) on *"Working mothers and the wellbeing of children"*. Gordon Neufeld is a developmental psychologist, and his presentation gave the focus group a psychologist's perspective on child wellbeing and child development: the spotlight of this final presentation was the concept of child attachment.

Main topics discussed and contributions from stakeholders

The presentation from Marietta Pongrácz, Hungarian Demographic Research Institute, was the starting point for a first brief discussion on the outcomes of the report. Participants made some general suggestions:

The report reflects the overemphasis of research on the division of work among heterosexual couple families and gives insufficient attention to other family types, therefore failing to reflect the variability and changing nature of family management patterns in European families. For example, single parent families, foster families and families caring for disabled persons must not be completely left out.

Another suggestion was that marginal groups such as migrant families and minorities should be included (participants considered that there is little information on family management among migrant families and raised the question of whether they tend to have traditional orientations and values).

The role of children in family management was also mentioned, given the fact that in some European families children are an important element in family management, such as being responsible for some household duties as well as for the care of young siblings. There was also a discussion on the age children should start to participate in domestic work at home. The importance of early socialisation regarding the (gender-biased) division of domestic work was pointed out.

Discussants also pointed out that existing research at the national level is not always in the English language. Comparative cross-national research needs to be reviewed and continued on a regular basis.

The overall debate within this focus group was very much centred on family management for families with children (heterosexual couples) and particularly their daily life after having children. The most relevant subjects discussed included the challenges of caring during the life-course - childrearing but also teenagers' and grandparents' care - negotiation relationships within the family (including all family members as well as the role of children in family management) and the connections between paid and unpaid work. With regard to childcare the debate covered the points of view of parents, mothers, fathers and children, the labour market perspective, and issues of gender equality. It became evident that these perspectives are different and not always reconcilable; while they may sometimes be complementary, at other times they conflict. The following points summarise this discussion.

Childcare, mother's or parents' care in the early years of a child's life? The child's perspective

The early years of a child's life were considered extremely important for the development of the future individual. Therefore, there was a discussion on the best arrangements for the care of children during this stage of life. On the one hand, the importance of giving more value to parental leave was underlined, not just in terms of increasing parental leave time but essentially to promote parental leave for mothers in order to motivate them to stay at home with their children as long as possible. However, this perspective discourages mothers from going back to work after childbirth. In general, mothers are viewed as being the crucial actors in developing and strengthening the emotional bonds with the child and the child's balanced development as a person.

As a psychologist of development Gordon Neufeld reinforced this perspective by bringing in the concept of the child's attachment. Although considering that gender is neutral in relation to children's attachment, Gordon Neufeld argued that after birth a child tends to be more attached to its mother than to its father. He raised the controversial issue of whether the focus on child wellbeing and development implies that child attachment is more important than gender equality, "*mothers are potentially more effective rearing children [...] children care nothing about gender equality*". According to his view, the discussion on childcare in the early years of a child's life must start from the child's perspective, from the centrality of the concept of child attachment: it is crucially important who the child is attached to, and that working attachment needs to be fully developed. Accordingly, day-care providers and teachers are considered to be a handicap, because children are not as attached to them as they are to parents. Parental separation affects children profoundly in their future development as adolescents and later as adults. On the other hand, the deeper the child attaches, the easier it is to be physically apart. Therefore Gordon Neufeld questions the model of early separation, believing that the child should be given more time in order to enable the full development of attachment.

In the ensuing discussion, other participants felt that the so-called child's perspective was strongly mother-centred (the stay at home mother), in contrast with other perspectives highlighting the social construction of biological bonds, the gender equality issue, the increasing participation of women in the labour market and the positive impact of high-quality childcare on parents and children's fulfilment. It was also stressed that there are studies showing that high-quality institutional care has no negative outcomes in terms of child wellbeing. The main challenges for research that resulted from this discussion are that research should address this issue of children's attachment and that more research is needed on the impact of early childcare on children's wellbeing.

For Gordon Neufeld one of most critical issues is that "*today we cannot go back to the parent in a home situation*". Therefore, and given that maternal employment is likely to increase, "*one of the most important questions we need to face today is how can mothers work outside the home and still cultivate the attachment required to raise their children; how can we best cultivate the attachment to the other adults involved in raising children, their relatives, parents, teachers, how can we mobilise grandparents as supporting parents?"* From a policy point of view the question is: what policy measures do we need to keep child attachment intact?

Men's/fathers' perspectives

The main discussion of this subject followed Michael Meuser's presentation. There was general agreement on the fact that during the last two or three decades a new cultural idea of fatherhood has developed in western European countries, centred around the notions of the *'new father'*, the *'active father'*, or the *'involved father'*. However, there is no widely held consensus on what this means, and we still have little research on what this new ideal of the 'involved father' means in terms of duties and participation in family management. *"Our knowledge on how fathers put fatherhood into practice is still limited and incomplete; we know more about the changing culture of fatherhood, on what is expected from fathers and how fathers themselves think about fatherhood, but concerning the conduct of fatherhood, the practice of fatherhood, we must be satisfied with some spotlights, and the little data we can rely on is not consistent".*

What role do men play in family management?

Father's participation in family management differs from Member State to Member State. Employment patterns show that men are not the sole bread-winners (both parents often work full-time), but men and women do not contribute in the same way; also patterns of employment (both parents working full-time) do not match the patterns of domestic work (women still do the majority of domestic work), and this relationship requires better understanding. The main point stressed by Michael Meuser was that *"there is a huge gap between the culture of fatherhood (that focuses on fathers' involvement in family management) and the conduct of fatherhood that is still affected by traditional patterns of the male breadwinner. If on the one hand men wish to participate more in family life, as some surveys indicate, on the other hand they only fulfil these wishes to a low degree".*

There was a discussion on the need to carry out further research on this gap, namely the need to link family research and gender research. According to Michael Meuser, *"until now fathers have being studied as a uniform group by comparing fathers' practices and attitudes with mothers' practices and attitudes, but we need more data on specific groups of fathers, more data on class, ethnicity, and educational background of fathers; working-class fathers usually*

155

do not participate in the discourse of involved fatherhood as educated middle-class fathers do. However working-class fathers are involved in family manage-ment on a very pragmatic basis, they do it but they do not talk about it; middle-class fathers regularly talk about it but seldom do it; therefore research should focus more on practices than on discourse".

Quality time concept of fathers' caring does not necessarily create the father's sharing

According to Michael Meuser, several images of fathers coexist: the tradi-tional breadwinner; the modern breadwinner; the holistic father. Qualitative and quantitative studies show that the modern breadwinner father is the most common pattern among contemporary men (the father sees himself as the main breadwinner, while the mother is responsible for domestic work, childcare and family life, but the division of work is not very strict); the modern breadwinner assists his wife in domestic work; identity is both work and family-centred; his presence within the family is relatively high during pregnancy and after childbirth, but decreases afterwards.

Another point stressed by Michael Meuser is that we cannot talk about a father's contribution to family management without talking about the struc-ture of the labour market. Changes in family management and getting fathers more involved are not only caused by changing attitudes towards fatherhood but can also be caused by structural changes in the labour market and working conditions. These take place independently of fathers' decisions and intentions: *"in understanding changes of family life we must go outside the family and take the workplace more into account".* The major question is: how is it possible to combine paternal engagement and family management with an occupational career?

The family perspective. Are there qualitative studies on the subjective perspective of family members?

The main idea stressed in this discussion was that the wellbeing of families is related to families' choices. *"How do they create and plan their family life? What are families' real needs today?" "What do they think about gender gaps in family management, task allocation and work-family balance?"* Participants agreed that although there are some studies at the national level, there is a gap in comparable cross-national studies, both qualitative and longitudinal.

Family management and the life-course perspective

It was often mentioned that when studying family management there is a need to consider the transitions in the life-course (to parenthood, children entering school, children leaving home, caring for elderly), and in particular the aspect of caring during the life-course: for children, teenagers, grandparents. It was also stressed that it is important to include in research the role children play in the allocation of tasks.

There was also a discussion on how economic pressures in a time of economic crisis impact on family management and affect family decisions (one participant mentioned the case of mothers in Romania who take parental leave even though they leave their children with family relatives and go to work due to financial constraints).

Gender equality perspective and family management of unpaid work

The idea of a gap between men and women's discourses and their daily family management was again raised. It was recognised that there are some quantitative studies which show that men want to participate more in family life, but it is important to research further why there is still a huge gap between rhetoric and practice. The importance of a qualitative approach to this issue was stressed, in order to have a better understanding of gender interactions within families. Regulation was underlined as a key concept for understanding this type of negotiation.

> "The patterns and trends of family management in the European Union show that female participation in the labour market is increasing across the EU in each Member State. The male breadwinner model is being replaced by alternative models, with variations between and within countries. Characteristics of welfare policy have been found to be responsible for cross-country variations. Good quality childcare services with a generous parental leave system can be major tools in reshaping female employment patterns. Yet, women still spend less time in the labour market, are more likely to take part-time jobs, and have more career breaks than men do. At the same time they are still primarily responsible for housework, as well as for child rearing, spending on average twice as many hours on these activities as men do. Very little qualitative research has been carried out to assess this phenomenon".

Another issue raised is that family management is completely different whether there is a child or not. After the birth of the first child both parents increase family time, but there is a growing discrepancy over time: men increase their working hours and women increase the amount of time spent with the family. It was suggested that it would be interesting to compare the division of domestic work in childless couples, who tend to share household tasks, but less so after the birth of a child.

The possible long-term effects of policies on gender equality was also mentioned, given the example of Swedish men, who seem to participate more in family management, with greater gender equality: *"to what extent is there a kind of long-term policy impact on the development of such participation?"*

How to value unpaid work? Are there policies that value unpaid work? How to value parental leave more? What would happen in Europe without all this unpaid work?

The issue of the value of unpaid work was raised. Two possible ways of valuing parenting work were mentioned: recognition and remuneration. Participants agreed that there are different psychological effects of paid and unpaid work on the individual; *"if you get paid for work you feel you get appreciation, unpaid work is valued in a different way".* A suggestion was made to include unpaid work related to childrearing in pension calculations as well as in GDP.

Major gaps and challenges for research

- More research is needed on the interactions/negotiations between parents regarding the division of paid and unpaid work (their practices, perceptions, justifications, preferences, factors that influence work sharing). Looking at everyday aspects of family management and negotiation processes between father and mother.
- More research on best practices for valuing unpaid work should be carried out.
- More comparative research on the subjective perspective of family members: what they really want, what their needs are.
- More research on the impact of structural constraints, cultural factors and welfare policies on family management.
- It is important to include children's contribution to domestic/paid labour in research, to study family management according to children's age, and to take into consideration children's views and opinions regarding their wishes in family management.

- Linking family research and research on the labour market (particularly regarding choices in family management and structural constraints set by the labour market and career orientation).
- The impact of the economic crisis on family decisions and family policies should be better researched.
- There is a huge need to include the male perspective on family management since there is a lack of quantitative and qualitative comparable studies on men's practices and perspectives of family management.
- There is a lack of research on the images of fatherhood, the conditions and obstacles for realising these models, and little attention has been given to the constraints on family change caused by labour market demands.
- More research on how new adolescence patterns, substance abuse, violence and insecurity affect family allocation of task management and involvement in the work force; the importance of studies dealing with work/family conflict among employed mothers of adolescents with high risk factors for substance use.
- More research on families with high stress levels (also identifying the major stress factors, what promotes stress and what diminishes it).
- Research should take into account the family management of marginal groups (minorities, migrants, families with disabled persons, families affected by poverty, etc.).
- Research should take into account the diversity of families (heterosexual, same sex, blended, single parents, families living together apart), in family management.
- The need for more research on quality time parents spend with children (primary/secondary childcare time) in order to get to know the best type of educational attitudes parents have towards children, with regard to setting limits, teaching, listening skills, educational security, sharing a good time together, etc.
- More research on best practices in work-family balance, which allow children to develop a secure attachment to their parents and reduce stress within the family during the early years of a child.
- Impact of early high quality childcare on child's wellbeing and development; long-term effects of early life experiences of maternal deprivation; the benefits of parental leave from the perspective of the child's wellbeing; impact of affordable high quality childcare on women's participation in the labour market; understanding the conditions which are required to preserve a child's attachment to parents/mothers when they work outside home; understanding the family-friendly actions that employers can take to preserve attachments

between children and parents (collection of best practices in order to promote and defend attachment).

2.1.6 Existential Field 6 - Social care and social services

Organisation of the focus group and keynote speeches

The focus group started with a presentation by Marjo Kuronen (University of Jyväskylä, Finland), who together with Kimmo Jokinen and Teppo Kröger authored the report "*Existential Field 6: Social care and social services*". Marjo gave a summary of the main findings of this report, which reviewed most of European comparative research carried out since the mid-1990s on social care and social services. Marjo's presentation was followed by a keynote speech by Anneli Anttonen (University of Tampere, Finland) on "*Care policies in transition*".

Anneli Anttonen commented on the report by discussing issues and questions which are currently at the heart of comparative research on social care. She stated that social care is of growing importance due to ageing and the related increase of care needs, but also due to the adult worker model which requires both parents of young children to work and which has gained popularity within EU employment policies. According to her, a key question is what happens to informal care – because it is currently the major source of care and will remain so in the future. For example, the tendency to expect workers to extend their careers in paid work (working longer hours and working longer over the life-course) can represent a kind of a threat to care and informal care, as it can create difficulties for spousal and other carers and therefore give rise to new tensions between paid and unpaid work. She stated that in the context of labour market relations and changes in employment we need to look at care as real work, because care is work and an activity somebody has to do: "*care is a labour-intensive activity*".

There is a continuous need for more and better care resulting from the expectations of the ageing middle classes: this is a big challenge for care services and policies. Good quality care is particularly important for the future, as people develop more consciousness of social care. On the other hand, there are major inequalities (care and social capital are needed to manage and negotiate complex systems of social policies); and there are significant differences between groups of people in terms of access to care services and informal resources. Anneli Anttonen identified an international tendency to move from services-in-kind to monetary benefits and the emergence of new hybrid forms of work and care. She also commented on the concept of defamilialisation. This is a problematic concept because it decreases the role of families as a source of care. However, she believes people still invest morally in families and informal care and that family responsibilities remain strong everywhere: *"the moral commitment to informal care is very strong"*. Defamilialisation may be related to social policy, and although there are more public policies, this does not mean that the idea of family is getting weaker. Even if people are moving into paid work, they still have a strong commitment to family members closest to them.

Anneli Anttonen also mentioned that transnational care is an emerging field that is becoming central in international care research. She was referring not only to immigrants as care workers or care workers in private houses but also to the different strategies migrant families have to develop in order to care for relatives living in another country or continent, and the importance of transnational relations of care and how care is organised. Finally, she raised the question of why the European Union has a European Employment Strategy but does not have a European Care Strategy: *"if the European Union wants to promote employment for everyone it must take into account care, what happens to care, they go hand in hand [...] if the European Union needs an employment strategy it also needs a care strategy"*.

A discussion followed the two keynote speeches, and four stakeholders presented statements. After the statements, the remaining time of the focus group was used for a general discussion about major gaps within comparative social care and social service research. The following paragraphs summarise the general discussion that took place in this focus group.

Main topics discussed and contributions from stakeholders

Some stakeholders' highlighted the need for more research on dynamics within families, particularly on unequal gendered power relations within the family and gender hierarchies that spread over different spheres of life. It was stated that without understanding changing gender inequalities within the family, it is not possible to reach gender equality in society. Attention should in particular be paid to domestic violence: it should be seen as gender-based and as a public concern, not as a private family affair. According to stakeholders, attaining gender equality requires a reform in values and gender role stereotypes as well as in the general social organisation of society. It was also stressed that although there are many prevention and protection programmes for victims of domestic violence all over the Europe, within the EU there are still no common standards on domestic violence. Social policies are considered to be crucial for promoting co-ordination between all the actors involved in the process of implementing the law. Public awareness campaigns, psychological support services and specific protection measures such as shelters were also mentioned as important policy measures in order to approach and protect victims of domestic violence.

Other stakeholders focussed on the wellbeing of children and young people, arguing that there have been dramatic increases in inequality across the EU, bringing greater marginalisation and pockets of disadvantaged communities, while the current economic crisis is plunging more families into poverty, governments are slashing budgets, and preventive and support services are under threat. As a consequence, a major problem is that of children and young people ending up in child protection and criminal justice systems. According to these stakeholders, governments should instead invest in high quality prevention, early intervention, and secure access to adequate services, including child and health-care (affordable services with universal access), and increase training and professional recognition of people working in the service and care sector.

Another point stressed by stakeholders was that families should be able to make choices in relation to what kind of education they want to give their children, and whether to choose if they want to care for their family members regardless of age. The example of Spain was given, where maternity leave is very short (only four months) and childcare services are limited. Due to the lack of childcare, some parents are forced to stay at home. In comparison, Nordic-style childcare services make it easier to achieve a balance, and the French system offers many opportunities as well. All in all, more flexibility and choice are needed within both childcare and eldercare services.

Stakeholders pointed out the growing demand for family support services (cleaning, cooking, etc.) due to ageing. Families' choice of social care arrangements depends on several factors such as existing formal care services, social networks and organisational cultures (employer perspective). In Europe there are huge national differences in the use of these services and too little comparative analysis: one of the barriers is the financial resources of families. Moreover, stakeholders stated that the links between migration and care are one of the main future challenges in the domain of social services. As a consequence, issues such as the qualifications and working conditions of migrant care workers ought to be studied. Additionally, the barriers that older people with a migrant background and/or with the Alzheimer's disease face in accessing care services also require more research attention.

Defamilialisation and familialisation of care

How can we address social work/social care and public services in order to strengthen families and keep them together? Who is giving the care is the crucial question for European countries: *"there is still a dual system of care: either family members are cared for within the family at home by the mother or by a middle-aged female who has given up her professional work for almost nothing in terms of financial compensation, or the care is done in an institution where the family cannot be; however there is a third option which is missing: a kind of intermediate care arrangement, a home care worker, or an institution for some hours of the day or some days of the week".*

It was pointed out that there should be more research on this mixed solution: *"the perspective of the care receiver, e.g. being an adult or a child is important, because whose voice is actually heard? For example, when we promote national care policies, whose voice is heard? Is it empowerment of users or is it empowerment of professionals and care workers?"*

What will happen to informal care in the near future? To what extent should informal care be regulated by the state?

It was stated that informal care is the major source of care and that informal care is one of the central questions in the field of care policies. One crucial question raised in the discussion was *"what happens to informal care? Will there be less informal care in the future?"* The discussion also pointed out that if, on the one hand, informal care is important because it allows families to stay together, on the other hand we do not know to what extent these families have the knowledge and skills to care. It was also stated that *"the best care is given within the family but also the worst",* thereby raising the question of how to control what happens in informal care within the family, in particular in the case of abuse in care relations: how to intervene?

Policies do impact on families

> *"If we look at public expenditure on families and children as a proportion of GDP we see that some countries have invested more money in childcare then others. If the government invests a lot of money in children and families (as the Nordic countries did in the 60s and 70s through child allowances and different types of benefits paid to families) this will have a positive outcome in the long term: child poverty, for example, is very low in Nordic countries. In the long run care policies and special childcare policies impact on the wellbeing of families and children. How to study this? By doing longitudinal comparisons between countries?"*

How to monitor and compare the quality of childcare services across European countries?

The availability of services is important but their quality is equally so. However, there is little data in European databases: *"if you want to do a critical comparison of childcare and look at differences and outcomes across European countries what you get out of EU databases is coverage rates and maybe how many people are working per child. This data is too limited: there is a need for more detailed data in order to compare quality of childcare and to identify the reasons for the different outcomes".*

What is the impact of access to and use of social services on reducing poverty and inequality? To what extent are available services reaching out to the most vulnerable groups?

Participants pointed out that there are new types of inequalities. There are vulnerable groups, vulnerable consumers and vulnerable managers of the complex policy care system, not only because of the lack of money or due to the traditional criteria associated with social class but also due to *"lack of knowledge or lack of language skills, specially when there are a lot of people suffering from memory diseases".* The importance of understanding the links between care and social capital was also emphasised.

A question was raised in connection with the social value of childcare provision. According to this view childcare services might have a role in *"achieving social cohesion and fundamental social democratic goals – making gender equality opportunities a reality, eliminating poverty, maximising life chances of all children irrespective of the parent's socio-economic background, reflecting the importance of high quality access and affordability of childcare services".* The contribution that early childcare services might make to breaking cycles of family deprivation, reducing inequalities and combating discrimination was also stressed. Ethnic minority children were also mentioned, particularly the fact that *"those whose native language is not the home country's benefit enormously from early childcare since they can get a start in language learning and improve their chances of integrating later on at school and within their communities".*

Families' perspective on care

"It is very important to look at care from the point of view of families and households". The discussion focussed on the issue of providing families with all the necessary conditions for making choices, considering not only those who

are in the labour market but also those outside it. Are families free to choose between full-time and part-time employment as well as between types of care services? And do they wish to use formal childcare? "*What does the shift from welfare government to welfare governance mean?*" and "*what is the role of the family in this shift?*"

Children's perspective

In the debate it was pointed out that there is an urgent need for research to focus more on children's perspectives and therefore also on their psychological and educational needs. It was also stated that the needs/interests of children are sometimes different from their parents' interests: "*what the children are saying and what the adults are saying is not the same, good services may not be what children want*".

Employers

The attitudes of employers were considered to be vital: if they are against female/maternal employment, then public policy measures like childcare and parental leave provisions are not sufficient to bring about change. Employers' interests influence flexible working arrangements, and there is a need today to promote more worker-friendly/family-friendly flexibility: "*Why should employers invest in family-friendly measures? The social responsibility of employers and private businesses needs to be restored, but there is also clear research evidence that proves that family-friendliness brings employers different economic benefits*".

Connections between childcare and eldercare

Linkages between elderly care and childcare policies, and between these and research, are often missing, because they are administered separately from each other. A life-course perspective is needed in policy and research. Participants felt that there are almost no reconciliation measures/studies on the family carers of older people. There are tensions and contradictions between the informal and formal economy/work as well as between childcare and eldercare, and research needs to highlight these: "*childcare seems to have a different status to elderly care*"; "*there are no special Europe-wide leave arrangements for the care of older people as there are in the area of parental leave*".

Migration and care

> *"Migration and transnational care will be crucial for policy and research in the future". "Policy should take into account the differing needs of different migrant groups". "Children who are left behind in the country of origin are in a very difficult situation: reunification of families is also an issue for care policy".*

Major gaps and challenges for research

- More comparative research on care leave arrangements, on state policies in this field and on company-level policies at a European level. There is more information on family-friendly company measures related to childcare but not so much information on those related to elderly care.
- More studies on informal care, including spousal care, mother and father care, different types of care in family relations.
- More research on organisational cultures and the employers' perspective on care (example of the project "Working Better"[5]).
- Research on how men are discriminated against in the labour market if they have to care for dependent relatives (elderly but also children).
- More comparative research on what the future generations are expecting from public care services to support the last stage of their life cycle in the long term. In what ways do they plan to resort to social services?
- Research on young people's opinions on the elderly.
- More research is needed on inequalities related to social care infrastructure, by looking at developments at global and regional levels, and the impact of accessing and using social services on inequalities in society. How do social services help to reduce poverty?
- The importance of incorporating the views of beneficiaries in research on care: the perspective of people in need of care/care receivers is still mainly missing (including children's perspective). Qualitative comparative research, the best way of understanding people's points of view and to explore how people experience care, is very useful here.
- There is a need for studies on new forms of dialogue between the generations (especially in families without grandparents).
- More research on the internationalisation of care and the different forms it takes (relations between care, gender and migration issues; global care chains and transnational care; different strategies for

---

[5] See *http://www.equalityhumanrights.com/key-projects/working-better/*.

caring for relatives living in other countries or continents; caring as an international business; care needs of migrant families; migrants as 'grey labour' in home-based care and formal care services).

- More research is needed on the dynamics (tensions and contradictions) of the changing relationship between formal and informal (family-based) care, and on changes in public policies over time (in-depth analysis of policy formation and the delivery process).
- There is a lack of knowledge and not enough data on care workers in private houses.
- There is very little research on children who receive institutional/foster care imposed by the government (experiences of different countries, different solutions).
- There is a need for more research on international adoption.
- There is a lack of research on children whose parents are no longer taking care of them, for example those whose parents are in prison or mentally ill. Likewise research is scarce on children who have previously been in institutional care and return home, as well as on the skills parents may require in order to take them back.
- Research on the best childcare solutions from the point of view of the child's interests and wellbeing: what are the best care arrangements to fit children's needs?
- Research is still scarce on the use of technology both in formal and informal care.
- Existing research concentrates on care for children and older people: care needs of other adult family members (e.g. people with disabilities) is missing.
- Quality (and not only quantity and availability) of formal care services should be studied in greater depth.

2.1.7 Existential Field 7 - Social inequality and diversity of families

Organisation of the focus group and keynote speeches

The session began with a presentation by Karin Wall (Institute of Social Sciences of the University of Lisbon and leader of Existential Field 7), of the report which summarises the state of the art of research on "Social inequality and diversity of families". Her presentation highlighted the main results in terms of major trends and research gaps in the four fields of analysis included in the general topic of "Social inequalities and diversity of families": migration, poverty, family violence and social inequalities of families. Following this presentation there was a brief discussion which was enhanced by two other presentations from experts as well as by the statements of stakeholders who took part in this session[6].

Claudine Attias-Donfut (Caisse Nationale d'Assurance Vieillesse) made a keynote speech on *"The social destiny of children of immigrant families – unchaining generations"*. She based her presentation on the results of research on intergenerational relations among immigrants. This study covered several aspects (family structures, living conditions, cultural norms, solidarities and conflicts) and examined a number of two-way relationships (between parents and their children; and between parents and their own parents). She addressed three main questions: 1 - Are inequalities in educational performance mainly determined by the socio-economic circumstances of the families in the country of immigration? 2 - Is there any influence of the social milieu of origin (in the country of birth)? 3 - Do ethnic origins (birth country) play a role? She mainly concluded that: family socio-economic circumstances and neighbourhood are stronger determinants than country of origin; the parents' social milieu of origin is more important than 'ethnicity' or country of origin; immigrants' daughters perform better and have fewer problems; only a small minority of immigrants have serious problems; the majority of children are on a path to success.

Maria das Dores Guerreiro (ISCTE – University Institute of Lisbon) made a presentation on *"Social inequalities and employment patterns"*. She highlighted the results of two surveys, *Quality of life in a changing Europe*[7] and *International European Values Survey*[8] which contain comparisons between European countries relating to people's overall work and life satisfaction. She concentrated on variations across countries, activity sectors, occupations, social class and gender. For example, in countries with very long hours in paid work, men and women are less satisfied with work and family. On the other hand, countries where people have a higher feeling of job insecurity also show lower degrees of life satisfaction.

The main idea stressed by Maria das Dores is that there are several factors determining family/work stress, such as: sex, marital status, age, having children at home, number of hours in paid and in unpaid work, sense of workload, occupation, cultural values, etc. It is important to take all these factors into account when trying to understand how families combine family and work and how they feel about it. She also emphasised that inequality in terms of families' wellbeing may be caused by families' internal configurations, such as: the age of family members, their care needs, and the way paid and unpaid work

---

[6] Collette Fagan was unable to be present, though she sent us her presentation and notes.
[7] See *http://www.projectquality.org/*.
[8] See *http://www.europeanvaluesstudy.eu/*.

is organised. According to Maria das Dores Guerreiro, there are specific groups which are still understudied: unemployed families, families affected by health problems (physical or mental disabilities), families whose children have been taken into foster care: *"all these families are known as dysfunctional families, but very little is known about their configurations, work-life balance, support networks, children's socialisation process"*. A major question is: how are policies supporting families not only with respect to financial resources but also in terms of skills and the empowerment they need?

Main topics discussed and contributions from stakeholders

The experts' presentations and the contributions by stakeholders underlined several key questions which were discussed by the group. The following topics summarise the debate within this focus group.

Topic 1 - Social inequalities and families

How unequal are European societies? How does this impact on families?

Participants agreed that social inequalities deriving from the unequal distribution of economic, social, educational, and cultural resources continue to impact strongly on family forms and dynamics, affecting families' opportunities and economic wellbeing. It was recognised that there is a lack of studies connecting social inequalities and family life at the national and particularly the cross-national levels; in the major databases - the European Social Survey, the International Social Survey Programme (ISSP) - a large amount of data has been examined in terms of gender equality across European societies, but social inequality and the linkages between social inequality and families have not.

How is social inequality produced and reproduced in families? Are policy and research only looking at the effects of social inequality or are they also trying to deal with the origins of social inequality?

> *"Researchers have moved away from the issue of social inequality and family life during the last few decades, and the focus has been much more on paradigms highlighting the concepts of agency, individualisation, choice and individual diversity [...] class analysis seems to be not so useful anymore [...] in democratic and individualised societies individuals and families have more options, they construct their families and their biographies with greater freedom and more opportunities; but social inequalities have been*

increasing in European societies, and their impact on family life must be taken into account. In this context the concept of class is probably still useful, even if it implies rethinking theoretical approaches based on these concepts".

The coexistence of old and new patterns of social inequalities in families. Research points in two different directions:

"On the one hand, social inequalities in family life seem to follow old and more traditional patterns of social inequality. On the other hand, these old patterns can coexist with the emergence of new patterns which need further research: for example: a) social inequalities linked to new types of conjugal homogamy; b) inequalities linked to differences between dual earner couples and male breadwinner couples; c) new forms of inequality which are emerging between upper and lower-class families: we are quite used to the trend according to which upper-class families spend more time helping their children with homework while lower-class families spend less time, but in fact what recent research seems to show is that both lower-class and upper-class parents spend the same amount of time helping their children with homework. Nevertheless, upper-class families provide other types of support to children. This is not being properly researched at present".

Cumulative aspects of social inequalities

"There are signs that there are cumulative processes occurring in families and individual lives, e.g. disabled people are more likely to be victims of rape; migrants have a higher probability of belonging to a lower class; the fact of belonging to a disadvantaged group might in turn be related to the likelihood of being disadvantaged in other aspects of life later on".

How to re-examine social inequality in Europe? Are there sufficient and effective indicators in international databases for measuring social inequalities?

"International databases have focussed on classical indicators. We need to go beyond them. If we only have indicators showing that European societies are unequal from the point of view of income (GDP), we do not really know how social inequality is being

produced, so we need to take various cultural, material and eco-nomic indicators into account".

The analysis of inequalities is still centred on certain types of families

"A subject like this - social inequalities and diversity of families - should include a broader spectrum of families: for example, fam-ily reunion is more difficult or even impossible for joint children of homosexual families. Same sex families are also discriminated against. The gender pay gap is higher in female-female families".

Major gaps and challenges for research

- There is very little research on new patterns of social inequality in fami-lies and on new forms of producing inequality. Some family forms and dynamics are very strongly related to class and others are not. It is impor-tant to carry out more national and cross-national research on social inequalities and how they impact on a variety of family indicators (e.g. living arrangements, interactions, division of labour, family formation and dissolution, patterns of fatherhood and motherhood, family networks, resource flows). It is also crucial to understand the process whereby fami-lies produce and reproduce material/social/cultural advantage and disad-vantage (e.g. the role of intergenerational resource flows); case studies are needed to analyse how families are transmitting and reproducing inequality and how they manage to improve their children's life chances.
- There is a need to know more about the cumulative aspects of social inequalities in order to understand the processes of cumulative disad-vantage that affect specific categories of families and people (e.g. the disabled, immigrants, minorities). More comparable data is required on families outside the labour market, the unemployed, the retired, the sick; families affected by health problems, physical or mental disability or some kind of addiction; families whose children have been taken into institutional care, families labelled as "families at risk". Very little is known in terms of cross-national studies on their configuration, age of family members, forms of interaction, organisation of paid work and unpaid work, support networks, children's socialisation, the way they balance different spheres of life.
- Research does not sufficiently cover the diversity of families with regard to lesbian and gay families' experiences. More research is needed on the gender pay gap in lesbian families as well as on other aspects of family life usually studied for families in general.

Topic 2 - Migration

Migration is a major challenge for European families, research, and policies: *"Migrant flows to Europe (as well as inside Europe) continue to be significant, with dual opportunities in the labour market (skilled and unskilled) as well as more diversity in family migration. Feminisation of migration and new types of family migration are emerging (e.g. women first migration). The number of foreign-born and mixed-born children (of couples of different nationalities) will increase over the coming years, thus representing a major challenge for families (for example, the need for families to negotiate cultural differences within schools and in local communities) as well for policies (e.g. the educational system) and for research. Not enough research has been carried out on how European societies are going to deal with this".*

Policies and attitudes to family reunion are becoming more restrictive: how is this going to affect immigrant families and their integration in the different European countries?

> *"In a context of restrictions on family reunion, and considering the emergence of new patterns of family migration (e.g. the feminisation of migration and mixed marriages): what happens if we have more and more couples who come to Europe and leave their children behind, in South America, in Africa? What does this mean from the point of view of parenting and from the point of view of integration in the host society?"*

The increasing importance of the concept of 'mobile families' – the need to consider all types of mobility. Analysing mobility and how it impacts on the reconciliation between work and family life

> *"Mobile families are likely to experience social isolation from kinship (as Jean Kellerhals said in his presentation in the plenary session); mobile families might have significant problems reconciling work and family life: caring for young children in the host country while caring at distance for children and other older relatives who were left behind in the sending country. How does this affect integration?"*

Mobile families and transnational care

"An issue that has recently made its way onto the research agenda (and is related to the feminisation of migration for the care sector) is the complexity of caring relationships and the 'transnationalisation' of care – the difficulties of taking care of children and (for example) other relatives who are left behind in the home country".

Mobility as a sense of Europe

"The concept of mobile families illustrates a kind of a European sense of family".

Mobility and the gender equality perspective

"The link between the concept of 'mobile families' and internal mobility within the European Union from a gender perspective: mobile families impact not only on working-class families but also on middle and upper-class families, particularly those who have highly skilled occupations, for example people involved in science careers. Women and men in highly skilled occupations have high expectations of mobility, but the ability to go abroad for a longer period is also largely related to men's and women's differing ability to cope with the demands of career progression".

Mobility and the life-course perspective

"It would be interesting to include the life-course perspective when studying mobility, because mobility seems to occur in specific life stages and may have different consequences for individuals and families according to whether it happens before or after having children".

What is the social destiny of children of immigrant families? Migrants from countries outside the EU face a greater risk of poverty, low integration and social mobility

"The importance of neighbour and family networks: this is a major challenge from the point of view of the integration of second and third generation immigrants".

"There is a need for more research on social mobility and the education-al success of children of immigrant families who have attended crèches and pre-school [...] are they doing better than previous generations?"

Major gaps and challenges for research

- European case-studies of international family migration tend to assume traditional paradigms of family organisation - the nuclear family above all - and have not fully explored the variety of family and household types which derive from home-country settings. There is a need to rethink the concept of families (male breadwinner versus many different types and forms of migrant families).
- It is also important to focus on changes within the family resulting from immigration: new types of family forms and organisation of gender roles (e.g. conflicts over women's roles, possible changes in the construction of masculinity which may affect both immigrant and non-immigrant populations).
- Further research on transnational families: the impact of national and cultural combinations on relationships, men's, women's and children's lives, host countries' attitudes; EU citizens travelling, studying and working abroad, etc.).
- Studies on students' migration are very recent and growing fast (examining social status, mobility and immigration policies).
- Need for research on mobile families according to a broader view of several types of mobility (see discussion).
- More research on the social mobility of children of migrant families. There is no data comparing cohorts of migrant children attending childcare in order to evaluate their social mobility.
- Little is known about undocumented immigrants or asylum seekers, those who are 'below the radar'. There is a need to improve the ways of reaching out to this group, to obtain data on illegal immigration, and to conduct further studies on aspects of health and social insurance for these immigrants, as well as on the impact of illegal immigrants' circumstances on their children's life chances.
- Studies on retirement migration of healthy north-western Europeans to southern Europe; but also within each European country, because more immigrant people will get older in the host countries, and there are no studies on this.
- There is still little knowledge on how cultural differences are being negotiated. How are host societies (and families in the host societies) responding to increased levels of immigration? What is going on in the schools?

- The effects of (limited) political participation on immigrants' integration and a deeper analysis of the reasons why naturalisation and dual citizenship are used (or not) by immigrants and their offspring.
- It is crucial to explore the positive aspects of immigration for families and individuals. More research is needed on 'success stories': for example, a better understanding of immigrants' entrepreneurship and related ethnic aspects of the economic benefit deriving from ethnic and social networks and transnational ties.
- More research on immigrants' fertility behaviour; very little is known about the differences between groups, or countries, if they are due to ethnic, cultural, socio-economic or political factors.

Topic 3 - Poverty

The persistence of poverty in European societies

> "How far in each country is there a persistence of poverty over the life-course of individuals, of men and women? In 2007 17 per cent of Europeans were considered to be at risk of poverty. The unemployed, immigrants from outside the EU, children in single parent households, those with low educational attainment levels, and elderly women are regarded as high risk in this context, as are the following types of household: single parents, large families, single persons".

Discussion on the narrow focus of the economic perspective of poverty which uses income as an indicator (Luxembourg Income Survey (LIS), European Community Household Panel (ECHP), European Union Statistics on Income and Living Conditions (EU-SILC))

> "Studies on poverty are based on low income, but indicators other than income should be used. Income is very rarely linked to other types of indicators, for example living conditions".

Limitations of the statistical approach: how can we achieve comparability in statistics on poverty in different countries?

> "Same statistics on poverty mean different things in different countries due to different definitions of concepts and their 'operationalisation'. Hence the importance of looking at households and not only at categories of people, and of combining both quantitative and qualitative approaches.

Poverty over the life-course

> *"The routes into poverty include accident, ill-health, unemployment, divorce, pregnancy, and lack of social and family networks. There is a need for more data on people who manage to get out of poverty, according to different life stages, and on social policy and its outcomes: the role of social policy as an incentive and as an opportunity for reducing the poverty gap. How are childcare facilities related to the prevention of poverty?"*

Major gaps and challenges for research

- More research on the life trajectories of poor people and routes into poverty, but with an emphasis on how to escape poverty;
- There are very few broader studies (both quantitative and qualitative) on the experience of poverty as well on the social patterns of poverty: there is a need to move beyond income indicators;
- Need for more data on the poverty of people who are caring or are cared for by family members;
- More studies on poor people/households in different urban and rural contexts.

Topic 4 - Family violence

Domestic violence continues to be significant

- Several types of family violence were identified: psychological, economic, physical, sexual. It is still largely gender-based (conjugal partners), but also occurs between parents and children, adults and elderly parents, boyfriend and girlfriend.

Domestic violence policies and legislation are still relatively new in many countries (1990s)

At present it is considered a public crime in several national legislations, on a par with other criminal offences.

Specific groups at risk

Low income households, those in which individuals have low educational attainment levels, children in large families and in families with alcohol problems, women with higher educational levels than their spouse, unemployed women with an employed partner, women in the process of separation, pregnant women, immigrant women of uncertain legal status, young women seeking abortion.

> *"Violence is not only about women. Most of the studies fail to take into account the fact that men are also victims of violence. Only 20-50 per cent of all the different forms of intimate partner violence are reported to the police, fewer relate to violence against men. Men seem to be more reluctant to report this violence".*

The problem of violence against disabled persons and elderly persons

> *"About 10 -13 per cent of women with disabilities reported having experienced abuse, a rate similar to that of women without disabilities. For all women, the abuser is often a partner or family member, but women with disabilities are more likely to be abused by health care providers or caretakers".*

Major gaps and challenges for research

- Very recent and little research looking in depth at families and violence, particularly variables and situations that encourage violence; lack of analysis using specific target samples of social categories of families to understand other forms of domestic violence.
- It is important to move beyond the gender unidirectional paradigm predominantly focussed on violence against women and to include violence against men.
- There is practically no research on what factors help people to break out of the cycle of violence.

2.1.8 Existential Field 8 - Family, media, family education and participation

Organisation of the focus group and keynote speeches

Sonia Livingston (London School of Economics) opened the focus group by presenting a brief overview of the main findings of the report on Existential Field 8 - *Family, media, family and education*, which is co-authored by herself and Ranjana Das (also London School of Economics). After this first presentation, the three keynote speakers provided their critical responses regarding the main research gaps and made some suggestions for the future Research Agenda.

Ann Phoenix (Institute of Education, University of London), made a presentation mainly focussed on implications for the family. She began by emphasising the importance of objective (economic factors) and subjective indicators when speaking about and measuring families' wellbeing: *"how people feel about their lives and how they are doing is key* [...] *subjective wellbeing is the key to understanding social policy terms on wellbeing".* She continued by reinforcing her belief in the importance and pertinence of the major trends and findings of the report authored by Sonia Livingstone and Ranjana Das and went on to focus on the major gaps in existing research on this field (see Major gaps and challenges for research).

José A. Simões (New University of Lisbon) made a critical response focussed on youth cultures research, media and family. He raised some questions: are youth cultures a product of media or is it the other way around? Are youth cultures homogenised or are they diversified? One ambiguous and complex relationship which needs further research, he believes, is how media plays a part in the construction of youth itself, in the way youth sees itself and in the way young people identify with what emerges from the media. He also stressed that there is a tension between two tendencies: individualisation (e.g. bedroom cultures, mobile phones) and mobility inside the home (media appropriation is in a complex relationship to space) on the one hand, and togetherness (family socialisation within the media) on the other. An important question for him is: what part does the family still play in media socialisation and socialisation in general?

Naureen Khan (Commission for Racial Equality, London) focused on stakeholders' perspectives on the future potential of this research, specifically regarding EU policy and legislation. Her focus was on the internet, mobile phones and associated technology as well as on the impact that the personalisation of media has on children and in the 'bedroom culture'. She pointed to the need for more research on the positive side of children's internet usage. Research usually focussed on the risks of children's exposure. She felt it would be interesting to know more about what goes on in the bedroom not only in terms of risks but also in terms of empowering children and their rights to privacy. Based on the report's findings, she stated that there is a significant children's usage of the internet: it would be interesting to know more about patterns of internet, mobile phone and other technologies' usage by children aged between six and eleven. She also mentioned the subject of parental mediation: there are various patterns in terms of mediation, but is parents' mediation effective? Is that the right angle to focus on in terms of children's usage? Shouldn't we know more in terms of children?

On the future Research Agenda she believes there are still gaps and challenges in understanding the importance of social networks. She gave the example of Facebook (for example, having thousands of friends on Facebook) and asked, "*What does that mean for friendship and relationships? What does it mean for that generation? What impact does it have on the development of family?*" Another interesting issue is the next generation parents who will be more confident and more aware of technology and how that will impact on their relationship with their children. She ended by emphasising that "*it is important to move away from that risk perspective and to be more proactive in working towards a more positive agenda*". The EU Institutions' approach to internet safety and media and technology is always "*a look in terms of risk perspective and too reactive*". In addition, although there are several EU strategies for media and use of technology, the impact of family research is very poorly covered in these strategies. Finally she pointed out that it is important to persuade decision-makers to carry out more comparative research on all 27 Member States, and not just on a few countries.

Main topics discussed and contributions from stakeholders

The discussion around the themes of safe use of internet, children's exposure to risks, their internet usage and parents' regulation of children's media usage (particularly internet and TV) dominated the overall debate. The following points summarise the discussion within this focus group:

Risks of children's exposure to the internet

"*New technologies such as the internet, mobile phones and video games have enormous potential in a positive as well as negative sense, and therefore we face new risks and opportunities that need to be identified and studied, for example how can parents be helped to develop their educational role at home by knowing both how their children use technology and learning to share that use with them, without abdicating their role. Is there any research on the effectiveness of different kinds of education that children can receive at school about how to be safe, how to participate online, how children connect their views and have a voice in participating?*" "*The challenge is to retain the notion of the child as an agent, but to recognise structure-constrained agency at the same time*".

Media and parenting

"*How can we reach out to parents? How are parents reaching each other? What are parents saying to each other? Where are parents going for advice when they need advice about parenting? They also use the internet (better-educated parents do so more frequently). How much research is there on this? How can the internet help parents' networks and how are they using it for parenting? Peer support, state support, online support – which works best? Social and economic differences do make a difference within families… we cannot have a general discourse".*

The importance of media in sustaining and shaping ethnic identities and transnational links - the example of global care chains - 'emotional transnationalism'

The example given referred to Philippine mothers who go to North America to work in households and cannot bring their children but still care for them at a distance using information and communication technologies as tools, like speaking and seeing through Skype and MSN (they see them every day, ask them for homework).

The impact of media on the subjective ideal of wellbeing

"There is a need for more research on the effects of media, not just the internet but also the effect that TV and the print media, including advertising, can have on shaping adults' values in consumer societies, the ideal family type, life-courses, ideal relations with children and within the family, ownership of property versus poverty and inequality; leisure life-styles, homogenised cultures versus individualism itself. Adults are also affected by media, and this can have an influence on their children. What do we get from ICT in terms of projection of identities and desires?"

The media as a cause of all social ills versus the potential of media in very different areas of life

"Important that the media are seen as neither good nor bad, but rather as a space, or as a resource which shapes all else; shift away from media effects and moral panic towards understanding the ways in which the media shapes identity, how everyday lives are mediated. For example, how can the media shape societal change towards more sustainable consumption?"

How can we use new technologies to support family relationships?

> *"A lot of research is oriented to the individual. It is difficult to find specific family-oriented impact research. What are the media doing in terms of family relations? Are they supporting or harming family relationships? These questions are not yet being researched. The example of online family mediation, courses on parenting, marriage preparation courses, education within schools in order to enhance family relations [...] there are lots of ways communication technologies can impact positively on family life, e.g. intergenerational interchanges and grandparenting through ICT".*

The impact information and communication technologies (ICT) technologies can have on reconciling work and family life

> *"Major impact on time management: the use of media to achieve harmonised management of time and family relations – is this sufficiently researched? Are we doing enough proactive work and trying to find the best solutions to help families reconcile work and family life? Existing research shows that technologies promise better adjustment of the work-life balance, but in fact their use tends to be directed by the workplace".*

Major gaps and challenges for research

- Focus on different types of families: there is a need to examine similarities and differences within and between households. Households are the site of reproduction of differences in ICT use by age, but also other variables. More research is needed on specific groups of families such as those with either disabled or dependent persons.
- More research on the process of how knowledge is transferred from the younger generations (who are better able to pick up new things) to the older generation in the household.
- There is very little research on the way ICT is used in mediating transnational family lives. Studies on mediated transnational, 'glocal' and hybrid identities need further development. In what ways are ICT the key to new modes of mothering and parenting for immigrant people, e.g. through 'global care chains' and 'emotional transnationalism'.
- The impact of new technologies on health, access to health, information on health; the linkages between the media, ageing and health support services.
- Research to support decision-making and monitoring of the amount of information (multiple messages) people receive; on how the same

message might be received differently by different members of the family.

- It is important to research the impact of media-transmitted biased messages on behaviour economics, both within and between households, and between the generations (some types of bias influence children more than the elderly).
- More research on media evaluation programmes – the evaluation of the effectiveness of programmes aimed at improving media competence and media literacy (in childcare, pre-schools and schools in general).
- Research is needed on how the media can shape families' attitudes towards more sustainable consumption (how one member influences the whole behaviour of the family e.g. mothers' purchasing decisions).
- More research on the bedroom culture and social networks of children, particularly internet usage among children aged between six and eleven.
- Need to refocus research on individuals' media use in terms of implications for family relationships, e.g. how ICT is used to mediate the making and breaking of relationships; the contribution of ICT to helping or hindering work-life balance e.g., via working from home – 'teleworking'.
- More research on how media can be a tool to assist parenthood, on social networks for parents; how are parents using the internet and talking to each other? How are they using the internet to help them in parenting? How they do advise each other? Support advice for parents in educating their children, etc.; research on parents' feelings that they have information needs: which parents, in which contexts, which information needs?
- Research on the impact of media in financial education, specially connected with the crisis, and the ability people actually have to manage the household budget.

Gaps in research and methodology. Three priorities for a future agenda

The media as content

More research is needed on how media content (on 'old' and 'new' platforms) supports or undermines family life, childhood and identities, and this should be available to guide parents, based on recognition of the fact there is a huge information need among parents (the 'sandwich generation').

The media as a tool

Diverse media platforms can be and are being used as tools to reach families and provide information, guidance and advice on various issues. Evaluation is needed to identify which approaches (messages, platforms, and contexts) are effective.

The media as infrastructure

Almost every aspect of family life - relationships, identities, health, education, values, work-life balance - is dependent in some way on media and information technologies. These bring opportunities and risks, and demand new critical and digital skills. Recognising this 'environmental' or 'infrastructural' aspect of media requires that media be considered a vital part of research projects on diverse aspects of family life.

2.2 Workshops on key policy issues

2.2.1 Workshop 1 - Transitions to adulthood

Organisation of the workshop and keynote speeches

The workshop began with an introductory report by Barbara Stauber (University of Tübingen) focusing on *"Transitions into parenthood, lessons from the expertise for the Family Platform"*.

According to Barbara Stauber, young people's entry into parenthood is tied in with other aspects of the complex process of transition, in particular the very important transition from school to work. After illustrating the concept of 'biographical transition', she focussed on the new problems that this process might involve for young people, drawing attention to changes over the last few decades (transitions that are fragmented and de-standardised, reversible and subject to risk and, above all, individualised). Within this framework, the adoption of public policies in support of young parenthood takes on particular importance, together with the deployment of policies aimed at facilitating the entry of young people into adult life. From a more theoretical perspective, Barbara Stauber stressed the importance of two phenomena with reciprocal tensions: the agency of young people, and the concept of capabilities.

The first term refers to *"the socially contextualised and temporally embedded ability to decide upon and perform the practices of everyday life"*; the second to *"the availability of opportunities – it is not enough to formally remove inequalities in resources. It is also necessary to actively facilitate access to them, creating real opportunities for (young) people to perceive their rights and transform them into claims".* In summary, the crucial question Barbara Stauber addressed is the importance of highlighting both the capacity of young people to act as protagonists in the processes of change that are taking place today as well as the constraints they have to cope with. From this perspective, the tensions between these two poles constitute the framework in which entry paths into adult life and parenthood unfold. Barbara Stauber concluded her presentation by identifying gaps that still persist in research (see *"Major gaps and challenges for research"*).

After Barbara Stauber's presentation, the discussion was opened to all participants including stakeholders' statements. The following paragraphs summarise the debate and discussion which took place.

Main topics discussed and contributions from stakeholders

The debate was polarised. Some of the participants shared the sociological perspective that considers transition to adulthood (and to parenthood) as a social construction – and as such, a phenomenon subject to variations at the historical and social level, influenced by political regimes, welfare contexts and so forth. Others, by contrast, expressed an individual (in the sense of extra-social) vision of the transition, relating it in an exclusive manner to the will of the individual/young person to confront his/her entry into the adult world. This latter perspective was focussed more on the concept of responsibility and "taking responsibilities" as an act on the part of individuals and the crucial marker of transition into adult life. In response to this position it was underlined that the way in which young people create their own cultures and give form to their own ways of life (and worlds) occurs within given social contexts and on the basis of specific (and unequal) economic, social, cultural and family resources: *"transition to adulthood is a social process which means it does not depend on the individual as a kind of non-social human-being".*

There are many paths for entering adulthood

The first issue that was emphasised in connection with the transition to adult life was the pluralisation of its forms and the growing social vulnerability that characterises them. More generally, the social and economic climate today, marked by a high level of uncertainty, has a negative effect both on the transition to adult life and the transition to parenthood. Here, the family of origin and the welfare policies in place play major roles in supporting young people. The role of the media in terms of recent available technologies was also mentioned as an important tool young people have for building their own expression and autonomy.

At what age, in the 21st Century, should we consider a person to be an adult?

A second issue that was discussed was age. How should we view age? Should it be considered as an exclusively biological phenomenon or does the meaning of age change in accordance with historical and social contexts? The age at which women have their first child, for example - today in the whole of Europe women have their first child at an increasingly advanced age - constitutes a clear indicator of social factors overriding biological ones (also related to the lengthening of the educational process which affects both young people and women). In the course of the debate, attention was also drawn to the importance of gender norms tied to age.

A question was raised concerning the consequences of these prolonged processes of becoming an adult - the fact that having children, a permanent partner, and a permanent job, and moving out from the parents' home are all taking place later in life - all contextual factors that are becoming more common in shaping the experience of being or not being an adult.

"Definition of adulthood is responsibility", "Responsibility is also potentially a political issue and a social issue"

A third important issue that emerged in the debate was that of responsibility, and as mentioned above debate was polarised on this issue. A number of participants insisted that it was vitally important to consider the assumption of responsibility - conceived as an act on the part of individuals - as the essential marker of entry into adult life. From this point of view the social conditions under which the transition takes place would appear to be of limited importance: *"becoming adult is becoming responsible for one's choices; the choice people have to make independently of economic and social circumstances"*.

In response there emerged another point of view, shared by other participants, according to which the assumption of responsibility itself - the possibility of conceiving of oneself as a responsible subject - possesses a social and political character. In other words, responsibility too has to be analysed in terms of a social framework and not as a simple act of individual liberty: *"decisions are taken according to resources that people have in their daily life, there are constraints and opportunities"*.

What would be the appropriate policies for enabling these transitions into adulthood? Should the state intervene in the process of transitions to adulthood by giving support to personal choices?

There were different views on the role of policies supporting personal choices aimed at achieving financial independence. A group of participants expressed doubts on the need to promote policies supporting transitions into adulthood. In their view public policies could even turn out to be counterproductive, acting in practice as a substitute for the free exercise of personal responsibility in the face of the tasks involved in transition.

Another group of participants agreed that facilitating transitions is a highly political issue and that all family policy is about these transitions. Two examples were given: one regarding a specific policy in Finland that promotes some autonomy of young people in terms of economic standards: *"staying in the parents' house until age 35 (as in Italy and some other countries) or at 22 (as in Finland) is related to policy decisions. In Finland every person who moves out of his/her parents home to study is given a housing allowance, which means they move out very early, at the age of 18; this gives them a sense of responsibility for being on their own".*

The second example refers to the lack of autonomy women might have in relation to maternity benefits, which are still linked to and dependent on employment and salary: *"policies support moving out from the parents' home but do not support becoming a parent until the person has a permanent job and salary-related benefits [...] a person has to be employed in order to get maternity benefits".* In general all the participants in the working group were in agreement in underlining the need for a strategic policy towards eligibility to maternity benefit regardless of the economic background of the mother.

Finally it was also underlined by many participants that while social policies *"can support transitions (to adulthood), they cannot design them".*

Major gaps and challenges for research

There is a need for comparative studies at the European level regarding transitions into parenthood. In particular:

- At the micro level: in what way do young people, women and men, negotiate their roles as mothers and fathers and try to reconcile them with their experiences as young people engaged in the transition to adult life.
- At the meso level: what social resources (institutional and informal) are available to support them in this trajectory, and what are the corresponding constraints?

- At the macro level: it is necessary to take into consideration the different transition regimes at the European level and the different degrees of sensitivity towards the tasks associated with parenthood and, more generally, towards the gender differences involved in the experience of parenthood.
- It is important to carry out further research on the strategies of young parents with reference to gender (gendering and de-gendering strategies): for example, a return to the traditional gender-based division of labour in the couple or, instead, a restructuring of gender roles after becoming parents for the first time.
- Expectations and young people's needs - "*subjective expectations and experiences of youth and adulthood - were considered a key question for research, since it was stressed that young people today expect different things from life/society compared to what their parents expected before them*".
- Also important and needing further research are transitions towards parenthood on the part of young migrants (and, in general, understanding this process in terms of transnational labour markets and the demand for labour).
- There is a need to explore the process of transition to parenthood in conditions of poverty and in the presence of housing problems.
- It was also suggested that it is important to explore dependency interactions between young and older generations as well as their impact on the autonomy of young people; research should also take into account cultural differences between and within countries.

2.2.2 Workshop 2 - Motherhood and fatherhood in Europe

Organisation of the workshop and keynote speeches

Margaret O'Brien (University of East Anglia) opened the session of this focus group with a presentation on "*Fathers in Europe: the negotiation of caring and earning?*" According to Margaret O'Brien, although there is a long legacy of research on father's work and family reconciliation in the European Union, fatherhood has not been a central issue in family policy developments in Europe. She addressed two questions: 1 - To what extent are European fathers becoming more involved in family life? 2 - How can we engage fathers in the work and care solutions of the future?

Starting with the first question, Margaret O'Brien presented some quantitative longitudinal data and concluded that European fathers are becoming more involved in family life. In fact, not only they are doing more and sharing more household tasks with their partners, but, and most noteworthy, they are increasingly involved in primary and active caring for small children, promoting (since the mid-1980s) the model of a 'new father' - in other words, a father who, besides being the main provider, is also a hands-on and loving one. However, there are considerable differences not only between European countries (with Nordic fathers spending more time in caring), but also within countries, when macro-social variables such as educational attainment levels, working hours, or even full-time/part-time activity of mothers are taken into account.

She also stressed that there are significant and diverse family contexts for becoming a father in contemporary Europe. In fact, fewer men are having children (voluntarily or involuntarily) and when they do, they do it later in life, in a wide range of family formations and sharing the financial responsibility with their partners. This leads us to the second question: how can fathers (as well mothers) work, care for their children and achieve personal wellbeing? According to Margaret O'Brien, the models of contemporary fathers, such as the active, the caring, or the nurturing father, which have corresponding images on television and advertising, seem to be in contradiction with the father of everyday life in terms of the availability of time to care and to involve oneself in family life: *"this mismatch may be a problem* [...] *particularly now that we are living times of economic insecurity and instability* [...] *the active father might be contested, men may feel less security in arguing for more time with their children in their working environment"*. As an example of these contradictions between father cultures and the conduct/behaviour of fatherhood she mentioned the fact that in the UK men who are employed for less than 26 weeks in the same workplace are not eligible to take the paternity leave of 15 days which has been available since 2003.

Given the fact that infant and child care is no longer a private 'mother only' family matter and that governments are becoming more involved in developing policies towards work and family reconciliation, Margaret O'Brien emphasised what she considers as a key policy issue: policies that promote choices and give parents freedom to choose between the available leave arrangements; if the parental leave is not well-paid or difficult to take, it does not become a real option.

In conclusion, Margaret O'Brien's presentation emphasised that there should be a connection between policies, labour market perspectives (employers) and fathers' and mothers' wishes, in order to find creative ways that include fathers and not only mothers in the care of children.

Main topics discussed and contributions from stakeholders

The discussion was very lively and focussed on the subject of politicising fatherhood and motherhood. Participants' positions were polarised around two different perspectives of two major recent trends in the EU: the regulation of early childhood through childcare services and leave policies directly tied to gender equality. In fact, inclusion in the political agenda of tools seeking to bring men more closely into childcare was seen by some participants as essential in order to accomplish gender equality in work life and family life; for others, it was regarded as dangerous social engineering which challenges the natural bonds and expertise within the family. However, it is important to note that both perspectives underlined the wellbeing of children as the major reference point.

Gender inequalities in childcare persist

> "As economic providers, mothers and fathers are becoming more equal; in childcare, inequality remains pronounced. How to 'equalise' the social and economic rights of women and men as parents (bearing in mind the interests of the child)?" This question is considered to be a challenge for welfare states: "fathers should be encouraged to do more housework and care and mothers should also be encouraged to let fathers do so".

What are the political drivers (both at local or national level) that might have an influence on the changing roles of fathers and mothers?

Concerning the drivers for more engagement of fathers in childcare, major research trends reveal that there are several macro and micro variables that might promote more involvement of fathers in caring for their children, namely educational attainment levels ("highly educated men are more likely to spend more time with their kids"); employment patterns, for example, full-time employment of mothers ("there is a link between mothers' employment and men's care time") and men's working hours ("the more paid work men do, the less time they spend with their children"; "although men's working hours are declining in Europe, fathers work more hours in comparison to men without children"); level of payment when taking paternity and parental leave: "men take leave when there is a high level of replacement".

However, the importance of getting to know more about men's wishes regarding the reconciliation of work and family life was also mentioned: "we know about the amount of time men and women spend with their chil-

dren (fathers' involvement in unpaid work - childcare, core domestic and non-routine domestic work - has increased) but we know less about what they feel about that time, their satisfaction, and the negotiations that take place in the home"

How much 'social engineering' do we accept in order to achieve gender equality? Nature and biology versus polices of social engineering

Some participants expressed the view that policies may seek to implement a kind of 'social engineering' which aims to promote the same amount of equality for both men and women in connection with childcare. This was considered as 'de-maternalising childhood'; it was considered that achieving complete gender equality might not always be in the best interest of the child. The example of breast-feeding was mentioned: *"you cannot replace the mother by the father if you are breast-feeding your child".* It was also argued that there are natural bonds between mothers and their children, and fathers are not as needed in the first years of a child's life as mothers are: *"mothers feel the needs of a child better than fathers".*

Another example that was very much discussed was a proposal which seems to be currently under discussion in Sweden concerning the division of the 16 months of well-paid parental leave into equal and non-transferable shares for each parent: *"what is in question is the right of mothers to take a long leave or the right of fathers to share part of that leave".*

Polarisation became evident once again, because for some participants to take away parents' right to choose who uses the parental leave and to make fathers take half would be devastating for breast-feeding as well as for the child's wellbeing; while for others fathers' involvement in childcare is a precondition for a fair balance between work and family life in dual-earner families, as well as being extremely important from the perspective of the child who experiences parental involvement and not only the mother's commitment. However, given the fact that time spent on unpaid work is significantly higher for women/mothers than for men/fathers, it was also suggested that a good model of gender equality should remunerate the unpaid childcare work which is mostly done by women/mothers.

Policy does not allow for free choice between genders

On the other hand, another group of participants stressed the fundamental role of policies in creating conditions for parents to choose. In this group it was considered that children benefit most when both parents are engaged in the first years of a child's life. It was stressed that it is neither the mother nor the father, it is 'both' mother and father. The example of Iceland was given, where high levels of breast-feeding seem to be combined with high take-up rates of parental leave by fathers. Several aspects were underlined in relation to the role of policies regarding men's involvement in childcare and household tasks.

Policies to promote parenting – role of the media

There was also a concern regarding policies which promote parenting not only among men but also amongst women. Given divorce rates, the decline of fertility (fewer children) and new fertility patterns such as the postponement of child-birth, some participants raised the question of promoting parenting as a benefit for people's lives by emphasising *"the joy versus the burden, a signal of commitment, family togetherness and personal identity for younger cohorts"* in order to encourage them to become mothers and fathers. The important role the media may play in promoting the notion of parenting as an exciting and positive aspect of life was also mentioned, because role models are also supported by the media.

Research and policies do not reflect diversity in families, with particular reference to same sex families

Another point raised in the debate was that both laws and research have been homophobic regarding same sex families, which are still invisible in the statistics. *"For example, the gender pay gap affects women, but how does this affect lesbian couples? Are these women having a double pay gap? What impact does this have on the children? On the other hand, men earn more, but how is it in gay couples? What about gay or lesbian parenting?"* The need for further research on these subjects was pointed out.

Another discussion relating to motherhood and fatherhood in same sex families focussed on the possibility of same sex families adopting a child. Some participants felt that there is a huge gap at policy level in rela-tion to adoption and fertility treatments in lesbian or gay families. Even when national laws recognise marriage between same sex partners, they exclude fertility and adoption. Laws also do not recognise rights and ties between gay stepfathers and lesbian stepmothers towards their step-children, for example when a biological father or mother dies. For some participants, representations of fatherhood and motherhood should have nothing to do with sexual orientation, and this independence should be carried over to the political level (*"unlink sexual orientation from being a mother or a father"*); however for others children's rights come before parents' rights.

Do we need a unique parental system throughout the EU?

There was general agreement that some basic rights should be required for all Member States and regulated under EU Directives. As an example, it was mentioned that there is no regulation on entitlement to paternity leave at the European level, and that many countries still do not provide it. Participants agreed that a global European Directive is needed to regulate either father's entitlements or the reconciliation of work and family life; breast feeding regulations were also mentioned but considered to be included in the Directive on Maternal Employment Protection.

The role of employers in promoting parenthood and family wellbeing

Finally, all participants agreed that employers must be brought into the discussion; there is a crucial need to engage employers in future conferences since they also have a fundamental role to play in promoting parenting and family wellbeing. "*State family policy can regulate some part of family life, but it is very much the work life that influences families, we have to build bridges between companies and families in family policy*".

Promoting parental leave over the family life-course

There was a proposal to include parental leave into life-course policies so that it is not just centred on the short period after birth; the possibility of taking parental leave at other stages of the family life-course such as, for example, when children become adolescents. This proposal was also seen as an alternative to the father's involvement in childcare: "*paternal discussion is very important, but we have to develop a parental leave over the life-course in other stages of children's life when they most value the presence of the father [...] how can we encourage fathers to take parental leave or time off when the children are much older, for example, when they are teenagers?*"

It was suggested there might be a family leave focussed on family care and not just children's care. For example, a family leave to care for other dependent relatives and not just centred on mothers or sisters as the main carers (as usually happens) but on other family members (such as fathers and brothers), who should be motivated to care during the family cycle of caring.

Major gaps and challenges for research

There was general agreement that research and policies have focussed on women as mothers and that fathering and fatherhood is mostly perceived from women's and children's points of view. Therefore research gaps are mostly related to the lack of reliable data on men's attitudes towards becoming/being a father. The following summarises the major research suggestions from participants:

- Need for research on the drivers that can influence fathers to be more involved in family life, particularly childcare and unpaid work in general.
- Further research on why men delay or miss out on fatherhood or want fewer children than their partners.
- Data is also needed on (potential) parents/young adults' feelings (of security or insecurity) about becoming a parent and raising and educating a child.
- Further research on parenting in same sex families in order to make these groups visible and mainstreaming the research.

2.2.3 Workshop 3 - Ageing, families and social policy

Organisation of the workshop and keynote speeches

In this workshop there were two keynote speakers: Claude Martin (CNRS/ EHESP University of Rennes), and Claudine Attias-Donfut (CNAV, France).

Claude Martin's presentation focussed on the impact of ageing at the EU level in relation to how care needs are evolving, as well as future care arrangements. Long-term care policies and welfare regimes were also mentioned, as well as the impact of those care arrangements on the family, introducing the subjective dimension of pressure and also the necessity of thinking more in terms of reconciling work and care for elderly persons. According to Claude Martin, "*ageing is one of the main challenges that most of our European countries are facing over coming decades*". He felt, however, that there had been (in some European countries) a kind of a split between family policy and social policy (particularly elderly care policies) as they are related to different interest groups, different research and decision-making fields with different administrative organisations. Considering that "*family does not stop with the ageing process*", there is a need to join together these two fields of research and policy - family and the vulnerable elderly - in care policies. Although the balance between state, market and family has changed dramatically since the eighties, with developments in welfare regimes as well as developments related to local authorities and collective insurances, *"the major part of caring responsibility and burden is still on the shoulders of family care-givers - spouses, daughters, daughters in law and of course some sons and male spouses, but this is a gender issue for all of our countries".*

Claude Martin stressed three main challenges for coming decades: ageing and the decrease in the EU population; the financial equilibrium of pension schemes; and the care deficit hypothesis as expressed in the reduction of the availability of *"free of charge services of women in the household".* According to him, the main future question for social care is not so much welfare state regimes and the differences between countries but how the reforms are to be carried out: *"we are all confronted with the same challenges and solutions: the combination of paid and unpaid, formal and informal care solutions".* Among the main future trends and needs, he highlighted the need for more flexible solutions developed at the local level (the regulation of care management on a local basis); the challenge of combining health care and social care; and the reinforcement of home-based care. He also stressed the importance of knowing more about care-givers' feelings and the meanings of pressure. As we will be confronted in the future with increasing numbers of people in the labour market combining elderly care and work (we usually think of work and family reconciliation in terms of childcare and not so much in terms of elderly care), a key policy question is how to manage the constraints of time, on the one hand, and the way people are feeling pressure or not, on the other: *"it is not only a question of the need for time, but also of the need to reduce pressure for these people".*

The second keynote address, by Claudine Attias-Donfut, focussed on family support, and outlined some results of the SHARE study (large European comparative longitudinal survey *"Survey of Health, Ageing and Retirement in Europe"*[9]): how family support is influenced by numerous factors (from the financial situation to the health status of the care-givers) and how, even though there are differences between countries, this support is mainly occasional, activated and present in situations of emergency and crisis, with the family then playing an insurance role. In fact, as already mentioned for childcare, family informal support and formal support (professional help) are complementary rather than in competition.

Summarising the research results, she stressed the important contribution elderly people make to family life, family solidarity and the economy, with the elderly being one of the most consistent providers of support to several family members, including other elderly persons. There are also significant

[9] See *http://www.share-project.org/.*

inequalities among families: *"the more social and financial resources, the more help is given"*; as well as significant gender inequalities, because men (when they are the main care-givers) are more likely to rely on professional support.

Main topics discussed and contributions from stakeholders

The following paragraphs summarise some of the main points of discussion:

Elderly care is mainly provided by family

> *"About 80 per cent of hours of care are provided by unpaid carers, mostly family carers; these carers have important sets of relationships, for example, the relationship between care and formal providers; their relationships with other family members; relationships with governments, but also increasingly relationships with policy areas".*

Implications of demographic trends for the future of care

The decrease in fertility rates also implies that in the future often a single child will have to care for his/her parents alone, and this means an increased burden. On the other hand, growing numbers of elderly persons imply (potentially) increasing caring needs. However, in the context of a parallel decrease in the number of young people, the question of a potential 'care deficit' arises – that is, a decline in the availability of unpaid/informal carers, at the same time as needs are increasing.

Intergenerational solidarity as a key issue

In connection with elderly persons, it was stressed that they are not only care receivers, but also care-givers who provide care for their grandchildren and also make transfers of money. Similarly, elderly persons should not only be regarded as a potential burden, but also as a resource – for family and for society. The importance of active ageing was also mentioned, in relation to their role in society: *"grandparents provide practical, emotional and financial support for their grandchildren [...] the birth of a first grandchild is often the moment when parents and grandparents find each other again [...] intergenerational solidarity can play a key role in developing fairer and more sustainable responses to the major economic and social challenges that the EU is facing today [...] public authorities should develop holistic and sustainable policies supporting all generations, and foster exchange of good practice and mutual learning between different generations".*

How are elderly people represented within society?

"How are they represented in terms of institutions and non-gov-ernmental organisations at national and European level? How can they let society know what their needs and their situation are?"

Family care is less and less considered as natural but rather as a choice

"There is increasing social demand for a full recognition of informal carers, women and men, who freely choose to dedicate themselves to their dependent family members [...] this will also have consequences for public support in the context of the links between formal and in-formal care, the supply of which should be locally provided and flex-ible, institutional but also home-based, and both affordable".

Sustainability of family care

Another element mentioned was the question of the sustainability of family care. The risk of a burden on carers was also mentioned, as well as the consequences in terms of wellbeing (feelings of pressure); the need for various forms of support, and for respite care was stressed (with provision of services such as day-care centres that would take care of dependent persons during the holiday season, so that carers may have a holiday as well).

A form of 'elderly sitting' was mentioned (that is the possibility of asking somebody to come to the home and stay with the elderly person during the day or in the evening, while the carer goes out).

The linkages between two major demographic trends: ageing and migration

Several questions were raised in connection with this topic: *"is migration slowing down the process of ageing? Is ageing changing the forms of migration, since the increasing needs of the elderly are attracting new types of care workers? Is there a new care sector mainly occupied by female migrants? Migrants them-selves are getting old and have specific needs that have not yet been studied".*

The perverse effects of some of the most flexible care solutions (*"Badanti"* in Italy, almost exclusively Romanian women), which might be leading to the development of a black market in migrant care workers, should be understood and researched.

Major gaps and challenges for research

- Lack of research on the subjective aspects of care arrangements (how is the caring arrangement experienced by carers?); the impact of this care on the carers' wellbeing, namely on the subjective feeling of pressure; research on some obvious key causes of problems for carers: managing incontinence; managing and living with someone who combines dependency with mental illness or depression.
- More information on sustainable family care: how and why people begin, maintain and decide to stop providing care (carer perspective). What works for carers in relation to training, respite, cash benefits, social security, and services support?
- It is important to focus on the contribution of spouses, who often do not consider what they are doing to be the provision of care and might be underestimated in the statistics on carers and caring.
- More information on what works in terms of building capacity – what kind of support really works for carers? When does information and training work best, how is it best provided, and who should provide it? Research should also look at good practices and how to provide a far better exchange of information on good practices in the domain of support to carers.
- There is also a lack of information on the challenges involved in reconciling work and care from the perspective of elderly care; more research should be carried out on the policy measures developed for those carers.
- More research is needed on the economic aspects of being a carer. What have the consequences of the current financial crisis been for carers? What happens to the carers who give up their jobs? What is going to happen to carers' pensions in later life?
- A lack of research on migrant care workers was also mentioned, including the effects of global distance caring chains for family members left behind in the country of origin; the specific needs of ageing and returning migrants.

2.2.4 Workshop 4 - Changes in conjugal life

Organisation of the workshop and keynote speeches

There were two presentations in this policy workshop. The first, by Eric Widmer (University of Geneva) was on *"The future of partnerships and family configurations"*, and the second, by Brian Heaphy (University of Manchester), was on *"Developments in conjugal life: same sex partnerships and lesbian and gay families".*

It is very important to understand what happens within conjugal ties, and Eric Widmer's presentation was focussed on how these conjugal ties are embedded in a larger set of relationships. What Eric Widmer emphasised was that family configurations (that is the larger structure of family ties, which might include grandparents, aunts, uncles, friends, colleagues, etc.) plays an important role in partnerships in late modernity, meaning that there is a variety of ties that can function as a backup to conjugal relationships. According to this family configuration, it is impossible to understand the conjugal relationship without referring to these larger sets of ties that support couples: *"no couple is an island; no couple can be understood in itself".*

In order to illustrate the importance of these ties beyond the husband and wife partnerships, Eric Widmer presented some results of the International Social Survey Programme (ISSP)[10] on the measurement of social networks, namely the persons relied upon when in need. Results revealed that partnerships are of major importance; the cohabiting partner is the first person called upon for support. However, there are a large number of alternative ties that play an important role in supporting partnership - the mother, the daughter, the sister, the brothers in law, etc., - *"if you take the sum of them into account, it becomes clear that conjugal life is not the only form of support within families in late modernity, particularly when considering the second person to be called upon for emotional support. Mothers in particular, but daughters too, play a major role in providing support to individuals across all countries within Europe".*

Accordingly, Eric Widmer suggested three patterns of relatedness: 1 - 'multiple ties-oriented' (less emphasis on partnerships and more on mother, father, sister); 2 - 'emphasis on conjugal relationships'; 3 - 'children-oriented' (more emphasis on son and daughter). He raised two questions: *"do family configurations matter for partnerships?"* and *"can we establish a link between the way configurations are structured and the wellbeing of couples?".* He put forward two hypotheses: 1 - firstly, that *"family matters beyond partnerships and nuclear families, there are ties between adults and parents and siblings that are really important for individual development and for conjugal life but also for the education of children";* 2 - secondly, that *"configurations and partnerships are interrelated; couples with more support interdependencies with relatives and friends will report higher conjugal quality than those with less supportive interdependencies".* An important point highlighted by research is that family resources exist beyond partnerships and nuclear families and

[10] See *http://www.issp.org/.*

that they can be used as social capital, *"something that individuals can use in order to advance in their own life, both in their intimate life, professionally, and in the education of their children".*

For this reason, Eric Widmer concluded that policy-makers should not only focus on marriage and nuclear families, *"because families are much richer than that"*, but should take into account this diversity of ties beyond the nuclear family; *"this will help us to promote partnerships without being entrenched in normative models of families which probably will be less and less present in the near future".*

Brian Heaphy's presentation was entitled *"Developments in conjugal life: same sex partnerships and lesbian and gay families".* He focussed on *"what is exceptional in same sex relationships and what is very ordinary?"* According to him, we are dealing with a population which is partially invisible in statistics and research: *"same sex, lesbian and gay families are a hard to reach population, particularly if looking for formalised couples".* Research on same sex relationships as well as on the changing legal contexts in which these families must be understood, tends to be based on small and *ad hoc* qualitative studies. Therefore, one of the main points stressed by Brian Heaphy is the absolute need for a more systematic review of the existing research on 'legitimate' and 'illegitimate' (not yet legally formalised) same sex partner relationships, in order to give feedback to research as well as to law and policies.

Implications of policy and legal developments regarding same sex couples were considered by Brian Heaphy as a key policy issue. According to him, talking about same sex families means talking about uneven developments: *"on the one hand it seems we are moving towards a broader legal recognition of partnerships, but those legal developments are uneven, they range from what might be seen as more formal marriage to what some people call 'marriage light'".* He also stated that there are not only uneven developments in terms of law and recognition of partnerships but there are also uneven developments in terms of the implications of those recognitions, for example, on the level of service provision: *"social policy is often underpinned by gender assumptions, by gender care, and gender responsibilities that don't fully account for same sex relationships [...] is it possible to conceive gender-neutral policies?"*

Another point stressed by Brian Heaphy was the challenges that same sex partners face in illegitimate contexts in terms of marginalisation and hostility, due to the way heterosexual norms are imposed or supported or actively pursued. Research suggests that the risks and threats that can emerge from this include violence, harassment, depleted social capital and social isolation. All these have implications in terms of a couple's wellbeing and resilience. On the other hand, recent research also points to the fact that same sex partners also feel unprepared to ask for family-supported services when things go wrong, for example in case of abusive relations and dissolution of the couple. In response to these illegitimate contexts (where same sex couples might experience highly stressful situations) 'families of choice' appear as creative responses to marginalisation. Families of choice include same sex relationships but tend not to be biological or legally formalised; it is not the biological relationship that matters, it is more the social relationship: *"Can policy capture these kinds of more dynamic relationships?"*

An interesting fact needing further research is that cultural guidelines, particularly on gender, are no longer applicable to same sex families, which tend to have highly negotiated relationships and also tend to be more equal because they are based on gender sameness; this area could benefit from further research. On the other hand, research should also focus on the gender pay gap that might be reinforced in same sex lesbian couples compared with same sex gay couples.

Complexity increases with the presence of children. Although there are new choices to become parents in same sex families (access to technology, informal parents' agreements, adoption, children from previous heterosexual relationships, etc.), a general perception still persists that children are more exposed to risks when living in same sex families, that the wellbeing of a child might be compromised by the nature of the same sex relationship. Brian Heaphy emphasises, however, that a key finding from research is that there is no discernible long-term impact on children's wellbeing within same sex relations compared to heterosexual ones. He also refers to his recent work on relationships among young couples in civil partnerships (which became legally possible in the UK in 2005), where he found notable continuities and similarities to young heterosexual marriages such as the focus on love, commitment, security, a tendency towards monogamous couple commitments, connections with family and cultural traditions, and secure and stable environments for children.

Main topics discussed and contributions from stakeholders

Wellbeing of families is associated with the existence of extended ties outside the nuclear family

> *"Women with bi-centric families feel much better curtailing their careers than women who do have not this kind of network; networks help to cope with the consequences of decreasing work participation".*

"Young adults who are in transition have a huge amount of friends cited as family members; the same happens with later-years families with small children, especially in families where there is no divorce; on the other hand, vulnerable individuals (those with psychiatric problems and incapacities) seem to have a very small family configuration based on blood ties, but might also include professionals as their family members".

What are the criteria for defining bi-centric families?

"Need to consider a series of indicators: frequency of interactions with friends, support provided by friends and family members, financial and emotional support, frequency of interactions with family members".

Why are there differences between countries in international comparison?

"Conjugal ties-oriented countries are to be found in countries with strong welfare systems; and multiple ties-oriented countries are more to be found in liberal non-interventionist family policy; this is not very clear, however, and needs further research".

Is there any relation between types of conjugal interactions and network configurations related to network ties?

"Types of conjugal interactions and configurations will be further researched. However, studies on recomposed families, step families and blended families reveal that there are very interesting signs that the two aspects - types of conjugal interaction and configurations - are very much interconnected".

Social policy and new forms of family

"Policies are addressing this issue mainly by recognising same sex marriages or same sex partnerships, but they are not dealing with other issues, namely social parenting in the context of same sex couples or blended families; there is a strong movement in Europe towards the recognition of same sex marriages, but there is little discussion of the real challenges for same sex families, which include how they are going to care and parent, including recognition of parental rights which are essential for the wellbeing of children".

"There is a need for more research on the gaps in policies dealing with new family situations. For example, in post-divorce families, who receives family benefits? It is usually the mother, even where there is joint custody of children; some couples negotiate, but there is no regulation on it".

Conflicting tendencies

"If you are living in a same sex partnership and do not live in a context of recognition there are implications for daily life and emotional roles, and this can also have an impact on children's wellbeing".

"There are political assumptions in care, service provision and family support services that support gender inequalities".

Children's wellbeing in same sex couples

Recent research shows that children do not suffer from having same sex parents:"*they suffer most from conflicting negotiations arising from their parents' divorce; however, they can experience discrimination at school. A child's wellbeing depends more on the environment than on the same sex nature of the couple*".

Major gaps and challenges for research

- More cross-national research on the internal dynamics of families across European societies is required, together with further research on types of conjugal interaction and their linkages with family configurations.
- Longitudinal studies on couples and conjugal life across the life-course: how do they build their relationship? When do they decide to get married and when do they decide to have children? Transitions to conjugal life and transitions into parenthood: how do couples manage transitions, and what are the factors that make some couples succeed and continue with their relationship? What factors influence couples to give up their relationship and divorce? More comparative work on routes into and out of partnerships; routes into parenting and post-dissolution arrangements.
- Further research on definition of the family, looking at how the notion of family is being built up across Europe.
- Look more at minority families such as immigrant, Roma families.
- Increasing cohabitation and decrease of marriage: reasons why young people are choosing to cohabit rather than get married (common trend

in people already married before, never married and new relationship where never married).

- Research possible linkages between marriage and participation in society (voluntary, political, etc.).
- More systematic review of the existing research on same sex families: how same sex partnerships in lesbian and gay families are (re)configured in different contexts of 'legitimacy' and marginalisation; in what ways are they and the challenges they face more or less ordinary and exceptional? Gender roles; parenting; child's wellbeing; gender gap, etc.; how are these families structured through their practices and also what are their problems? Do same sex couples have new choices in becoming parents? Who is the biological parent?

2.2.5 Workshop 5 - Family relationships and wellbeing

Organisation of the workshop and keynote speeches

The workshop consisted of one keynote presentation and four statements, followed by a brief discussion on main research points and key policy issues.

As a developmental psychologist, the keynote speaker Gordon Neufeld (University of British Columbia) focussed on *"Family relationships and the wellbeing of the children both as today's children and as tomorrow's adults"*. According to Gordon Neufeld, when the literature on this subject is reviewed, one theme stands out from the others: the effect of separation on children. Often the conclusion is *"that separation from parents - whether physical and emotional - adversely impacts a child more than any other single experience. The impact of separation can be far reaching: behaviour, development and personality"*.

One major research question which therefore arises, in Gordon Neufeld's view, is: *"how do we take children from their families to care for them and educate them, yet provide sufficient connection so that they do not experience the deleterious effect of separation?"* One central concept is attachment as well as maturation: *"if deep attachment enables a child to preserve a sense of connection, then we should be looking at the conditions that are required to cultivate this kind of attachment [...] maturation, not schooling or socialising, is the primary process rendering children fit for adult society"*. In other words: the more a child is attached, the more he will be able to adapt to society, be resilient and be emotionally fit for society. If parents and various institutions are aware of and sensitive to this attachment, both will find solutions which minimise the impact of separation. The solution must be focussed on *"the development of a child's capacity for relationship and the resulting ability to preserve a sense of connection even when physically separated"*. Therefore a final message is that *"the wellbeing of today's children, tomorrow's adults, and our future society, will depend upon our ability to support the family as the womb of psychological maturation"; "how can we support families to cultivate the kind of attachments that will give birth to the realisation of human potential?"*

Main topics discussed and contributions from stakeholders

Among the stakeholders' contributions was a statement outlining the conclusions of several recent international studies that agree on the negative effects that marital breakdown has on the happiness of children and parents involved, and on national economies as well. These conclusions have been summarised in the *"2009 The Family Watch Annual Report – The Sustainable Family"*. In the light of these findings, this statement reinforced the idea that some prerequisites exist that enable a family to be 'sustainable', according to the definition coined by the Brundtland Report[11] in 1987.

Another statement stressed the importance of positive parenting and empowering parents in their educational role. It was argued that the fight against child poverty in Europe has become a top political priority and that a strong focus has been placed on promoting the quality of life and the wellbeing of children. A 'strength-based approach' should be taken: an approach which values parents' empowerment. To create a good environment for children, there is a need to support families in their parental role. Actions that remove barriers to positive parenting should be further promoted, raising awareness and increasing recognition of the social value of parental roles.

A recent survey among ethnic minority groups in Bulgaria was quoted in order to reinforce the idea of solidarity between generations as one very important aspect of family wellbeing in those groups.

In the general discussion there was some controversy about the attachment theory presented by Gordon Neufeld. The main reactions highlighted the fact that there is a professional debate on this topic, in which there is disagreement with the model presented, and that existing empirical research shows there are other more important threats to the child's wellbeing and future development as adults, such as violence and emotional threats. There are other possible alternatives in terms of attachment to parents: adoption was presented as an example of the possible re-attachment of children.

---

[11] See *http://www.un-documents.net/wced-ocf.htm*.

Quality of life and wellbeing of children

"It is strongly determined by their family situation and the quality and accessibility of services; more attention should be paid to ensure families' access to appropriate material resources but also psychological and social support for parents' empowerment"

Reconciling employment and family life: links to child poverty and wellbeing

"A good work/life balance for parents is critical to the wellbeing of children and society, as both income poverty and time poverty can harm child development. Children whose parents are not in paid work are more likely to be poor, while mothers who have interrupted their careers to care for their children are at higher risk of poverty in later life".

Fathers' involvement

"Solo caretaking by fathers is associated with their continued caretaking of older children and grandchildren. Research shows that early active involvement of fathers can lead to a range of positive outcomes for children and young people. These include better peer relationships, fewer behavioural problems, lower criminality and substance abuse, higher educational and occupational mobility relative to their parents' employment, and higher self-esteem. Conversely, low involvement of fathers is linked to negative outcomes for children, and the links tend to be stronger for vulnerable children".

Family structures and the psychological aspects of a family

"A bridge was established between family structures and the psychological aspects of a family. The Lisbon Conference has focussed very much on family structures and on how to adapt society to new family structures from a sociological point of view. Of course this perspective is very important, but we think it would also be most interesting to establish bridges with a psychological perspective: to take into account the impact of structural changes on individuals, on their personal development and wellbeing".

Family wellbeing, cohesion and care

> *"Since many of us work in the EU institutional environment, we realise that most of the fundamental EU policy texts refer to social cohesion (alongside economic and territorial cohesion) as a way out of the crisis and as an instrument of dynamic growth (see for instance "Commission Work Paper 2010"). Family is the initial model of wellbeing. It is the first laboratory for social models, for social cohesion. We wonder how we can implement cohesion and wellbeing in society if citizens do not have that cultural model implanted in them by education and experience developed from a family context".*

Major gaps and challenges for research

- Research that helps to understand what leads to stable families (sociology in connection with the psychological perspective).
- Research that helps to understand how better to educate and train parents on parenting and couple life.
- Research that helps to understand what families (father, mother, children) actually want.
- Research on how policies can support families in cultivating the kind of attachments that ensure the development of human potential.
- Research on the impact of joint custody (which is becoming more frequent after divorce) on fathers' and mothers' professional careers, for example in the case of qualified parents, the main obstacles which arise, and also negotiations within couples.
- It is also important to research the impact of joint custody on children's wellbeing in comparison to other forms of custody.

2.2.6 Workshop 6 - Gender equality and families

Organisation of the workshop and keynote speeches

In this session there were two keynote speeches, one by Ilona Ostner (University of Göttingen) and the other by Shirley Dex (University of London). Both presented a number of important aspects to map the 'state of the art' on issues related to gender equality in contemporary Europe whilst pointing out major problems and gaps in both research and policy making.

Monitoring gender equality (at the EU level) led to the production of statistics producing a high level of linkage between policies and the production of gender indicators that have enabled measurement of gender developments. In her presentation on *"Gender equality and families"*, Shirley Dex presented an overview of major trends on gender equality in a number of key areas. For example, there has been increasing equality in employment, particularly among younger women. However, the same cannot be said for older women and women with small children. Departing from a cross-national perspective it is possible to track major trends in models of family and work balance, which show that Europe is not homogeneous and that policies at supra-national level have not led all Member States to the same gender policy solutions. According to Shirley Dex, it is crucial to establish a new framework for thinking about gender equality issues in families, in order to make comparisons between European countries. Clear differences emerge between countries when examining part-time labour, pay issues and part-time pay penalty, pay gaps between men and women, and women's education.

Major questions concerning gender equality and families relate to: *"how to solve the simultaneous need for money/labour and caring time?"* and *"Are there conflicts between gender equality objectives and needs of families?"* Shirley Dex concluded that there is a need to rethink the importance and role of flexibility, childcare services, division of labour between mother and father and unpaid work, as well as potential time off. Flexibility has been seen as a solution, but might also have ambivalent outcomes, since men and women have different problems dealing with new flexible forms of employment; on the other hand, although part-time work is increasing in EU countries, it is found in low-paid and gender-segregated jobs, while skilled jobs are not adaptable to part-time arrangements; this increases employment inequalities between men and women. Shirley Dex also stressed that childcare coverage rates are quite uneven across the EU. Public coverage for children below the age of three is still lacking, and this creates problems in reconciling work and family. Regarding the division of labour between mother and father and unpaid work, although the overall amount of hours of paid and unpaid work has become more equal between genders, the distribution of time is still unequal, women do more, and this must be taken into account. However, it is important to recognise that overall gender segregation has improved, which is an important conclusion when looking at key data on gender.

From a policy-making perspective, Ilona Ostner focussed on the degree of success attained at the EU level in terms of gender equality policies. The starting point for her presentation on "the success and surprise story of EU gender policies" was the following question: *"why have gender policies been successful to a certain extent?"* According to Ilona Ostner, if we start out from a historical perspective, the success of gender policies in the EU is not yet fully understood. In order to address this question we need to take into account the real ways in which gender equality policies are built up, so as to further understand the complex causalities underlying the somewhat surprising pathways of gender equality policies in the EU. Therefore we need not only to map what policies exist today, but to be aware of the specific agenda that lies behind policy-making.

A main reason for focusing on these complex causalities, from a political standpoint, is related to what Ilona Ostner considers to be the element of surprise. Why have gender policies been so successful? The fact is that EU Member States had not anticipated gender policies and the inroads they made. The EU can be considered a weak state, its institutions are weak, but nonetheless influence policies undertaken at the national level. However, this process of 'Europeanisation' has often resulted in the forced or unwilling compliance with gender equality policies that were not priorities at the national level. There is, however, some 'success', which is partly due to the feminist debates which have marked the political agenda since the 1990s. Why is this happening? How do political analysts explain change? According to Ilona Ostner there are complex factors and causalities that rest on societal and political explanations. For instance, political analysts would emphasise political explanation and institutional constraints. In this perspective, gender policies are developed through the appearance of some windows of opportunity and then *"you need actors who speak and act in terms that can be sold to those who make the public decisions"* (in this sense the role of epistemic communities is of the utmost importance).

Gender policies need a window of opportunity, e.g. the EU and OCDE building coalitions that bring together transnational and national actors (the top bureaucrats); and the selective inclusion of experts (in this case certain leading feminists, amongst other epistemic communities). In every single Member State the idea and the perception of what is important may be different. But these perceptions at the hands of leading epistemic communities open avenues for the rise of lobbying groups, who have a role in deciding how gender equality should be addressed. Nevertheless, any gender-related policies at the EU level must pass through two 'eyes of the needle' in order to be discussed, adopted, and implemented: first at the level of the Union, with its narrow conception of equal opportunities in terms of equal treatment and its stringent requirement for consensus in the Council; and secondly in the variable implementation of EU legislation in the 'gender order' of each individual Member State.

These processes can be viewed in two ways. Gender policies are a result of negative integration, starting from the problems that have arisen as a result of the need for the free movement of workers in Europe. It was not expected that these policies would also promote positive integration. The most important surprise factor is the success of gender policies at the supranational level, with a historical movement from concerns with "equal pay and equal opportunities for men and women" to a more generalised focus on general anti-discrimination legislation (with a whole set of targets which were brought in with the Treaty of Amsterdam and other measures particularly related to mothers' employability and child care targets), which led to enhancement of regulation on matters of gender equality. The process has, however, been one of vertical integration linked to supra-national forms of regulation. Some regulatory inconsistencies are unresolved, stemming from what appears to be a "ping-pong game" between the national and European levels of regulation.

For Ilona Ostner what is important is to see how the process of institution-alisation of gender equality has evolved and resulted in positive integration. There is new 'constitutionalised' legislation that has extended the meaning of gender equality, as a key part of the whole process of developing anti-discrimination targets and policy measures. However this important trend is also an ambivalent one, in spite of its success in regulating gender equality and constructing the whole debate around gender issues as an equivalent of gender equality: on the one hand, employment has been a very impor-tant catalyst, but on the other, fertility policies are also of major importance for arguments in favour of gender equality policies.

Why and how has this happened? From the 1990s onwards new social risks have had to be dealt with. Declining fertility is important because we have labour shortages. It is not the number of children but the quality of children that matters (a functionalist argument). In conclusion, European gender policies are successful, yet:

- Today they are not the most important, if considered *per se*.
- Gender policies have never been an issue *per se*, but rather are linked to other issues (labour shortages, demographic ageing, for instance).
- Gender policies have been highly dependent on the building up of coalitions. This is how politics actually works, and it is a problem that has to be further addressed and monitored in the future.

Main topics discussed and contributions from stakeholders

Reconciliation of work and family life

There was general agreement that one of the most important subjects for debate in terms of policies and policy-making was the problem of recon-ciling work and family life.

Gender equality policies are not a neutral subject, as they presuppose ideological conceptions of the ideal family with a 'gender contract'. The 'ideal arrangement' generated heated debate and some disagreement.

Private choice versus public regulation of gender and families

For some participants, gender models belong to the private sphere and must be freely chosen and not imposed by public regulation. The state should not impose the adult worker model upon women, but rather respect men's, women's and couples' freedom of choice. This is an important debate, in which there are opposing views of gender relations, reflected in differing visions of what gender policies should be.

It was also stated that although the focus on individual choice must be taken into account when considering gender equality, it has to be addressed in different terms. As Shirley Dex noted, mothers and fathers do not have to work full-time. Part-time can be seen as a solution, but only if it is considered in equal terms for both men and women. There is, however, a pay differential that has to be taken into account. Gender equality targets do not recognise the potential variability of choice, e.g. part-time work is undervalued, and the rights of part-time workers remain a problematic issue.

New solutions to reconcile family and work should be put forward. What is the financial value of housework and childcare? What role should the state have in transferring money to families in order to keep mothers/child-rearers at home?

It was also argued that new solutions to reconcile family and work should be put forward (e.g. pension credits for homemakers, whether female or male). One suggested solution was women's self-employment as baby-sitters for their own children. This leads to a key question: should care arrangements be paid for by the government? This was also considered an important gap in research.

Care

Another issue raised in the discussion was care. It was argued that the right to participate in care has to be implemented for men and for women, and that the model of the adult full-time worker has to be rethought from the point of view of gender equality and family life. However, two major problems arise here: 1 - How to pay for care and how to implement policies that support care arrangements? This was considered to be the main challenge for the future policy agenda. 2 - The problem of fertility as a backdrop for gender equality policies must be taken into consideration in this context.

Labour market. How might the changing structure of the labour market affect gender equality policies and gender arrangements in family life?

It was recognised that people today face new risks (job insecurity) and that *"we cannot recommend the breadwinner model because of the risks it involves"*. New policies are needed to deal with new individualised risks (divorce, unemployment, etc.). There is also an economic issue in connection with sustainability: *"without contributions, the state is unable to pay"*; *"as some researchers have shown, the more insecure jobs are, the more hours people work"*.

Integrating gender and family policies

Another important issue discussed by participants is the need to integrate family and gender equality policies. Family policies are less advanced at the EU level than gender equality policies.

Major gaps and challenges for research

- Rethinking the models for equality and linking them with the demands of the labour market.
- More research on how to regulate care arrangements and gender equality.
- More research on the consequences of different care arrangements for gender equality.
- The need to research men and fatherhood (e.g. statement on paternity leave).
- Lack of data on gender equality in couples and on parents as a couple and on family forms in general.

2.2.7 Workshop 7 - Reconciling work and care for young children: parental leaves

Organisation of the workshop and keynote speeches

There were two keynote speeches in this session. The first keynote speech was by Fred Deven (Kenniscentrum WVG, Belgium) who made a presentation on parental leave policies across Europe.

Parental leave across Europe is a kind of umbrella concept covering an increasingly complex reality of policies and practices. There are two critical factors related to parental leave. The first is replacement payment, which ranges from an earnings-related payment, up to a maximum of 100 per cent (some countries put a ceiling on this), to a flat-rate payment or even no payment at all. There are many countries which can be very generous in the length of leave but do not provide replacement payment. The second factor is eligibility, which has been *"disregarded a little bit".* What Fred Deven wanted to emphasise was that eligibility is not as widespread as some people might think it is, even in those countries which are known to be the most generous in terms of paid parental leave arrangements (like Sweden, for example).

Eligibility is employment-related in most countries (including Sweden). Therefore there are several categories of employed persons who are not eligible, for example those who are self-employed, those who have temporary contracts, or those who work in small companies. There are also significant differences between public sector organisations and private sector companies. On the other hand, if a person is eligible, the replacement payment might be *"very conditional on your prior working history, so if you have built up rights you may be eligible for the generous earnings-related payment, but if you have just started or entered the country you receive a flat-rate payment".* Hence the importance of framing leave policies in the context of the issues of inequality and also democracy.

Among the main ideas presented by Fred Deven three proposals stand out. One is the need to contextualise research as well as policy leave arrangements within a broader context in order to understand them properly. Leave policies are only one instrument for European public authorities to facilitate the reconciliation of work and care for young children. There are other tools such as early childhood and education services, cash benefits, and flexible working conditions. When trying to understand leave policies it is important to bear in mind the different perspectives of the diverse actors involved: families, public authorities, stakeholders (social partners, NGO), and the media (in terms of their images of what is good parenting, for example). It is also important to take into account that there might differences of opinion between family members because of potentially conflicting interests, for example over the length of the leave period from the child's perspective, which may be different from the interest of the working parents.

Another idea stressed by Fred Deven is the importance of having a research approach to the collection of data on parental leave take-up rates. Data is still collected by the administrative departments responsible for the payment. He mentioned two sources of information on comparative data on leave policies in European countries: one is the data in the annual review

---

[12] See *http://www.leavenetwork.org/.*

of the *International Network on Leave Policy and Research*[12] (which includes about 30 countries, most of them European and some transatlantic). The other is the data collected through the recent Council of Europe question-naire on family policies (of which a significant part was on leave policies) involving 40 countries, and which resulted in a database on European family policy[13]. Fred Deven made some final suggestions for further reflection: how to frame parental leave and leave policies within broader issues, and how to conceptualise and how to implement types of leave in terms of care for dependent persons in a broader perspective which looks at the family life-course. He emphasised the need to frame leave policies within broader issues such as the different stages of the life-course and gave the example of Belgium's 'time credit system'.

Daniel Erler (Familienservice GmbH) focussed on the German parental leave system and highlighted what he considers an important issue, *"freedom of choice"*, which has dominated (West) German family policy discourse for a long time. His presentation highlighted the impact of leave policies on childcare-related family behaviour and going back to work after leave, which also has an impact on the increased numbers of women in the labour market (as in the case of recent leave policy change in Germany).

In 1986, when parental leave was introduced in Germany, the whole political debate centred on enabling parents to *"freely choose"* to stay at home and care for their children, because the main concern was the well-being of children and their emotional development. Between 1986 and 1992 parental leave in Germany gradually increased to three years, two of which were paid (not very well), while at same time female employment decreased. The main point Daniel Erler wanted to stress is that talking about *"free choice"* means offering a number of alternatives, thus also including childcare facilities in the scenario of options for parents: *"if you look at a leave scheme and it offers no childcare services, then there is no free choice"*. In the case of Germany there was an incentive to leave the labour market but there was not really a choice for families because there were almost no childcare services for children between the ages of zero and three. Only recently, in the late nineties, was there a discussion on the relevance of such long leave periods for mothers (leave was also for fathers, but fathers' take-up rates were very low, at around two to five per cent), because prolonged labour market absences were also seen as having negative repercussions on future career prospects for women. There was a political discussion on the need for adapting parental leave, giving parents greater freedom, and engaging

---

[13]   See *http://www.coe.int/t/dg3/familypolicy/Source/Family_Policy_in_Council_of_Europe_member_ states_en.pdf.*

fathers in childcare. A new law in 2007 represented a radical shift away from the previous basic idea that enabled parents to stay at home during the first three years after birth. The new law reduced the parental leave period, though it increased the payment with the introduction of 12 months' parental leave paid at 66 per cent (33 per cent for 24 months) of prior income plus two months of 'fathers only' leave, not transferable to mothers. The new law is based on two principles: one is to increase women's participation in the labour market (mothers' returning to work after one year of well-paid parental leave); the other is to motivate fathers towards more participation in childcare. Overall the intention is to reduce parents' absence from the labour market.

Daniel Erler also stressed the importance of involving fathers in parental leave time without penalising parental leave time for mothers. The principle of extending leave on the condition that leave is shared was conceived in order not to penalise parental leave time. Instead of reducing the 12 months' leave to ten months' leave if fathers did not take the two additional months, policy makers decided to keep the 12 months and give two months additional paid parental leave in the case of fathers/other spouse sharing. This policy led to an increase in fathers' take-up of leave (20 per cent of fathers take the two months leave) and to a decrease in mothers' period of leave. A group of highly skilled women increased their leave period, however, by taking the whole one-year paid leave (previously they would return to the labour market after a short period of leave). At the same time, childcare services are being developed. This will eventually give so-called freedom of choice to those parents who want to combine work and childcare services. Employers have been receptive to the fact that women tend to stay out of labour market for one year instead of three years.

Main topics discussed and contributions from stakeholders

Successful work-family reconciliation strategies (including father's involvement in parental leave) require an integrated approach/multi-dimensionality of reconciliation between work and family

From the workshop presentations and discussions it became clear that parental leave schemes are only one aspect of successful work-family reconciliation strategies, which require a multi-dimensional and integrated approach to the issues of time, care and money if they are effectively to enable mothers and fathers to combine their work and family life. What has been emerging quite clearly from research on parental leave schemes across Europe is that fathers will only start to use leave entitlements if they are well-paid and at least partly non-transferable. However, the effects of leave schemes are also strongly mediated by interactions with other social policy aspects, e.g. childcare services, child allowances, and pension entitlements. Hence a comprehensive understanding of leave policy effects needs to control for numerous intervening factors, necessitating a holistic research and policy approach: *"parental leave is the end result of policy considerations in the following areas: maternal health, health of the fœtus, fertility policy, labour market policies, gender equality, children's rights, family policies, etc.".*

Diversity and complexity of national leave schemes

"While European Union directives and regulations have led to some tentative convergence, leave schemes across Europe remain highly diverse, reflecting different historical policy legacies as well as cultural preferences. It is important to respect and allow for differences because one cannot simply impose one system on countries with very different socio-economic contexts".

"It is especially difficult to conduct comparative research, because there is very little comparable data available, and data is mostly collected directly by the institutions responsible for administering leave benefits. These institutions are not necessarily concerned with the collection of comparable data".

Parental leave schemes and parents' freedom of choice

All workshop participants seemed to agree that one of the crucial questions regarding parental leave schemes is parents' freedom of choice, i.e. enabling parents to choose between staying at home to care for their children, for a longer or shorter period of time. However, it also emerged very clearly that parental leave schemes only foster true freedom of choice if they are complemented by a sufficient supply of external childcare solutions, offering affordable quality services.

"Benefits need to be income-related. For if they are not, many parents, but particularly fathers, who usually contribute more to the family income, will not be able to take up their entitlements, because the related income loss is unsustainable".

"Yet real freedom of choice also necessitates flexible working options for parents as well as a family-friendly working culture within companies. If working parents are not offered flexibility that

suits the needs of their family, or if they fear that taking leave will compromise their future earnings and career prospects, they are unlikely to use their entitlements".

Parental leave and involvement of fathers

Current proposals to include fathers in parental leave by means of 'father only' (compulsory) leave (meaning that time it is not transferable to mothers) were also discussed: *"Fathers' involvement in parental leave, and the gender sharing of parental leave, is closely related to well-paid individual entitlement to parental leave. At present this entitlement exists in only a few European countries, which means that fathers' involvement in parental leave is not being encouraged, and is not high on the political agenda".*

Parental leave and social inequalities

"Income has an impact on the take-up of parental leave. Earnings-related parental benefits have advantages in relation to parental leave producing social inequalities; for example, in countries where salaries are very low, people still do not have the free option of staying at home or returning to work, so they must work. On the other hand, highly skilled and better paid persons are more likely to take parental leave, and this also applies to fathers' take-up rates".

Employers' perspectives

Unsurprisingly, all workshop participants agreed that employers need to participate in the consultation processes on future family policy strategies, because they are crucial stakeholders, and without their collaboration all policy initiatives are likely to have only limited effects.

As employer perspectives and attitudes appear to be crucial, it might be useful to gain some deeper cross-national insights into their positions. It may therefore be useful to commission a cross-national survey on employer attitudes, for example, to the perceived costs and benefits of leave or care policies. Some insights in this respect can be gleaned from existing survey data, e.g. the European Working Conditions Survey[14] or the European Company Survey[15]. However, none of these allow for an in-depth investigation of employer attitudes to the very specific issues of work-family reconciliation policies.

[14] See *http://www.eurofound.europa.eu/ewco/surveys/index.htm.*
[15] See *http://www.eurofound.europa.eu/surveys/companysurvey/2009/index.htm.*

Parental leave policies and parental leave take-up in contexts of economic crisis

> "On the one hand it is known that the effects of leave entitlements are strongly influenced by the economic performance of a country, because in times of uncertainty people tend be more careful in taking leave. On the other hand it is known that various countries are reviewing the costs of leave schemes and are considering cutbacks in this area. If benefit levels are reduced, this is likely to have repercussions on leave take-up patterns. There is an acute need for more cross-national, comparative research on the impact of leave policies, especially with regard to the labour market behaviour of mothers and fathers".

Parental leave policies over the life-course

> "A major question is how to conceptualise and how to implement types of leave in terms of care for dependent persons in a broader perspective over the family life-course".

A good example is the 'career break system'/'time credit system' (Belgium) "which goes beyond the narrow part of the first three months or the first year. The idea is that over the life-course of all your professional career you can drop out for a time, to provide care, certainly, but you do not have to be specific about your reasons. You retain your rights to go back to your job".

Major gaps and challenges for research

- Grasping the complexity of leaves and options and also understanding this in the context of class, gender, companies for which people work, regional differences, and different cultures.
- Improve research on mothers' and fathers' take-up rates (there is a lack of information not only on take-up rates but also on the educational and socio-economic backgrounds of parents who take leave). There is also a significant lack of information on total or partially unpaid leave arrangements. Statistics are driven by the administrative department responsible for the payment, which does not have a research approach to data collection).
- More research on young women's and men's family planning, namely their prospects for the transition to parenthood and their expectations regarding available leave arrangements and childcare facilities and labour market participation.

- Further research on the impact of 20 weeks' maternal entitlement to job protection, on longer breast-feeding periods, and children's immunisation rates.
- Further research on the connection between long-term parental leave at low rates of pay and the decline in female participation in the labour market, and some negative effects on female career prospects (also taking into account different levels of women's education and qualifications).
- Commission a cross-national survey on employer attitudes, for example to the perceived costs and benefits of leave or care policies. In-depth investigation of employer attitudes to the very specific issues of work-family reconciliation policies.
- It would be also important to explore how fathering and work are seen by society (what does it mean for a man who wants to care not just for one week or one month, but really wants to care in the long-term?).

2.2.8 Workshop 8 - Reaching out to families: the role of family associations and other institutions

Organisation of the workshop and keynote speeches

This session consisted mainly of statements (both prepared and informal) made by family associations present at the FAMILYPLATFORM Conference. There was also a brief discussion on common points. However, there were unfortunately no representatives of research or policy stakeholder groups. Because most of the session was devoted to statements, this topic *"Organisation of the workshop and keynote speeches"* provides a synopsis of the different kinds of family associations which took part in the workshop.

Gezinsbond (Flanders, Belgium)

The first statement was presented by Luk de Smet (Director General of the Gezinsbond). Gezinsbond has one guiding principle: care for the material and immaterial quality of family life and the principle of solidarity and justice where the family and its members are concerned. It has three aims: 1) to promote solidarity between families; 2) to protect the interests of all families, with special concern for large families; 3) to work towards a family and child-friendly climate.

Gezinsbond was formed in 1921 as the League of Large Families of Belgium shortly after the First World War. At that time it was not unusual for families to have ten or more children. It did not originate from previously existing family associations but instead was formed by a small group of people who launched the idea of a family association which would call upon local people to recruit families as members. It went on to be a co-founder of the International Union of Family Associations (the IUFO – now the World Family Organisation) in 1948. Following adoption of the Treaty of Rome (1958), the European Region IUFO entered into a dialogue with the newly-founded European institutions and, as a result of this discussion and the new treaty, helped to form COFACE. Today, Gezinsbond is a large non-profit making association with 280,000 members in Flanders and Brussels, all of whom are individual families. Members are primarily middle-class, but Gezinsbond is reaching out to families with 'an ethnic or culturally diverse background', and to those in poverty.

Eric de Wasch (Member of the Administrative Council of Gezinsbond) added a short statement on a number of additional areas. The first was the 'Family Impact Report', which examines the impact of all policies on different aspects of family life: these reports, Gezinsbond argues, should be entrusted to the person(s) in charge of family policy and monitor all policy formation. The second was family modulation, which is direct government support to families. The third was good practice in consultation between employees and employers so as to tackle the challenge of reconciling work and family life.

Associations Familiales Catholiques (France)

The National Confederation of Catholic Family Associations (CNAFC) was founded in 1905 and currently has 35,000 member families throughout France. They have been a member of COFACE since 1958, and they founded the Federation of Catholic Family Associations in Europe (FAFCE) together with Familienbund der Katholiken 20 years ago.

Their representative at the Conference, Françoise Meauze, argued that family associations are only effective when they are representative and when their members are volunteers. Family associations should put forward ideas that correspond to families' needs, and react to legal and political developments that have an impact on families. They are also important for their political lobbying work, and in promoting family mainstreaming. We were reminded that Article 16 of the Lisbon Treaty enhances dialogue with civil society, and that family associations should search for increased recognition. Mention was made of COFACE, FAFCE, and the World Movement of Mothers as organisations promoting families, as well as the Family Intergroup within the European Parliament, and the Commission on Social Issues, Health and Family of the Parliamentary Assembly of the Council of Europe. These actors rely on instruments such as the Universal Declaration on Human Rights (Articles 3 and 16), the European Social Charter, and the European Treaties. In conclusion, Françoise Meauze stated that family associations should be afforded legal recognition in all European countries, so as to enhance family mainstreaming. An emphasis was put on subsidiarity and also on the difference between family policy (which is preventive) and social policy (which is reparatory).

UNAF (Union Nationale des Associations Familiales, France)

In France, the government has created a state body for family associations, UNAF. UNAF receives funding from the state and consists of around 8,000 associations, representing 800,000 family members. It has four missions: 1) to provide public authorities with opinions on family-related issues and put forward measures in all aspects of family policy; 2) to be the official representative of all families in dealings with public authorities; 3) to deliver family services entrusted to it by the State; 4) to uphold the material and moral interests of families in accordance with the law.

Forum Europeén des Femmes (Brussels, Belgium)

A much younger organisation, though no less active, is the Forum Europeén des Femmes. It has been in existence for six years and is based in Brussels. Most of its members are expatriate professionals in Brussels. It is active in reconciling work, family and private life and aims to promote a more healthy work-life balance, in the belief that strong families are the beginning of a cohesive society and that wellbeing and families start with care in the family. They say it is impossible to work for a cohesive society without the presence of strong families.

Cana Movement (Malta)

Cana Movement developed in a strongly Catholic country, providing services to members. In a country of only 400,000 people, it has 1,000 volunteers who help organise activities and sustain the movement. They organise marriage preparation courses and counselling services that the Maltese Government now relies on Cana Movement to provide.

The Ombudsman for Parents' Rights (Poland)

In Poland, one of the main institutions mentioned in the constitution is the family, but there is almost no family policy. Government's experience of the family is predominantly negative, its only form of contact with families being through social services (who may face problems such as alcoholism or domestic violence). The Ombudsman for Parents Rights' in Poland, only recently established, is looking to change this perception and to call for parents to be involved in dialogue with government on family issues. Run with minimal resources and 100 active volunteers, it is nevertheless able to organise street demonstrations of more than 30,000 people on the internet.

Main topics discussed and contributions from stakeholders

Family organisations in the new Member States are facing particularly harsh times. Families were previously supported by the communist regimes, but have since had such support withdrawn. Families in the new Member States have only had a few years to build up family associations to represent them, and the current crisis and the relatively undeveloped civil society places them in additional need of support. In Latvia, for example, the Family Ministry was closed and family support was cut.

Leonids Mucenieks, of the Union of Latvian Large Family Associations, called for the following kinds of support: a) financial support from national governments and the EU to provide stronger support to families during this time of crisis; b) to see greater progress at the EU level in the field of practical consolidation of family rights and family-friendly policies; c) development of European grant programmes, which could help family associations to organise activities without co-payments.

Family organisations from the older Member States are well placed to offer some support to younger family organisations in terms of information sharing and capacity building, but increased support may be needed at an EU level. The point was also made in discussion and during the plenary session that family organisations can exert pressure on national, regional, or local governments by taking concerted action at an EU level. This can push national governments to take action. The European Alliance of Families helps promote EU level co-operation and should be strengthened.

Major gaps and challenges for research

- Evaluation of effective practices for reaching out to different families (development and interchange of good practices).

- Research on the constituencies of family associations: which groups are represented in family associations (by age, ethnicity, religion, geographical region, etc.) and which groups are not.
- The role of family organisations in influencing policy.
- More research on the role of local government in monitoring quality of education and teaching, and on connections between schools, family, local government and neighbourhood.
- The introduction of a family impact report/assessment. With this family impact report it might be possible to assess the impact of certain policy measures on opportunities for families.
- Analysis of consultation processes with family organisations. Understanding consultation processes between employees and management may be the key to addressing the challenges of reconciling work and family.

2.3 Methodological issues identified in focus groups and workshop sessions

With respect to the main methodological gaps and challenges which were common to all 16 working groups, participants debated and identified the following issues:

The perspective of national statistical offices and most of the data collection approaches as well as the interpretations of existing data were regarded as highly problematic, emphasising the need for valid and synchronised definitions and concepts used in research (concepts and sources differ widely between Member States). There was particular concern over the need to harmonise at a comparative level variables and categories describing family life, providing valid information on all kinds of family forms (e.g. patchwork families, same sex couples, percentages of separated couples, civil partnerships, number of consecutive marriages), including the (biological and social) status of children. It was also mentioned that official statistics of the European Member States still take the nuclear family as their reference model. Hence, they no longer reflect the variety of family life and relationships today, as family life is understood and lived differentially (e.g. family of choice, family as a network, etc.). Thus, data addressing relationships and kinship in more detail are needed. It was also mentioned that national statistics should include data on the number of men and women caring for their dependent family members at home.

The importance of including under-researched countries, especially the new Member States, covering all 27 members and/or those which do not belong to the OECD, was also mentioned by participants. The need for current data on regional and local levels was also emphasised, because differences within countries are also significant.

The comparability of European family databases with other important topical areas such as national labour markets and educational systems was also seen as an important methodological approach.

There was a particular concern with the **harmonisation of concepts and indicators defining and measuring several fields of family research**. Some examples refer to domestic violence, childcare, housework, parental leave entitlements and take-up, 'substantial childcare' (when measuring father's involvement in childcare). Participants also underlined **the need for more and new comparable indicators (both qualitative and quantitative)** in connection with the study of social inequality and poverty, **in particular indicators which go beyond income**. The dominant focus on income poverty provides a very specific outlook centred on the notion of "poor people" rather than on the experience of poverty and how this affects family life and individuals within families. Few (both quantitative and qualitative) studies, at least with a comparative focus, **highlight the experience and social patterns of poverty and families in poverty**. These include loss of dignity, choice, and control, limited access to social capital and to assets of other kinds, poor health, few opportunities and an uncertain future. Social analysis of families and poverty would also benefit from a **reinforcement of the household/family as a significant unit of analysis**.

The need for new and more comparable indicators (both qualitative and quantitative) also applies to further research on family wellbeing, satisfaction with living environments, involvement of fathers in family life, quality of childcare and elderly care, and the daily and biographical processes of doing family. Vital information on ethnicity was mentioned as still not being available in European-level studies. The harmonisation of migration statistics was also considered to be of great importance.

There was general agreement on the **lack of cross-national comparative research, both quantitative and qualitative, as well on the importance of the life-course approach and the longitudinal survey design as well as panel research** in order to get a deeper insight into the development of family forms, structures and development over time. There is a need to study family management and decision-making processes, taking into account the life-course perspective, which examines different periods of transition in the family life-course (early childcare, child entering school, adolescence, children leaving home, elderly care, divorce, remarriage); best arrangements in terms of work and family life balance and the needs of fathers and mothers to have more time to spend with their children; family values and ideals, their behaviour and attitudes; gaps between the theory and practice of gender roles; development of family policy in order to assess the impact of policy changes on overall support for families and on the outcomes of family-oriented policies. It was mentioned that there is a need to **evaluate longitudinal data sets and designs across the EU**, namely existing data sets and their design and content, in order to draw conclusions on their suitability/usefulness for current research questions as well as on their potential for future cross-national research.

Most studies and research projects are based on aggregate national data instead of setting the focus on variations between the different social, cultural and regional backgrounds of families within the different states. **Cross-national comparisons would provide a better understanding** of the differences and similarities between families with different social, cultural and regional backgrounds.

More qualitative insights are required on men's attitudes to having children (for example their reproductive behaviour, with information on the number of children they fathered or expect to father), on men's experiences as fathers, on partners' attitudes and interactions, and on their feelings and wishes in relation to their role in reconciliation. There is also a need for more **qualitative research on care arrangements**, in particular on the time devoted to care tasks and the constraints on them, but also on what carers feel about the time they spend on caring, their feelings of pressure, and the meaning of pressure.

The lack of data on specific cohorts of people was also underlined, for example, the importance of special data sets for the study of migrant populations and their spatial concentration as well as comparative data on migration across Europe which would provide a European perspective; there is a lack of in-depth analysis using specific targeted samples of social categories/families in order to understand diverse forms of domestic violence; new family forms (same sex families, LAT, patchwork families); and designing research for specific groups of families: some groups of families are unlikely to get into sampling frames and/or are by nature more fluctuating.

The importance of a family/family member perspective: ensuring that several and not only one representative of the family answers the ques-tionnaire and speaks on behalf of the whole family.

In discussion on **the importance of focusing on impact analysis studies**, it was stressed that there should be more cross-national (qualita-tive and quantitative) comparative research on the impact of family poli-cies within countries (regional differences, for example) as well as between countries. Participants emphasised the need for more in-depth, qualita-tive comparisons to understand and explain family policy reforms across countries, with a need for a better and up-to-date typology of "family policy systems" which takes into account national variety, developments in institutional forms, changes over time and available financial resources.

The importance of developing a multi-method approach by supplementing quantitative and statistical approaches with qualitative approaches and case studies. It would be important to have greater financial investment in methodological advances linking qualitative and quantitative studies.

Finally, it was considered important to ensure that **sociological research establishes bridges with a psychological approach**, in order to take into account the impact of structural changes on individuals' personal development and wellbeing. The lack of integration between different fields (intersectionality) was also mentioned.

2.4 Final comments: selected elements on the research and policy agenda

Drawing on the discussions, statements, keynote speeches, and other written documents and notes produced or reviewed during the Conference, our aim in this final section is to pinpoint some of the main research topics/themes and issues which were suggested or argued for by participants in the three-day Conference in Lisbon. Given the wide range and number of suggestions, the main objective here is to record and summarise these proposals, with a view to future debate, rather than to set out overall recommendations. The selected elements are based on the overlaps and broad emphases which emerge from the Conference and the previous sections of this chapter.

Selected topics and issues for the European Research Agenda

Topics and issues identified as important for the future Research Agenda include the following:

1. **Contemporary parenthood, motherhood and fatherhood**. The need for a deeper understanding of parenthood and parenting is a topic which emerges repeatedly as a key issue for future research. *The future of parenthood among young people in Europe and across the new plurality of families is seen as a major interest for both research and policy.* This implies focusing on a wide range of themes and issues, such as: examining the new models of motherhood and fatherhood (including legal aspects and their implications as well as the values and practices of parenting types); understanding how young people plan and envisage parenthood; seeing how the new models relate to gender and social

inequalities, as well as to different family forms and conjugal divisions of paid and unpaid work; analysing the social processes that promote or hinder fathers' involvement in parenting practices; understanding how parents deal with illness and disability in children; seeing how media can be a tool to help parents in parenthood and how they incorporate media into their daily lives; analysing dissemination of these models across Europe and capturing the roles played by different family polices in promoting these changes/models.

Suggestions for Project Topics:

Negotiating parenthood: understanding the decision-making processes of the transition to parenthood (first or second child). Example of study design: in depth interviews with 20-30 couples with young children, covering different social and economic back-grounds. Case comparisons across EU countries.

Fathers taking leave and working flexibility: understanding family experiences of fathers on leave and their use of flexible work. Example of study design: qualitative household level study; 20-30 dual earner couples with young children, with fathers in a variety of leave and employment situations; in-depth interviews to under-stand how fathers (and/or mothers) manage tension between time, money, services and care. Case comparisons across EU countries.

Strongly connected to the topic of contemporary childhood and family well-being, the necessary research on contemporary parenthood is also closely intertwined with the need for a greater understanding of how children's lives and outcomes are currently affected by both motherhood and father-hood forms and how these have changed (see Topic 2).

2. **Children's experiences, trajectories and outcomes**. Another major trend in the discussion on the gaps and challenges for future research focussed on children in families, in particular on the need for a better understanding of the *experiences of children and of how their lives and outcomes are affected by different elements of their family lives* (e.g. the effects on children of living in different family structures or within diverse parental and educational models); the effects on children living with parents who either both work full-time, or where one works full-time and one part-time, or only one

works; or with parents with atypical timetables; the experiences of parent's partnership breakdown; the experience of living in poverty for short or long periods in institutional settings and in families with different educational, financial and social resources; the effects of experiencing different types of childcare – at different ages and with different amounts or hours of care; understanding how media, in particular internet usage (but also other kinds of media, such as television, advertising, etc.) are shaping children's lives.

Moreover, the broad issue of children's lives and outcomes came across frequently as a cross-cutting pathway into research on family life, in particular since it could encourage:

a. research projects tying together different fields and aspects of family life: parenthood, working couples/mothers, schooling, child development and outcomes, social inequality and poverty, the impact of new technologies and changing living environments or communities;

b. research projects focusing simultaneously on various family issues which may be seen as tensions or dilemmas of contemporary families with children (e.g. how to combine the interests of children, working parents and the labour market; finding quality care solutions for young children below the age of three; time use and quality time with children; the meaning of choice in family life; family management of media, schooling and parenting; positive or negative effects of different types of childcare).

Suggestions for Project Topics:

Children and maternal employment: how different types of maternal employment and care (working full-time, part-time; using different types of care) influence children's lives and development, in the context of diverse welfare and gender equality regimes. Example of study design: cross-national survey of mothers with young children in a variety of employment situations, in different EU countries or birth cohort study.

Children's experiences and outcomes in families outside the labour market (unemployed, retired, sick) and/or "at risk" families (suffering physical, mental disability or some kind of addiction).

Other proposals for cross-cutting research programmes stressed the need for studying the effects of organisational change in the world of work and daily life (hours of work, workload, geographical mobility, multilocality in daily life, etc.) on the life rhythms of children according to age groups (pre-school, 6-12 years, adolescents).

In summary, more knowledge on the evaluation and impact/effects of different forms of child care, as well as their linkages to maternal and paternal employment, labour market constraints, parental leave systems and changing gender equalities/inequalities within the family, was generally considered as an important challenge for both research and policy-making.

3. **Changing family composition, structures and networks**. *A better understanding of old and new family forms and their development over time, of why differences in family composition and structures occur and why their extent differs across EU countries* was identified as an important issue for research and policy. Discussions on research gaps in this field pointed repeatedly to the following methodological problems: the lack of longitudinal and cohort data; difficulties in dealing with the concepts and indicators of family living arrangements, particularly those addressing the existing plurality of family life and relationships; an overly strong focus on the nuclear family model and the household unit and on aggregate national data rather than variations between different social and regional groups; problems regarding the comparability of European family databases with other databases in order to understand the influence, at the cross-national level, of welfare, labour market and educational systems.

   Four interrelated research topics within this fundamental field of research on family composition over time and across social groups and national contexts may be underlined:

   1. The need for further and improved data on *family composition and structures*, their plurality within national contexts and across Europe and the main factors shaping variation and diversity. Deeper understanding of new family and conjugal living arrangements (e.g. blended families, same sex unions, families separated by migration, lone fathers, joint custody families) and of the differences between social, cultural and regional groups are seen as major challenges for future research on this topic.
   2. Moving beyond the focus on the household unit and the standard nuclear family model, the need for research to grasp the diverse meanings and new notions of family and family relationships in

228

late modernity, in particular the *sets or configurations of close relationships*, which may include a variety of important alternative ties providing support and resources (e.g. friends, relatives beyond the household unit or from other generations, colleagues).

3. Drawing on a life-course perspective, greater *understanding of family formation, transitions and trajectories*, including decision-making processes and reasons underlying or delaying family transitions (such as the transition to parenthood, to conjugal life or to divorce), as well as the linkages between different types of life trajectories, in particular between career and family trajectories; family transitions and decision-making processes must be understood in the context of specific historical, social, normative, institutional and generational contrasts.

4. Understanding the *differential effects of major demographic trends* (e.g. rising life expectancy, low fertility, increasing geographical mobility and immigration) on family forms, intergenerational relations and networks over the life-course.

Suggestions for Project Topics:

Family forms across Europe: obtaining further and improved data on family composition and structures.

Families as networks: mapping the resources that exist beyond the nuclear family, their effects on family relationships (gender and intergenerational) and care.

Changing meanings of "family": grasping the diverse meanings and new notions of family and family bonds (including a variety of relatives and non-relatives providing attachment, support and resources).

5. **Post-divorce family forms and relationships**. Analysis of post-divorce situations is another major issue for future research and policy-making pinpointed by discussions, presentations and documents. After divorce, 'joint custody' is becoming more and more frequent due to changes in legislation in most European countries. There is a need for research on the diverse patterns of these post-divorce family forms and how couples negotiate and decide on the new living arrangements. But there is also need to further the analysis of their impact on mothers' and fathers' professional life/careers, on child care arrangements, and on children's experiences and outcomes.

Suggestions for Project Topics:

Post-divorce living arrangements: how parents structure daily life and negotiate parenting within shared residence arrangements ("joint custody") and consequences of this on children. Example of study design: in-depth interviews with divorced couples (each member separately) with children under 16. EU countries with "joint custody".

Family legislation after divorce: comparative research on how policies across Europe are dealing with new post-divorce family situations such as "joint custody", tax deductions, and who receives family benefits.

6. **Families, social inequalities and living environments**. *Families and social inequalities* also emerge as a cross-cutting issue, mainly due to the fact that research on families during the past few decades has tended to neglect analysis of social, cultural, spatial, environmental and regional differentiation and its consequences on family life and experiences. Four interrelated research topics within this fundamental field of research were highlighted:

- The need for a *deeper understanding of social inequalities between families*: for example, how long families/different types of families spend in disadvantage or poverty; how and why some types of families accumulate advantages (e.g. well-paid dual career couples) or disadvantages; what the experiences and effects on family members of living in disadvantaged families or environments (or in difficult housing situations) are; how and why the extent of social inequality between families and its effects on family outcomes differs across European countries.

- The need to understand more about the *role of families in reproducing social inequality across the generations*, thus affecting children's life chances. Transmission of social advantage and disadvantage via the family may take place both at material and socio-cultural levels: for example, how do unequal endowments of 'cultural capital' in families influence children's acquisition of social and educational skills, and how do differences in income levels or social capital affect the living conditions of children and the inheritance of economic capital and material advantage over the life-course.

- The need for greater understanding of the *linkages between policies and inequalities between and within families*, by examining not only how policies help to check the worst inequalities produced by differential access to resources and living environments, but also in what ways policies are likely to challenge the entrenched advantages some families have and pass on to their children. Research on the causes and consequences of social inequalities and how policies tackle them is key to understanding the relative position of disadvantaged families and families at risk of failing.

- The need for research on specific types of families which may be more vulnerable to disadvantage, poverty or difficult living or housing conditions. Given the increase in immigration, as well as increased mobility in general within the EU, research on *immigrant families* and on families from minority ethnic groups was considered by participants a major and urgent challenge for research and policy-making. Research is still scarce on immigrant families and on the positive or negative changes resulting from migration. Relevant issues for research which deserve more attention include: the role of families in promoting the integration of their members; types of spatial concentration or dispersion and the way this affects how immigrant families settle and how the host city copes; immigration and care (how immigrant families manage work and care, the effects of transnational care practices on family life, and the crucial role that immigrants play today as care workers for dependent people in Europe); developments in immigration policies, in particular restrictions on family reunion, and their impact on family life; understanding how subsequent generations are coping, who succeeds and who fails to thrive in different local, national and cross-national contexts.

Suggestions for Project Topics:

The complex connections between social inequality and family life: how are families transmitting and reproducing inequalities and how does this affect children's and young people's life chances.

Spatial concentration and immigrant families: how is spatial concentration affecting access to resources (education, health, and integration)? And how do host cities and families cope with migrant groups?

Social inequalities and school underachievement and drop-out: the family and social trajectories of children with low achievement and how it affects their life experiences.

Sustainability and family dynamics: what is the impact of family dynamics on the environment? Examining the role of families (mother, fathers, children) in daily purchasing, use of energy, use of transport, etc.

7. **Doing family: family interactions and processes over the life-course**. Another major trend of the discussion on gaps in research stressed the importance of focusing on *the interactions between people within the family and on their practices in everyday life and over the life-course*. From this perspective families are seen to be constructed through multiple forms of interaction (from physical and emotional interactions to cognitive, social, spatial and media-related interactions; from interactions involving cohesion and solidarity as well as conflicts, demands, stress, and even violence). A major challenge for research within this approach is therefore to understand the daily and biographical shaping of common life as a family, built upon the interactions and daily life of the different members of the family (of conjugal partners, children, fathers and mothers, siblings) and of the couple/family in relation to significant others and wider societal contexts.

This approach to family studies points to a variety of potential, interrelated themes which are important for both research and policy-making. For example: examining family practices and negotiations of paid and unpaid work and the existing gap between attitudes and practices regarding gender roles; understanding the diverse procedures and models of negotiating and practicing parenthood and partnership, and also of specific events or family transitions (illness, death, leaving home, birth of a child, etc.); understanding the interactional dimensions (emotional, physical, cognitive, social) of motherhood and fatherhood; understanding support practices and mutual care between family members and different generations; studying the effects of different interactional factors on the wellbeing of families and couples (for example, the extent to which a rich relational environment, implying support practices and ties beyond the nuclear family, are important factors for conjugal and family wellbeing, even beyond divorce); understanding how families, but also in particular children, deal with high-conflict situations; comparing practices and daily life over the life-course in diverse types of families, such as blended families, large families, lone parent families, migrant families or same sex couples.

Suggestions for Project Topics:

Family interactions and wellbeing: the effect of different types of interaction on the wellbeing of families and couples (e.g. to what extent is a rich relational environment outside the family an important factor in conjugal and family wellbeing?)

The life rhythms of children: how does organisational change in the world of work, schooling and daily life (hours of work, workload, geographical mobility, multilocality in daily life, media, etc.) affect the life rhythms of children according to age group (pre-school, 6-12 years, adolescents)?

Media as a tool for parenthood: how are parents using the internet to help them in parenting?

Family practices and negotiation of unpaid work: "opening the black box" of the gender gap between discourse and practice.

Transitions to adulthood in European societies: mapping the extent of de-standardisation.

8. **Ageing, families and social policies**. Ageing was recognised as one of the main challenges that European societies and families will be facing over coming decades. Research issues discussed and suggested during the Conference cut across a variety of questions and topics. *Understanding changes in the life trajectories and transitions of people aged fifty and over* (e.g. transition to new partnerships, to postponed or anticipated retirement, to grandparenthood, to dependency on others in daily life), in the context of different labour market and welfare contexts, was emphasised as a first important topic for research.

    Other key issues for research included the following: understanding intergenerational support and solidarity, from the perspective of elderly persons both as *care receivers and as care-givers*; understanding *how active ageing is impacting on support for dependent persons*; identifying the values, practices and important contributions of grandparenthood; understanding the connections between *ageing and migration* (immigration as a factor which slows down the process of ageing; the relationship between the growing needs of elderly care in ageing societies and the immigration of female care workers); examining *the sustainability of different care arrangements* (e.g. carer's needs for training, respite, cash benefits, services, support for reconciling work and family life, use of new technologies); *understanding the subjective dimensions of care* (how care and the problems of caring are experienced by care-givers and by care receivers); a deeper understanding of the *new trends in social care for the elderly*, whereby flexibility and complementarity (between state, market and family, between paid and unpaid care, and between solutions developed at local or national levels,) are being highlighted and developed in most European societies.

Suggestions for Project Topics:

The transition to elderly life: examining family and work trajectories of ageing men and women and their effects on personal wellbeing.

Socially innovative forms of care: mapping best practices (including community and neighbourhood networks, information and communication technology in health support services, elderly care-friendly companies).

Welfare states, migration and care: the politics of care and the role of immigrant women in formal/informal care services across different European countries.

The experience of caring and being cared for: incorporating the views of the people in need of care/care-givers – subjective aspects and financial implications of being a carer and a care receiver.

8. **Family policies**. Analysis of family policies and of the intersections between family policy and other policies (e.g. gender equality, labour market, educational, social security, immigration), both at local, national and cross-national levels, is a cross-cutting issue which was raised in all the sessions of the Conference. Many proposals and thoughts on the challenges and research gaps may therefore be found in the summaries of the focus groups and workshops presented in the earlier sections of this chapter. Overall, more analysis and comparison of family policy trends in Europe were recommended. The following selected elements seek to highlight some of the more specific topics, aspects or gaps in policy research which were identified as important:

   • A deeper understanding of how family policy is culturally, institutionally, politically and historically embedded in each country; in particular, the need for more research on how the development of national policy measures is being shaped by differences in socio-political pathways, regulatory frameworks and financing possibilities

   • Understanding changes in family policy measures and priorities as a response to contemporary societal challenges and difficulties, in particular the economic crisis.

   • The need to improve and renew existing typologies of family policies in general, as well as the typologies related to specific fields of family policy, such as institutional frameworks, parental leave systems, social care patterns, cash and tax benefit systems. In this field, the need to move beyond dichotomic concepts such as familialisation/defamilialisation, formal/informal, choice/no choice, north/south divide, etc., towards a better understanding of the on-going complexities of family policy developments (for example, the complex ways in which policies are currently mixing and balancing formal care, informal care and immigrant worker care in order to provide care for older persons more effectively).

- A better understanding of the rationales and consequences of some of the more recent and sometimes controversial developments in policies, such as: cash for home care versus day-care for children under three; increases in maternity leave versus increases in paternity leave and measures to promote gender sharing of parental leave; universal versus selective family allowances.

- Greater understanding of the linkages between policy measures/entitlements and family ideals and practices. For example, the need for further data on the practices and consequences of parental leaves (coverage rates and uses of different leaves, decision-making processes and strategies underlying use, parents' and other actors' perspectives on different types and consequences of leave).

- Compensating for long-standing gaps in research on family policies. There is less research on care services for the elderly and the reconciliation of work and caring for elderly persons than on child care; not enough attention given to the importance of a life-course perspective for the framing of policies; inadequate data on tax benefits; less attention given to the quality (and the quality standards) of services than to the quantity; less attention given to the perspectives and measures implemented at local or regional levels and by employers; not enough attention given to the evaluation of existing policy measures and the need for developments in the tools (new types of services, leaves, etc.) of family policies; little attention given to the perspectives of policy makers and to how and why evidence-based policies are being developed; there are countries which are systematically under-researched.

- The need for greater understanding of the role and contributions that different types of NGO and family associations are making today and could make in the future (in the context of different national and cross-national frameworks) to the building up of support for families and policy-making.

Suggestions for Project Topics:

The politics of family policies: understanding how national policy measures are being shaped by differences in socio-political pathways, regulatory frameworks and financing possibilities.

Evaluating recent changes in policies for parents in Europe and how they affect families: for example, cash for care, leave for fathers, developments in care services, family benefits (e.g. cross-national comparison of the politics, uses and consequences of extended home-care leave in different European countries, such as Norway, Finland, Austria, etc.).

The impact of EU policies on family reunion: cases of recent restrictions on family reunion of non-EU immigrants and their consequences on family life.

In summary, in this chapter we have given a general overview of the dynamics and outcomes of the critical review process, a key event of FAMILYPLATFORM, which took place during a three-day Conference in Lisbon in May 2010.

Various conclusions may be drawn from this process, two of which are of particular importance. The first is that there are some major concerns regarding the future research and policy agenda. Against a backdrop of growing inequalities, the economic crisis, and new dilemmas and risks facing families, in particular young families and children, key concerns are the challenges of parenthood and parental negotiation, care for young children and elderly persons, difficult life transitions and work-family balance, and the changing forms, meanings and practices of contemporary families in Europe.

Family research is also of vital concern in the policy context, with a demand for in-depth analysis of policy processes and effects, both at national and cross-national levels, in order to generate evidence-based awareness and policy developments. The second conclusion is related to the impact of a Conference which involved a plurality of perspectives by bringing together around 140 participants from various sectors of society. The discussions and controversies which emerged from the Conference took the debate to a higher level of mutual understanding and recognition, as well as helping to contribute to a clearer awareness of the diversity of family actors and agendas at national and European levels. This awareness may be seen as an essential driver of dialogue and democracy in Europe.

Chapter 3: Facets and Preconditions of Wellbeing of Families - Results of Future Scenarios

Olaf Kapella, Anne-Claire de Liedekerke & Julie de Bergeyck

3.1 Introduction

What living arrangements and family forms will people choose in the future? Will families remain the central place where individuals' needs are fulfilled? Who will care for our children and elderly relatives and how will they do so? What are the uncertainties which could affect the wellbeing of families? What factors or policies improve the wellbeing of families, and how can they be implemented or encouraged?

If we think about the future of the family, these questions and many more spring to mind. One of FAMILYPLATFORM's key activities was to examine future societal challenges, factors and policies which will have a strong impact on families. To fulfil this goal, "Future Scenarios" were outlined to describe possible futures of families in Europe in 2035 as well as key policy and research issues.

From the outset it should be made clear that in developing these Future Scenarios and key policy issues, a very creative technique has been used – known as the Foresight Approach. This method does not claim to be a scientific simulation, nor does it claim to foresee the true future and challenges of families. It is, however, a unique and creative method of bringing to life possible futures for families. Our aim was to describe possible scenarios and families in the form of narratives and research issues, as well as key policy issues and social innovations for researchers, policy makers, NGOs and all those others involved in work for and with families.

Even though this exercise was very creative, its results are firmly grounded on the findings of the work in the FAMILYPLATFORM. It was one of several steps leading up to the Research Agenda on Families and Family Policy – which is of course, FAMILYPLATFORM's ultimate goal.

The Future Scenarios are the result of wide-ranging discussions between the members of the Consortium and the Advisory Board of the FAMILY-PLATFORM. Credit is due to each member for their creativity, ideas and willingness to maintain the process of discussion for close to a year. The Consortium encompassed different scientific disciplines as well as different organisations, such as universities, NGOs and policy makers. To acknowledge the valuable work of each one, we present the names of all those who contributed to the work.

Consortium

- Technical University of Dortmund, Germany: Kim-Patrick Sabla, Matthias Euteneuer, Uwe Uhlendorff
- State Institute for Family Research, University of Bamberg, Germany: Loreen Beier, Dirk Hofäcker, Elisa Marchese, Marina Rupp
- Family Research Centre, University of Jyväskylä, Finland: Kimmo Jokinen, Marjo Kuronen
- Austrian Institute for Family Studies, University of Vienna, Austria: Sonja Blum, Olaf Kapella, Christiane Rille-Pfeiffer
- Demographic Research Institute Budapest, Hungary: Szuzsa Blaskó, Zsolt Spéder
- Institute of Social Science, University of Lisbon, Portugal: Mafalda Leitão, Vasco Ramos, Karin Wall
- Department of Sociology and Social Research, University of Milan-Bicocca, Italy: Carmen Leccardi, Sveva Magaraggia, Miriam Perego
- Institute of International and Social Studies, Tallinn University, Estonia: Epp Reiska, Ellu Saar
- Department of Media and Communication, London School of Economics, United Kingdom: Ranjana Das, Sonja Livingstone
- Confederation of Family Organisations in the European Union (COFACE), Belgium: Linden Farrer, William Lay
- Forum delle Associazioni Familiari, Italy: Francesco Belletti, Lorenza Rebuzzini
- Mouvement Mondial des Mères Europe, Brussels, Belgium: Julie de Bergeyck, Anne-Claire de Liedekerke, Joan Stevens

Advisory Board

- European Commission, Brussels, Belgium: Krzysztof Iszkowski, Emanuela Tassa
- Haro, Sweden: Jonas Himmelstrand, Madeleine Wallin
- National Family Planning and Parenting Institute, London, United Kingdom: Clem Henricson
- Mouvement Mondial des Mères Europe, Brussels, Belgium: Owen Stevens
- University of Jyväskylä, Finland: Lea Pulkkinen

Guests

- Bundesforum Familie, Berlin, Germany: Katherine Bird
- European Commission, DG Research: Elie Faroult

3.1.1 Scientific background for the work

In our view, it was important to construct Future Scenarios and the key policy and research issues for the wellbeing of families in the year 2035 on a firm scientific foundation. It should be borne in mind that FAMILYPLATFORM and its different working stages are part of a process and not just isolated steps. FAMILYPLATFORM's first step was to describe the state of the art and major trends in family research (Kuronen, 2010). Work on the Future Scenarios was therefore constructed on this firm scientific foundation.

To give an impression of the foundation on which work on the Future Scenarios was built, a summary of the major trends described in the first stage of the project are presented first, followed by a summary of the results of a structured brainstorming exercise to identify the key aspects and facets of the term "wellbeing of families".

3.1.2 Major trends in the Existential Fields

Demographic topics and trends such as declining fertility rates and post-ponement of first childbirth, or the influence of economic crises on individual living arrangements and family forms, have a long tradition of in-depth and detailed study across Europe. In the work of FAMILYPLATFORM, the various research areas in family studies and family policy were divided into eight Existential Fields. Each Existential Field produced a report[1] on its topic. The report covers the state of the art of scientific studies across Europe as well as describing major trends in each field of research. The construction of the Future Scenarios for the wellbeing of families in 2035 was based on and rooted in this work. Future trends that are visible today were central to the construction of the Future Scenarios.

A brief summary of the major trends from the Existential Fields will be given as an introduction, because family life in the future will be influenced by these trends.

Ageing populations across Europe

The proportion of people over the age of 60 in Western Europe will rise dramati-cally, from 21 per cent in 2008 to 33 per cent in 2035. In Eastern Europe the increase is virtually identical: from 19 per cent in 2008 to 32 per cent in 2035 (Table 1).

---

[1] For details see Working Reports: Beier, L.; Hofäcker, D., Marchese, E.; Rupp, M. (2010). Belletti, F.; Rebuzzini, L. (2010). Blaskó, Z.; Herche, V. (2010). Blum, S.; Rille-Pfeiffer, C. (2010). Kuronen, M.; Jokinen, K.; Kröger, T. (2010). Leccardi, C.; Perego, M. (2010). Livingstone, S.; Das, R. (2010). Reiska, E.; Saar, E.; Viilmann, K. (2010). Wall, K.; Leitão, M.; Ramos, V. (2010).

Table 1. Demographic changes between 2008 and 2035

		2008		2035	
		Western Europe	Eastern Europe	Western Europe	Eastern Europe
Life Expectancy	Female	81.49	77.87	86.75	83.42
	Male	75.26	70.00	80.55	75.55
Total Fertility Rate		1,51	1,31	1,56	1,48
Proportion below age 20		23%	22%	18 %	16%
Proportion above age 60		21%	19%	33%	32%
Proportion above age 80		4%	3%	7%	6%

Source: Lutz/Sanderson/Scherbov (2008).

Postponement of marriage and first childbirth, and generally decreasing number of children. Decline in the number of children per women, even though fertility aspirations are still at a comparatively high level

This trend is found all over Europe, but differs between regions: the age at first marriage in the Scandinavian countries is the highest in Europe, with the average in Sweden at 31.1 years, and lowest in southern Europe, for example, in Portugal at 26.4 years. The average age of women at first birth is currently lowest in eastern Europe, for example in Bulgaria (24.9 years) and in Hungary (26.9 years). In contrast, the average age of women at first birth is highest in Switzerland (29.5 years) and in the United Kingdom (29.8 years).

Increasing number of out-of-wedlock births

Later marriages are also reflected in an increase in out-of-wedlock births. Being married has lost its central role as a precondition for family formation. However, cohabiting relationships are often a transition phase to a later marriage.

Decreasing marriage rates, increasing divorce rates and increasing rates of re-marriage

Increasing diversity of living arrangements/increase in new types of family life

The "nuclear" family, often also referred to as the "classical" family, remains the dominant family type across Europe, but its numbers are decreasing, while the number of families of other forms is increasing. In particular, growth has been observed in the number of lone parents, stepfamilies and cohabiting couples, but also in "new" or "rare" forms such as foster and adoptive families, rainbow families, multi-generation households, and families with more than one common household, such as "living apart together" and commuter families.

The prolonged presence of young people within the family of origin

All across Europe it can be observed that young men and women stay longer in their family of origin. The highest rates are found in central and southern Europe. For example, young people leave home after the age of 28 in Belgium, Slovakia, Italy and Malta and between the ages of 22 and 23 in Finland.

The new role of grandparents

Increasing life expectancy and better health across the whole of the life-course have led to grandparents playing a more important role in families. On the one hand, they are becoming an important resource for their children and their children's families (e.g. care work) and, on the other hand, grandparents themselves are choosing more frequently to become active subjects in their own lives e.g. deciding autonomously how to spend free time and money.

The field of family policy has gained in importance and expanded

After decades of insignificance and low prestige, the field of family policy has expanded and gained in importance. In general, it can be said that the Nordic and Anglo-American countries have less explicit family policies than the conservative, Mediterranean and post-socialist ones, and they do not protect the family as a social unit in their constitutions.

In recent years, defamilialisation has been more pronounced in national family policies than re-familialisation

There are ongoing trends of re-familialisation (where the family is responsible for the welfare of its members) and defamilialisation (where social policy takes over welfare and care responsibilities) in Europe, which may be positive (e.g. increases in parental leave benefits), or negative (e.g. reduction of family allowances). This leads to a greater mixing of re-familialising and defamilialising measures, such as Nordic countries introducing re-familialising childcare expansion.

The most important family policy issue has been reform of childcare services

Childcare for children over the age of three (up until school entry) is well developed across Europe, with care rates of at least 90 per cent. Childcare for children aged under three is particularly well developed in the Nordic countries, the Netherlands and Belgium. In Austria, the Czech Republic, Hungary, Lithuania, Malta, Poland and Slovakia the care rates are less than 10 per cent of all children under three. In several countries the last pre-school year has been made compulsory.

Polarisation between families with very low and very high incomes

The risk of poverty increases in line with the number of dependent children in the household, but also for lone-parent and single-adult households. In Mediterranean countries and in most of central and eastern Europe, the risk of poverty in families with two children is higher than if the family had only one child. For example, in Greece, Italy, Portugal, Spain, Latvia, Lithuania, Poland and Hungary at least one-third of households with three or more children have an income below the poverty line.

Growing diversity and instability of work

Employment patterns are shifting away from full-time, non-temporary employment for men and women. This is not just due to employee choice, but also to employers' preferences and the deregulation of labour markets.

Differentiation in education levels along the urban/rural dimension

Differences between European countries in terms of educational attainment levels are not as clear cut as the differences between urban and rural populations within the vast majority of European countries. Since educational facilities are harder to reach in rural areas, and parents' educational levels as well as the family's financial situation are related to school performance and the future educational level attained by their children, children in rural areas face multiple obstacles to achieving their full potential in education.

Mismatch between diversification of the life-course and housing market developments

Europe is on the whole characterised by a reduction in the rental housing stock. There is a wide variation in housing ownership status across Europe: home ownership rates are generally higher in the new EU Member States (e.g. Estonia, Hungary, Lithuania, and Slovenia) but also in Spain, Greece and Italy. There is a tendency for lower levels of ownership in Germany, Austria, the Netherlands and Poland. The number of people per household tends to be lower in northern Europe and higher in Mediterranean countries and the new Member States. In a comparison of living space and number of rooms per dweller, homes in central and eastern Europe are smaller and more often overcrowded.

Growing responsibilities of local governments

Local authorities (such as regions and municipalities) are taking on more and more independent responsibility for political issues, such as schooling, childcare services and taxes. Global solutions have to be appropriate to local realities for families and individuals. This trend is also known as "glocalisation".

Different actors are working together at the local level to reshape realities for families

To adapt global solutions and challenges to local realities and family life there is increased networking at the local and regional level between different actors such as NGOs, the public sector, private companies, trade unions, and family associations. The intercultural dimension is also gaining in importance at a local level as well.

Gendered division of paid and unpaid work

Even if the level of female employment has increased (from the 1960s onwards), men still spend more time in paid work and women in unpaid work (such as domestic tasks and childrearing). Women spend less time in the labour market and are more likely to take part-time jobs and have more career breaks than men – on average women spend twice as many hours a week in unpaid work. But statistics show that the gender gap in the level of labour market activity is decreasing – the difference between men and women in the level of labour market activity fell from 18.6 per cent in 1997 to 13.7 per cent in 2008 in the EU27 countries.

There are significant differences in the gendered division of time spent caring for children. In general, women do the majority of childcare. In the Netherlands and Nordic countries the division of time spent on childcare between men and women tends to be most equal – women do twice as much childcare (around 16 hours per week) as men (seven to eight hours per week). In all other countries, men spend on average only four to five hours per week caring for their children. The largest gap is noted in Anglo-Saxon countries, where women spend 14.2 hours per week and men just 4.1.

Social care is going public

The trend towards the institutionalisation and professionalisation of care work and services for families will continue. This does not mean that social care is provided as a public service, but rather that it will be provided as a blend of public and private market-based services. Most researchers agree that the main differences in social care arrangements are to be found between southern and northern Europe.

Social care still remains a combination of formal and informal care. In particular, the role of women in providing care for children, the elderly and other dependent family members is remarkable.

Childcare remains the main focus of social care policy at the national and international level. This is not only related to the needs of the economy, the labour market and gender equality policy, but also expressed through authorities paying more attention to the quality of childcare services and the educational aims and content of formal services.

The globalisation and internationalisation of care and care work will increase. The further development of cross-national care relations, global care chains and transnational care will lead to an international market for care services. Furthermore, the number of migrant care workers in formal and informal care work will increase.

Extreme vulnerability of migrant families and their children, particularly of non-EU immigrant families in comparison with other families and EU migrant families

There are continued and significant migration flows into Europe, and an increasing feminisation of migration. The vulnerability of migrant families is apparent in several areas of everyday life: they work in lower-paid and lower-skilled jobs, they have atypical working hours, are more frequently exposed to poverty and unemployment, they often have weak family networks and consequently major problems in reconciling work and family life with young children, and they mainly live in segregated urban areas. Consequently, the integration and wellbeing of second and third generation migrants is a major challenge. Immigration policies tend to be restrictive or ambiguous with regard to family reunion and the legalisation of immigrants and their children.

Higher risk of exposure to poverty of some social groups and types of households

The average risk of exposure to poverty for households in the EU27 is 17%. The following groups have a higher risk of exposure to poverty: unemployed households (43%), immigrants from outside the EU (30-45%), children in lone-parent households (34%), people with low educational attainment levels (23%), elderly women (22%), young adults aged 16-24 (20%), children (20%), lone-parent families (34%), large families (25%), and single person households (25%).

More and more areas of our social life rely on new information and communication technology

New, interactive, individualised and personalised media technologies are rapidly contributing to a diverse media environment in Europe. For example, educational systems across Europe, from school through to university, increasingly rely on classrooms enhanced with technologies. Health, ageing support and other care and support services also rely increasingly on new technologies, especially within the home.

3.1.3 Key aspects of the wellbeing of the family

Before starting work on the possible Future Scenarios for family wellbeing in 2035, the different facets and preconditions of "wellbeing of the family" had to be defined. The aim was not to establish criteria for a definition of wellbeing of families in general (that would have been a project on its own), but to identify the key facets and preconditions of wellbeing for families that could then be used for constructing the Future Scenarios.

To ensure that the points of view of all the different members and dialogue groups of the Consortium and Advisory Board of FAMILYPLATFORM (researcher, stakeholder, policy maker, NGO) were taken into account, a systematic method of brainstorming was used. The participants were divided into small groups and asked to suggest key aspects/issues to define wellbeing of the family. After sharing the results of the group work and refining them over several rounds of feedback, ten key aspects were identified.

1. **Security for individual members of the family and for the family itself.** On the one hand, security applies to general and material conditions, such as social and economic support for the family. On the other hand, when using security as an indicator of the wellbeing of the family we are also talking about emotional (immaterial) security for individual family members, such as a child's right to be raised without violence, or to feel secure. Indeed, a part of wellbeing is feeling secure in personal and family life today, and maintaining that feeling in the future by seeing opportunities and knowing that there will be a place in society for families with different living arrangements in the future.

2. **Individual self-fulfilment.** Self-fulfilment is an issue that cuts across different areas of life for every human being, including work-life balance, partnership, and involvement in society. It is strongly related to freedom of choice at an individual level but also at the level of living arrangements and family forms. It encompasses being able to plan and realise the choice of a specific family form or living arrangement at a specific time. Self-fulfilment can be achieved through individual resources, but also through support by family members and society.

3. **Health.** Health is a multi-dimensional precondition for wellbeing encompassing, for example, the way the health-care system is organised, access to different health-care services, and the range of health-care services on offer. Additional aspects include the environment and how it helps the individual to be and stay healthy, as well as information and communication technologies. The issue of health has to be differentiated by gender and social group or status, since access can be very variable. There is not just an objective dimension to health, but also an individual, personal one: *"How healthy do I personally feel? What options do I have to influence my own health and the environment I live in?"*

4. **Involvement in society (citizenship/participation).** In general, involvement in society can be described along lines of inclusion or exclusion. This may be different for women and men, for individuals and for families, as well as for different social groups. It should be understood not only in terms of active personal engagement in society, but also in terms of how a person is integrated into or excluded from society, which can lead to poverty or exclusion for that person or family or living form. Involvement in society may also be visible in different ways, such as engagement in NGOs, social associations and voluntary work.

5. **Love, respect and tolerance.** Basic human needs are a central aspect of the wellbeing of families and individuals. Needs such as love, respect and tolerance have an impact on the emotional and physical health of each family member and can prevent crises or problems within families

and between their members. On the other hand, we have to be aware that if these needs are not present or cannot be satisfied by the family, then family may be a harmful or even violent environment. Furthermore, the intergenerational aspect of fulfilling these basic needs within the family should also be stressed.

6. **Balance**. With wellbeing in mind, it is important to point out the individual and the societal dimensions of balance. At the individual level, we are talking about the personal balance one finds in life, the ability to manage ambivalent situations, and being able to negotiate in order to establish or maintain balance. Over the life-course this personal balance is likely to change and may have to be re-established. At the societal level, it refers to work-life balance and opportunities in society.

7. **Time**. Modern society is composed of many different 'worlds', each with its own time structure, and this makes negotiation and co-ordination necessary. In today's society, time is not only of central importance for individuals but also for living arrangements and family forms which are becoming increasingly difficult to co-ordinate. Families and individuals should be empowered to spend time together as a family and as a couple, to maintain family relations and to be able to manage family and personal tasks.

8. **Equality**. Equality is the central aim of modern society in Europe and intersects with virtually all other issues. It can be understood in a very general sense as each person having the same opportunities to participate in society, in every possible way, and at every possible level. Equality refers to many aspects, such as gender, sexual orientation, social position, religious or ethical beliefs and cultural heritage and traditions.

9. **Support for families**. Support for the family comes from both inside and outside the family. It can be provided by the state, or more locally by municipalities or regions through policies, regulations and services. "Family mainstreaming" should be introduced as a term meaning the attempt to integrate the family perspective into existing and new policies, even if those policies are not explicitly designed to affect families. A central aspect and starting point in support for families is the recognition of the work families and their members are doing for society and its individual members. That fact should be visible and acknowledged throughout the different fields of society.

10. **Living and environmental conditions**. Living and environmental conditions have a broad range of aspects that affect the wellbeing of families and individuals, for example educational opportunities and expectations, housing and urban development, intergenerational solidarity, the role of information and communication technologies and the

media, and the economic situation. Living and environmental conditions should be understood as the basic foundation for achieving the other preconditions for wellbeing mentioned here.

3.2 Methodological approach

3.2.1 The Foresight Approach

In order to debate the future wellbeing of the family in 2035, a participative approach involving the different groups of experts in FAMILYPLATFORM was required. With the Foresight Approach, a popular approach to describing the future was chosen. By using this approach, possible futures for families and living arrangements can be identified, and desirable and undesirable Future Scenarios imagined. This provides a basis for developing future strategies.

Foresight is a creative and systematic attempt to look into the longer-term future of society as a means of policy consultation, and therefore this method fits perfectly with the project design of FAMILYPLATFORM.

The "Foresight Approach is becoming increasingly attractive for governments, national research agencies and businesses in their efforts at coping with the increasing complexity of new technologies and decision environments, in an increased techno-economic competition world-wide... Since the 1990s quite a number of major foresight exercises have been launched in many European countries" (Kuhlmann, 2002).

Foresight is neither a prognosis or simulation, nor a plan for the future, but does require participants to take an active role throughout the entire process. In FAMILYPLATFORM it was used as a tool for creating narratives in order to bring to life possible family forms and living arrangements in the future, and to make these intelligible to policy makers, scientists and other stakeholders. Furthermore, it was used as a creative means of building a bridge between the knowledge of the individual experts who prepared the Existential Field reports, and the future agenda for family research and family policy.

> "The majority of experts consider foresight essentially as a collective and consultative process, with the process itself being equally or even more important than the outcome. Foresight exercises are ways of obtaining options, conflicting or otherwise, about future developments, most of which are already established. Foresight in this sense is an essential contributor to the creation, either collectively or individually, of models of the future. Such models are im-

portant because they are capable of creating synthesis, they are disruptive and interfere with current modes of thought, thus forming and shifting values" (Kuhlmann, 2002).

The Foresight Approach is especially useful for policy consulting. Turoff *et al.* (2002) describe the effect of a structured approach such as Foresight in generating the strongest possible opposing views on potential solutions for a major policy topic. Kuhlmann (2002) describes just some of the major purposes of Foresight in the context of policy making as being to discover what new demands and new possibilities will emerge, as well as new ideas, and to discuss desirable and undesirable futures, identify a choice of opportunities, and assess potential impacts and opportunities.

A central technique in the Foresight Approach is the Delphi method. This was developed in the USA by Gordon and Helmer in the 1960s (see Gordon/Helmer, 1964). FAMILYPLATFORM also used the Delphi method for its Foresight exercise. Delphi can be pictured as a systematic, interactive approach for looking into the future. It is based on a structured group discussion by a panel of experts. It makes use of the implicit and explicit knowledge of the experts as well as their experience. In FAMILYPLATFORM the expert panel was comprised of scientists from different disciplines, and stakeholders and policy makers from across Europe. The experts have to answer questions in different rounds. These questions have to be discussed, and a common result has to be found in the group. After the first round, feedback is given to the panel and new groups are formed to further discuss the subject and to comment on the results of the first groups. There can be several rounds of discussion, feedback being given to the panel after each round (see Turoff, 2002).

3.2.2 Methodological to constructing Future Scenarios

Delphi was put into practice at the very first meeting of FAMILYPLATFORM, when Consortium and Advisory Board members started discussing Future Scenarios for families and living arrangements in the year 2035. Work on constructing the Future Scenarios was shaped by three key questions:

- What challenges for the wellbeing of families and living arrangements might arise in the future?
- What are the key drivers[2] in these changes?
- What scenarios will be defined and how will the drivers work together – how to construct the scenarios?

---

[2] "Driver" is defined as something that will impact in a certain way (positively or negatively) on an aspect of the future wellbeing of the family. It may already be present today.

Figure 1. Cornerstones of the Future Scenarios

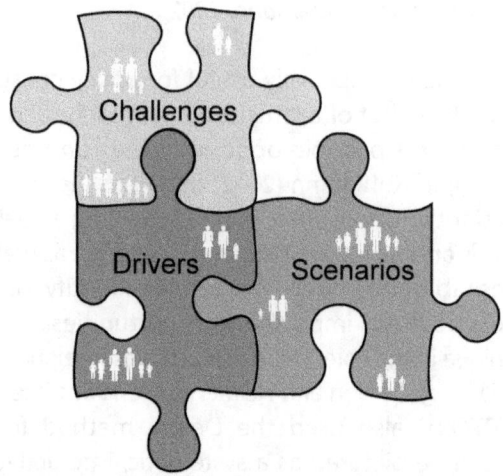

The future scenario narratives were based on the results of discussion of these three questions. As required in the Delphi method, work on the questions was structured and interactive, and involved several 'feedback loops'. Discussions between FAMILYPLATFORM experts took place at several meetings:

- The 'kick-off' meeting in Brussels, 26-28 October 2009
- Major trend meeting in Jyväskylä, Finland, 23-27 February 2010
- 1st Future Scenario meeting in Witten, Germany, 28-30 April 2010
- Critical Review meeting in Lisbon, Portugal, 25-27 May 2010
- 2nd Future Scenario meeting in Witten, Germany, 9-11 June 2010

The discussion continued between meetings through feedback loops via e-mail (participants were asked several times to give constant feedback on the results). In addition to discussion between the FAMILYPLATFORM Consortium and Advisory Board, all members of FAMILYPLATFORM were invited to participate and give their feedback on the Future Scenarios at the Lisbon conference and on the FAMILYPLATFORM website.

In the first step, the panel of experts discussed the question of the key challenges facing the family in the future. Small groups had brainstorming sessions to come to a consensus on identifying challenges and prioritising four of them. After the first round, the groups were mixed up, and new groups with one new member were formed. Each new expert presented the results from their first group which were further discussed in the new group. Then the new group agreed on and prioritised four key challenges.

Several of these feedback rounds were conducted, leading to prioritisation of the following challenges:

- Work-life balance and time management
- Changes in behaviour (family break-up, changing sexual morality, reduction in parenting skills, balancing individual fulfilment within families, individualism/selfishness)
- Ageing/demographic change
- Uncertainty
- Gender roles of father/mother, cultural representations of gender roles, gender responsibilities, denial of gender identity
- Diversity (of family models, gender, father/mother, cultural, etc.)
- Lack of mainstreaming, families not valued by society, public respect for parents and family values
- Economic crisis
- Immigration

On the basis of these challenges, the second step for constructing the Future Scenarios was to define the "key drivers". A similarly structured and interactive brainstorming process was employed. The plenum was again divided into small groups and was asked to collect the trends (drivers) from all Existential Fields that will have an impact on family in the future. The drivers had to be agreed upon and prioritised by the group. In further rounds, the individual members of the plenum were again involved in several new groups, which discussed the drivers, generating alternatives which were further discussed. After several feedback loops, each group identified and ranked the following drivers (1 being the most important):

Table 2. Drivers ranked by importance by working groups

	Group 1	Group 2	Group 3	Group 4
1	Demographic changes	Care Services	Migration	Demographic Change
2	Education/Values	Demographic Change	Care Systems	Migration
3	Gender	New Technologies	Education/Values	Inequalities
4	Inequalities	Gender Roles	Inequalities	Education/Values

Out of this collection of different drivers across all the Existential Fields, the plenum agreed on four drivers that will have a major impact of family life in the future:

- Inequalities (social, cultural, economic, gender, ethical, etc.)
- Migration
- Education and Values in Society
- Care Systems

Once the key drivers had been defined, it was possible to move on to the final step in the creative process: constructing the scenarios. It was agreed that four different possible Future Scenarios for the family and living arrangements would be described in a narrative fashion. This entailed, on the one hand, describing the outline of society on the basis of the drivers identified and, on the other, giving examples of different family forms and how they would live in the future society described in each scenario. As a starting point for the core of the Future Scenarios - the narrative descriptions - the following outlines of each scenario were agreed upon:

- **Scenario 1**: equal opportunities – open migration – diverse education and values – mix of private and public care systems
- **Scenario 2**: increasing inequalities – no migration (very select) – private education and extreme positions in values – privatisation of care systems
- **Scenario 3**: increasing inequalities – open limited migration – private education – accepted diverse values – privatisation of care systems
- **Scenario 4**: equal opportunities at a low level – restricted migration – rigid public education with very specific curricula – accepted diverse values – public care systems

3.3 Possible family and living forms in 2035

This chapter describes in detail the societal context for the four different scenarios chosen. For each of the scenarios, the group created four narratives of one or more pages, thus a total of 16 narratives (which are detailed in the Annex to this chapter).

3.3.1 Scenario 1: equal opportunities, open migration, diverse education and values, mix of private and public care systems

Basic societal context

A positive cultural attitude prevails in society. Integration and multiculturalism seem to be working. This is a confident society with few social and economic fears, a strong welfare system and high levels of social solidarity. There is full employment, and this scenario's vitality depends on continued economic growth.

Equal opportunities

Governments are responsible for the careful management of gender and social equalities. The state seeks to reduce gender inequality, which is an explicit common goal in Europe, not only by means of family policy, but through policies in other sectors as well. Choices are not linked to the gender or social status of an individual. There is a more equal sharing of child rearing and unpaid work. The parenting role is valued, both for fatherhood as well as for motherhood. Equality in gender roles ensures that men have the right to choose to stay at home, while women have the right to self-fulfilment in their professional careers. More flexible time schedules are possible, and families can choose the family model that suits them best.

There is a strong redistribution of resources. Equal opportunities exist in terms of education and employment. All children have equal life chances, and there is a high degree of social cohesion. Equality exists across the life span, from the pensioned and the elderly through to parents and young children. Lifelong learning ensures adult access to education throughout the course of the adult life.

In terms of culture, universal access to information and communication technologies ensures that there is no digital exclusion. Digital inclusion to help fight social exclusion is a policy priority. There is linguistic and religious diversity, with shared commonalities. There is a gradual assimilation of migrants into society without forced regulations on learning specific languages. The curriculum handles cultural diversity by incorporating these into the teaching itself.

Open migration

Europe is one big nation. People want to migrate within, to and from Europe but also to other parts of the world. There is no discrimination against migrants. They work on a legal basis and are able to obtain social security benefits. There is virtually no black market for labour and unauthorised work. Diversity in migrant family lives is seen to enrich the diversity of European societies.

Even in a society with open migration, migration rates may still be low and accompanied by some degree of regulation of open migration. An important aspect of migration is linked to a person's current position in the life-course. For instance, for educational reasons young people are a dynamic and mobile group. There are different forms of migration: for reasons of employment (more relevant for those coming into Europe from outside) or for educational reasons (more relevant for young people).

Another interesting point is that there is a cultural willingness to migrate amongst certain social groups. Well educated people with good jobs might wish to migrate; however, one notices less of this type of migration due to new technologies connecting employment opportunities and organisations across borders. New technologies allow people to stay in their own countries. There is also a rise in female migrants entering the continent as a result of job opportunities arising in the care and social services sector. Consequently, there is a feminisation of migration, with young women coming alone, often initially as lone mothers.

We observe an increasing diversity in Europe, creating specific challenges (potential reinforcement of traditional gender roles, for instance) as well as opportunities (changing attitudes to older people). New family forms emerge, for example cross-national families. If migrants leave their families behind, negative effects may arise because of the separation, but on the other hand, they are able to support their families financially in their home country.

Diverse education

Educational systems are grounded on a general scheme of public education supplemented by private and non-profit education. Basic standards of high-quality education are assured by common certification of the quality of institutions. The system affords equal opportunities and access to education for all. There is some freedom of choice, but a greater amount of public regulation. Education systems are well resourced and reflect individual abilities.

There is a rise in e-learning, and lifelong learning is accessible to all. It was a European policy decision to invest in education and make it a priority. The universal availability of e-learning is a move towards this goal. There is investment in digitally enhanced classrooms, and traditional learning skills (e.g. going to a library) are being replaced by new digital skills and literacy (greater peer collaboration, just-in-time learning, new media literacy). Lifelong learning is available for adults throughout their lives and helps combat social exclusion.

There is a move towards greater diversity in education, and every child has the opportunity for "open education" regardless of parents' decisions, because children's rights to make educational decisions or to spend time at home are respected. The high costs of education are no longer relevant, since public funding schemes for public, private and non-profit education ensure that costs are covered. The education system is flexible and can accommodate and support every individual child's abilities. There is a high degree of parental involvement in educational decisions and a linkage between educational institutions and communities. Schools are inclusive and provide for more than just learning, leading to the sharing of common experiences and the creation of shared reference points for culture, literature, art, sports, etc. Linguistic diversity is present in many schools, an important factor in the integration of children from other cultures. This scenario also sees a greater investment in more resources for education, smaller classes, the promotion of linguistic skills, more teacher training and higher degrees of parental involvement in school work.

Diverse values

This is a tolerant society founded on human rights that respects diversity and encompasses shared values. A variety of different life styles, gender roles and family models are widely accepted. There is an emphasis on equality and wellbeing, and education is a priority in society.

Mix of private and public care systems

Care systems are primarily organised by local government, and a pluralistic welfare system has developed. Local authorities provide money directly to families, and families can choose how they wish to use it (either directly for care, or to be paid to others for care services). Care is more sensitive to demands and is de-commodified.

In this mix of private and public systems, there is freedom to choose between familial care and outsourced care. In this context, it is necessary to differentiate between care for children and for elderly people. For children, the basic emphasis is on early childhood education, while for the elderly, public regulation and state initiatives promote care solutions. Co-operatives build up care services, and there is a diversity and flexibility of care systems funded by public money, with common quality standards. New communication technologies help support ageing and sheltered housing, which allows the elderly to enjoy the independence and comfort of their own homes. The priority is preventive care, which also recognises the rights of the elderly to stay at home.

How is life experienced at the individual level?

Families in this scenario have choices, but they need to be flexible and able to negotiate, which generates potential for stress. Individual choices are possible in this society, which leads to the question of what this means for children, and for whom and by whom decisions are made.

Individual orientations take the focus away from occasions for 'family togetherness'. This leads to uncertainties with regard to the creation, transmission and sustenance of common cultures, within and across families and generations.

The quality of life, measured by material standards, is fairly high, and there are hardly any pressing material needs. Hence other factors emerge as important – the quality of intra-generational and inter-generational relationships, which are increasingly complicated, and the question of continuous negotiation between individuals.

Individual choices imply additional responsibilities, for instance deciding to give more time to the family. Since material needs are taken care of with relative ease, the need to stick together (for example, to make sacrifices for mutual goals) is eliminated. These circumstances might do away with interdependence and might encourage a drifting apart of family members; there are emerging emotional needs for greater family solidarity and support as a result.

Family reunions and get-togethers may help in terms of conflict management and planning, and help the family to function as a unit. Housing arrangements will be an important topic – communal ways for generations to live together will be needed.

3.3.2 Scenario 2: increasing inequalities, no migration (or very select), private education and extreme values, privatisation of care systems

Basic societal context

Due to the vast debts built up in almost all European Member States during the 2010s, public provision of welfare (including care, health and education) has been sold off and privatised. In effect, the government has withdrawn from the provision of welfare, with the result that the state's power to shape society has been greatly reduced; partly because of this, the nation-state's legitimacy is increasingly questioned. To fill the gap in public support, two competing models of welfare provision have developed, each centred on a different ideal of who provides the necessary services:

1. Community-based support
2. Market-based support

Each family positions itself in relating to these sectors, although it may draw on one or the other for certain services. This positioning is a source of tension in some families.

In this scenario, society is deregulated, with practically no government intervention. Deregulation means that everybody is ultimately responsible for choosing how to maintain his or her own wellbeing. Inequalities between different social groups are high and have increasingly polarised society. Furthermore, rapid technological developments have taken place, which also have an effect on daily family life.

All social services (education, care, health, etc.,) are available either through the market (for paying customers) or within a community (for recognised members). As national governments were rolled back, they only retained responsibility for a limited range of policy areas (e.g. criminal law and enforcement, trade agreements, providing physical infrastructure). Geographically, boundaries between societies may be based more on self-defined, semi-autonomous regions than on currently existing national borders. These developments are supported by transnational corporations relocating their staff, but also cut across ethnic, religious and life-style lines. The EU's right to exist is questioned and continuously scrutinised. But there are still some regulations handled at the EU or national level, as well as international regulations by international institutions such as the UN and the WHO.

In this society there is hardly any mobility between social groups, and only limited freedom of choice.

Increasing Inequalities

Inequalities between social groups are continually increasing – and these inequalities are increasingly apparent not only in economic terms, but also in terms of values. The retreat to more extreme positions in values and the high degree of social segregation has led to marked inequalities between different social, ethnic, cultural and religious groups. Depending on their social, cultural and financial background, families participate differently in the education and care systems.

Gender equality or inequality varies according to social position. Whereas the market and some communities do not differentiate on the basis of gender at all, other communities enforce a rigid gendered division of labour and attempt to reproduce these gender hierarchies. Many communities are positioned between these two extremes.

Migration is severely restricted

Borders within Europe have closed. Mobility between countries is restricted to tourism (and therefore dependent on individuals' financial position), and there are exceptions only for wealthy people and those with higher levels of education.

Low-paid work is carried out increasingly by people born in Europe, rather than those who migrated from abroad when borders were more open, leading to shifts in the gaps between different social classes. A number of ex-migrants still live in Europe, and they still constitute a majority of the caring professions (e.g. nurses, care workers, social workers), but this number has decreased since Europe's borders began to close.

In spite of the closed borders, illegal migration from inside and outside of Europe still exists. Illegal immigrants do not receive any social support and have no officially recognised citizenship. They are organised into informal ethnic networks (often community-based), where support is provided, and this represents a third dimension of inequality. An increasing number of people are choosing to migrate from Europe to areas of the world which are more favourable to live in, though this trend is still at an early stage.

There is great potential for conflict between social and ethnic groups. A hierarchy of communities is developing, depending either on their proximity to the market-orientated section of society or along ethnic, religious or cultural lines. Different migrant and cultural groups try to maintain and defend their own traditions and values, which they seek to transfer to the next generation.

Due to the lack of social contact between groups, cultural segregation and inequalities are increasing. Further inequalities arise between those who can move within Europe and those who cannot, and those who are forced to move within Europe (e.g. illegal migrants). Social mobility within the community or the market-oriented sector of society is still possible, but due to social pressures to belong and identify with a certain set of values of a given community, this occurs infrequently.

National and regional states are also trying to cultivate a sense of 'belonging', with a number of competing nationalisms and regionalisms found within Europe, some harking back to various shared pasts (e.g. the nineteenth or twentieth centuries) and others attempting to create a narrative encompassing all the different communities as they exist in 2035. These attempts have met with some success, but overall the power of the state to shape identity has been reduced, and competing sets of identities are becoming increasingly entrenched – some of which are deeply antagonistic to the state itself, for ideological, practical, or ethnic or regionalist reasons.

Education is completely privatised

The state no longer has any responsibility for the education system, and parents decide where their children are educated. Due to technological advances and the relatively low number of children, virtual schooling has become standard. Children in different towns, regions and countries can all be present in the same virtual classroom. For this reason, every social or cultural group can offer schooling in accordance with their beliefs and values. Access is either for community members only or, for the commercially operated schools, regulated by fees. Very expensive face-to-face private schooling (e.g. Eton, Harrow, Salem, etc.) is still available and serves the purpose of creating international elite networks. Consequently, social, ethnic or cultural groups ensure that their children are educated and raised the way they choose, and this has led to high diversity and de-standardisation of educational systems and increasing social segregation. The social and financial background of families therefore affects the educational development of their children, and educational inequality between social groups is increasing.

Rapid advances in new technologies and media have had far-reaching effects on the way families organise their everyday life: the self-cleaning home is a reality, and a revolution in information technology has made it possible for much more paid work to be done at home. The result in some families is a spatial concentration in the private home. The way children use the internet and educational computer programs depends on their family's social, educational and financial status. Additionally, the way that children individually learn depends on their environment (noise, space, time, guided internet use, etc.). Alternative media and different types of internets (some free, others commercial) cater to some of the social needs of families with different abilities to pay. Families with greater financial resources can chose whether they want to use the newest technologies. For sections of the less well-off, virtual communities have become more important, and the relevance of direct face-to-face communication and personal relationships has decreased, leading to increased social isolation.

Values vary and extreme positions prevail

Values vary between different social groups, depending on their economic and cultural backgrounds. On the one hand, radical groups with very rigid values have arisen (e.g. various perceived forms of traditionalism vs. different kinds of anarchical or libertarian values). Less interaction between social groups and families means less exposure to alternative belief structures and different everyday life practices, resulting in less reflection and consideration of values. The segregated education system supports this trend and encourages the emergence of extremely varied value systems. On the other hand, in different sections of society, a new process of re-familialisation is taking place, which sees new forms of living arrangements, such as common childcare within the community, support by the community, or neighbourhoods providing care facilities, becoming more widely accepted. In these communities, intergenerational help is widely available and valued.

Care systems have been privatised

The state no longer provides any care whatsoever, and private insurance and market solutions for care have increased rapidly as a result. In addition, communities began early on to step in and provide care services themselves. In both cases, responsibility within families and intergenerational help are increasing.

The privatisation of childcare has resulted in the emergence of different care markets catering to different financial and cultural needs. Quality and price are directly correlated, resulting in a polarisation between highly-qualified and well-paid carers on the one hand, and poorly educated and badly paid carers on the other. Poor families and families with more than one child do not have access to high quality childcare on the market. Many families in this situation, who cannot draw on the support of a community, deliberately have only one child.

Each community organises childcare in its own way. In more traditionally focused communities, care is delegated to older women; in more egalitarian communities, everyone takes turns. Those living in communities with non-market values experience greater numbers of people being active and volunteering within the community, until they are forced to stop due to ill health. In many communities, volunteers are well-respected and play an important part in society. Care for the elderly is often provided by the "young-old" (mid-50s to mid-70s), supported by technical innovations. Robots, for example, play an important role in care for the elderly as do alarm and monitoring systems, which mean that older people can live longer on their own in their own homes.

With respect to pension systems, the market-oriented section of society invests more in private and company pensions. Many communities have developed their own community pension schemes that people can pay into, and pensioners are generally either looked after in their families or communities, or through a market-based solution. Many people work until they die, and early retirement is increasingly rare and generally confined to market-oriented families and communities. Some community insurance schemes are facing the problem of how to ration their funds, that is, deciding who receives which form of medical care and who does not.

How is life at the individual level?

There are pressures on families and their individual members to conform to the norms and values of their communities because being ostracised means losing access to care and other welfare services. This lack of freedom of choice is a source of tension between generations and within couples, and prompts discussions about the family's present and future location within the community or the market.

Intergenerational tension also exists because more generations have to live together. This is quite a new development, to which people have not yet become adjusted. Furthermore, tensions between neighbouring communities also exist, if one feels that the other is trying to attract members or create divisions: rumours and hostilities frequently arise. Many communities are responding to this by creating structures and projects which work to promote inter-community understanding and co-operation.

There is an emergence of "pragmatic parents" who juggle work and family obligations and duties and have hardly any leisure time or just plain fun. Their relationship with each other is often based on practical consid-erations rather than on love and affection, which affects their wellbeing. Tensions within the family are further increased by having to take in and care for older family members, especially when this situation is unexpected. In this scenario, employers' policies for supporting families are a decisive factor in reconciling work and family life.

Community leaders, or indeed the varied forms of community decision-making processes (authoritarian, inclusive, nepotistic, ecological, demo-cratic, authoritarian, etc.,) play an increased role in ensuring individual personal development and personal freedom in order to allow communi-ties to grow and change. The outcomes of these different styles of commu-nity governance are varied.

The complexity of this society and the different family forms and back-grounds may, on the one hand, be a challenge for children and, on the other hand, be enriching. The high degree of responsibility for care in the communities and the demographic structure (including more elderly people) mean that care is delivered by dedicated community members and that intergenerational living offers support to many families at many different levels.

3.3.3 Scenario 3: increasing inequality, open limited migration, private education, accepted diverse values, privatisation of care systems

Basic societal context

This society is a "contract society", in which individuals cannot rely on a universal welfare state and so enter into contracts for the provision and remuneration of these services. Civil society has a very important role – everybody is an active member of society, and associations and religious groups are very active. A "tribal society" has arisen, based strongly on affiliation to one or another group. In some areas, gated communities have emerged. Integration of individuals is dependent on being a member of a specific association or group (e.g. migrant, job-seekers, etc.). On the one hand, this has led to the segregation of young people into two groups, those who are integrated into communities and extended families, and those who are excluded, and this produces conflicts. On the other hand, it can help young people to be integrated into a community, and enables local authorities to help manage conflict.

Increasing inequalities

Inequalities exist in all areas (gender, economic, social, and ethnic) due to social segregation, which also produces social unrest. Knowledge gaps due to differential access to communication technologies also produce conflict.

Gender roles are not generally defined by society, they are negotiated. Negotiation plays an important role in society. Different gender roles are present in different social groups.

Open limited migration

There is limited open migration. People have different legal statuses and different citizenship rights, and migration has been privatised. It is possible to buy national citizenship by paying taxes and there is also European citizenship. For EU nationals, it is very easy to change citizenship to that of another Member State. The public services offered by Member States to their citizens differ significantly, as does taxation. As a result, people are free to choose a bundle of services (education, health-care, public pension, etc.,) based on the income taxes they are prepared to pay. Market mechanisms provide sufficient employment opportunities, and Europe is one big marketplace – something that the European Union is mostly concerned with regulating. Black markets for labour and segmentation of the labour market have developed. Migration is influenced much more by family ties than by economic factors.

Private education

There is only limited public education and no public kindergartens. Private schools are much better than public schools. The European Union guarantees a minimum educational standard, and there is basic schooling only up to the age of 14. All universities are private and frequently supported by foundations, such as the Nokia University or the Mercedes-Benz University. Communities have their own schools. E-learning and digital learning have been privatised, and knowledge gaps in society are growing. Transnational television (diasporic media) is widespread and brings migrant communities in different locations together in front of the TV.

Accepted diverse values

Values are very diverse. Social equality is not a priority and the most important value in society is freedom. Associations and firms (the market) play an important role in defining values. Social control and cohesion is provided by associations and extended families. Religious organisations also play a leading role. Values are shaped by the groups to which individuals belong, or else they choose to be members of a group that shares their set of values. Consequently, ways of life, attitudes and values vary between groups, but not greatly within them.

Privatisation of care systems

Care systems are private and provided by charities or financed by private insurance. There are minimum standards, but the private sector does achieve higher quality. There is a minimal level of health-care provided by the public sector, but there is no public childcare at all. The lack of public care is to some degree compensated for by strong and active neighbourhood ties and the growing voluntary sector. Different kinds of associations, religious movements and volunteer organisations step in to redress the lack of state welfare, sometimes with funding from private companies. There are strong obligations on families to provide for themselves. Childcare provision, for example, is a mix of company and association-funded (private interests) and informal care systems.

How will life be at the individual level?

In this society, people are increasingly insecure in their living conditions and relationships. The future is unclear. People demand more public responsibility and clamour for a more secure future for families.

There is a high degree of freedom of choice in this scenario, and this has its own consequences. There is greater ambivalence about freedom of choice and choices in general. There is more social inequality and consequent social unrest.

Social organisations and voluntary groups have increased in importance, and there is a growing dependence on these organisations. Consequently, there is potential for more neighbourhood support networks. Parental choices in connection with maintaining a work-life balance are crucial, as the state does not play a big role. This is especially important for children, their rights and their best interests.

3.3.4 Scenario 4: equal opportunities at a low level, restricted migration, rigid public education with very specific curricula, accepted diverse values, public care systems

Basic societal context

In this scenario Europe is very strong, yet isolated from the rest of the world. The climate has changed considerably over the last twenty years, leaving Europe more hospitable than many other parts of the world, though in many parts of Europe it is much less habitable than it was in the past. Because of resulting migratory movements, Europe has closed its borders, and now aims to radically control them.

The European Union has attained the status of an autonomous "megastate". National borders within the EU no longer exist, even though there are still elections for a European Government and elections at national levels. Identification with "Europe" is very widespread among the population. Every European still speaks their native language, but there is one central language in Europe which is spoken by every European. Due to these low barriers, mobility across Europe is high and travel is frequent.

Natural resources within Europe have become scarce. Minimal-impact and resource-light technologies have therefore become more important. Some of these are labour-intensive, which has helped alleviate the unemployment problems which plagued Europe in the first 25 years of the century.

European policies focus on the individual, and everybody is responsible for him or herself, regardless of the family constellation they live in. The state provides only basic support and services. There is no family-oriented policy, but the state supports special groups, for example children or older people, as individuals.

Major changes have occurred in reproductive technology. In addition to natural conception and childbirth, declining male fertility rates encouraged the development of alternative methods. These include:

- Surrogate motherhood
- In vitro fertilisation
- "Growing" babies outside the womb

One consequence is an increase in women/couples who produce large numbers of children, either to have a large family themselves or on behalf of other couples. A second consequence is the uncoupling of women's employment performance and achievements from their reproductive potential. Employers can no longer (implicitly) discriminate against women on the basis of their childbearing potential. Therefore, women are able to fully capitalise on their better performance in the education system (which began towards the end of the twentieth century) and they have taken over far more positions of power in all sectors of society and the economy.

The value of children is very high, because they are considered the foundation of the European population and economy. The state is pro-natalist and encourages the "natural" production of children by targeting benefits at children, supplemented by its own measures to "grow" children outside of the womb. The pro-natalist policy is a reaction to demographic change.

As a consequence of pro-natalist policies and advances in reproductive technologies there is currently a fierce debate on genetic engineering. Issues include:

- *"Should the state control which genes are more frequently reproduced (e.g. best suited to the changed climate, appearance, intelligence, etc.)?"*
- *"Should parents be able to choose the phenotypes (observable characteristics) of their in vitro baby?"*

Wild Card Scenario[3]: the state can "order" the production of more citizens, according to specific needs (such as military needs or a need for highly qualified people; genetic pooling facilitates the selection of specific characteristics).

---

[3]   Wild Card means that an unexpected and not very likely event occurs.

Energy costs are very high, so public transport is supported by the state, but private transport (cars) is very expensive and barely exists any longer. The world is divided into different blocks (e.g. South America, China, ASEAN, Indian subcontinent) that do still trade with each other (by sea), but only in high quality, high priced goods. Transport costs are very high (no more oil) and there is more re-use and recycling than in the past.

Due to the policy of supporting children directly and targeting benefits at them, the state empowers children from an early age. Vouchers are given for many of the things that children need (e.g. nappies, school books, etc.) and children learn very early to spend their vouchers themselves. Child benefits are not dependent on the family situation.

Equal opportunities at a low level

Since the European state provides basic public care services and education with equal access for all citizens from birth onwards, social mobility is in principle possible for all social or cultural groups. In fact, however, it is still dependent on the family of origin's social and financial resources. Due to the state's orientation towards the individual, it is up to each individual to take advantage of the facilities offered. Since these public services satisfy minimal needs, there is a baseline to inequality that no-one drops below. Inequalities nevertheless arise because some individuals start off at a higher level and strive to maintain a higher standard of living.

Due to climate change, the different regions of Europe are more or less attractive to live in. The state provides basic quality public housing, but often in unattractive regions. Nice homes in attractive regions are expensive. Young people may stay at home longer, until they can afford to move to their own place. On the other hand, the state may provide the individual with housing.

The state has little or no control over the labour market. There is a minimum wage, but no guaranteed job. The public sector is very large (childcare, education, health-care, etc.) and is regulated by the state. In the private sector, there is significant competition for jobs, and good education gives potential employees a competitive advantage.

Support during times of unemployment is no longer linked to the bene-ficiary's family constellation. Factors which today are taken into account for calculating welfare benefits, such as number of children or partner's income, no longer play a role. However, the state provides basic support in case of unemployment ("Nobody has to live on the street") and children in the households of the unemployed receive direct support.

In terms of gender inequalities, women's higher educational achieve-ments and female networking have brought women into leading and powerful positions in society. Gender inequalities have changed, and there is now a kind of female dominance of society, which makes reconciliation of work and family life possible for women. In this female-dominated society, motherhood is not a handicap. Since active fatherhood has been encouraged over recent decades, men are much more active in the household and in childcare. In this respect at least, there is gender equality.

Regarding social inequalities, the state's involvement guarantees low-level equality by meeting the needs of the individual. In the wild card scenario, sex and reproduction are no longer linked.

At the level of beliefs and culture, there are hardly any inequalities any longer. Different ethic and religious backgrounds are accepted equally throughout Europe. However, this does not include tolerance of foreigners from outside Europe. Therefore, illegal immigrants suffer from social and cultural inequalities, and racism towards non-Europeans has increased.

Migration is almost impossible

Europe's external borders are now closed for most purposes, and international trade has decreased. There are fewer opportunities to work outside Europe, though tourism is still possible. Due to the high level of border protection, militarism is on the rise in the EU, and Europe spends large sums on defence.

Within Europe, people can move and work freely. Inter-European trade is free, since there are no frontiers left. Because of the high level of identification with the 'mega-state EU' and separation from the outside world, people have developed a strong common identity, a kind of nationalism which makes them suspicious of everything outside their EU.

Public education is provided at a basic level

A rigid educational system is provided universally up to the age of 16. Schools teach a common basic curriculum, which is identical across Europe, promoting mobility (e.g. same language education, etc.). In each country, children learn components of national history and culture, including national anthems and national languages. Being tri-lingual is increasingly becoming the norm. This educational framework facilitates social mobility because everybody has the same opportunities right from the start. On the other hand, there are additional school programmes available on the market without public support, which only rich parents are able to afford for their children. These high-cost, supplementary, premium educational programmes reinforce social inequalities.

After the age of 16, young people either start an apprenticeship or continue their education at a privately managed school or college. In addition to the professional armed forces, all 18 year-olds (male and female) can serve for a year as a European Border Guard on a voluntary basis. As a "reward", participants can study at university for two years for free.

Diverse values are accepted

The "mega-state" Europe encompasses a high diversity of former country-specific values, identities and backgrounds. These differences within the European population are accepted, true to the motto "We are different and we love it", but only if you are European. Europeans are very suspicious of foreigners or migrants who moved to Europe before the frontiers were closed (because they could be here illegally or, in the worst case, terrorists or secret agents from outside Europe). Such fears are frequently whipped up in the popular press.

Because of general acceptance of different European values, the acceptance of different family forms is also very widespread. This acceptance is also favoured by the orientation of the state philosophy towards the individual: the family form is not relevant for public care or health and social services. The individual is responsible for his or her own life.

Care systems are public and cover basic needs

The entire care and social support system is targeted at the individual. The state offers basic facilities and people can choose to what degree they utilise this available public support. They have to decide individually how and if they manage to supplement basic quality care for children and/or older family members. As far as these services are concerned, the state guarantees basic care for everyone. These policies are not directed at families or designed to support families as an institution, but to support special groups (e.g. children, people with disabilities, older people, etc.). Since there is comprehensive care service coverage for very young children, parental leave is very short.

Alongside the basic public social support system a relatively good and functional market has developed, which provides additional services at higher cost.

How will life be at the individual level?

Relationships, kinship and networks are an important part of society and are highly valued. The energy crisis and closing of the borders have brought people together. Long-term and committed relationships are valued and easier to achieve with greater gender equality and greater personal freedom. But since men tend to be less well educated than women, it can be difficult for well educated women to find a partner. The past has shown clearly that in the selection of a partner, the choice is driven by a homogametic focus, which means that most people form relationships with someone of the same educational level as themselves. Greater acceptance of different family forms and living arrangements can lead to more same sex relationships (not necessarily sexual relations) and a greater variety of care and living relationships. However, high mobility may negatively impact wellbeing and the work-life balance as well as relationships.

There is more freedom and less pressure in personal relationships, since the state takes care of individuals' basic needs. The pressures and stresses of insecure jobs can, to a certain extent, be cushioned by the state, which provides childcare, schooling, housing, etc. This reduces pressure on mothers and fathers, as long as they are not employed in a high-pressure private sector job. With the minimum wage or welfare benefits in case of unemployment, a minimum standard of living is possible, so there is more time for family and personal interests.

The individual is empowered in this society. But masculinity has to be redefined, and a new lobby for men's advancement has arisen to work towards gender equality in pay and to campaign against the causes of the inequalities faced by boys in education.

Due the pressures on children (being responsible from an early age, constant observation by CCTV, limited leisure time activities because of the high cost of visiting friends, and a climate of fear), depression among children is more common. Within the family, relationships tend to be happy: it is the external structures of society and the growing fear in society that induce the stress.

3.4 Key policy issues and research questions

One of the main objectives of these "Future Scenarios" has been to outline key policy questions and research issues relating to the wellbeing of the family – as derived directly from the narratives.

The four scenarios and the 16 family narratives constitute the "possible futures" that were the basis of this discussion. We used them to identify the following:

- *"What are the uncertainties which could affect the wellbeing of the family?"*

- *"What research questions directly relate to the future wellbeing of the family? Therefore, what are the relevant research issues?"*
- *"What factors or policies improve the wellbeing of the family?"*
- *"How can they be implemented or encouraged?"*
- *"What are the 'dead-ends' or 'bottle-necks' that could lead to a breakdown in families' wellbeing, and what are the instruments to prevent it?" Therefore, identification of key policy issues and social innovations.*

Discussion of the narratives and scenarios led to the following specific issues being identified: the importance of intergenerational solidarity and communities, the importance of sufficient time for families, the issues of unpaid work and care arrangements, children's perspectives (rights, best interests, and impact on wellbeing), periods of family transition, family mainstreaming and individualisation, and the impact of technological advances on families. These are described in more detail below.

3.4.1 Importance of intergenerational solidarity and communities

A striking element emerging from the scenarios was the importance of intergenerational and community solidarity. Family and community solidarity remain important for families' wellbeing in all scenarios, and in some scenarios increase in importance. Indeed, in scenarios with a weak welfare system, they become crucial.

The scenarios showed that families as well as community networks provided care services for their members, thus freeing the state and the taxpayer from the costs. In a situation of financial crisis or depression, family and community resources proved vital.

In the vast majority of the narratives, families relied on support and help from grandparents, siblings, cousins, friends, and neighbours, as well as from local communities (based on ethnic, religious, social, and cultural cleavages). These offer alternative and reliable solutions for the provision of care.

In certain scenarios (especially 2 and 3), governments have almost completely withdrawn from providing welfare in the widest sense (care, health, education). Societies in these narratives have become deregulated, with practically no government intervention. To fill the gap in public support, strong community-based support has emerged. Societies without public welfare are greatly affected by social inequalities and tend to consist of segregated local communities focused around common ethnic, religious, class and income-related socio-cultural groups. As integration in these so-called "tribal societies" (Scenario 2) is only possible for members of a specific association or group, there is strong segregation in society between those people integrated into communities and extended families, and those who are excluded, a situation which is likely to produce a great deal of (possibly violent) conflict.

Quoting scenarios:

"Each community organises childcare in its own way. In more traditionally focused communities, care is delegated to older women; in more egalitarian communities, everyone takes turns. Those living in communities with non-market values experience greater numbers of people being active and volunteering within the community, until they are forced to stop due to ill health. In many communities, volunteers are well-respected and play an important part in society. Care for the elderly is often provided by the "young-old" (mid-50s to mid-70s), supported by technical innovations. Robots, for example, play an important role in care for the elderly as do alarm and monitoring systems, which mean that older people can live longer on their own in their own homes" (S2)[4].

"Childcare is largely organised in the family (siblings and cousins rather than grandparents) and the community. A rotation system was set up, in which a group of parents take turns to look after each other's children, one day a week for each family" (S2.1). It is there not only to provide care support, but also for the purposes of education, health, social control and cohesion: "community school[s]… [where] well educated community members teach for some hours a week as part of their contribution to the community" (S2.1). "Every social or cultural group can offer schooling in accordance with their beliefs and values. Access is either for community members only or, for the commercially-operated schools, regulated by fees" (S2).

Some of the results found in the scenarios and narratives include "*There is a great deal of pressure within the community… to conform*" (S2.1). Or, "*Because of the lack of public care systems, having children is very important for parents to ensure that they are cared for in old age*" (S2.2). Or, "*A 'tribal society' has arisen,*

---

[4] Please note the meaning of the following abbreviations: S2 means Scenario 2, S2.1 means Narrative 1 of Scenario 2, and so on.

based strongly on affiliation to one or another group. In some areas, gated communities have emerged. Integration of individuals is dependent on being a member of a specific association or group (e.g. migrant, job-seekers, etc.). On the one hand, this has led to the segregation of young people into two groups, those who are integrated into communities and extended families, and those who are excluded, and this produces conflicts" (S3). Or, "[Anna] *lives in a co-op-erative community, which provides some social cohesion. She is happy that her "work" not only brings in money, but also produces some kind of commonality and brings people together"* (S3.2). Or, "*Relationships, kinship and networks are an important part of society and are highly valued. The energy crisis and closing of the borders have brought people together. Long-term and committed relation-ships are valued and easier to achieve with greater gender equality and greater personal freedom"* (S4). Or, "*Many communities have developed their own community pension schemes that people can pay into, and pensioners are gener-ally either looked after in their families or communities, or through a market-based solution. Many people work until they die, and early retirement is increasingly rare and generally confined to market-oriented families and communities"* (S2).

It seems that intergenerational solidarity and community support have become the backbone of support for families in the narratives. But a society in which care and education are based on community support alone was also seen as being particularly prone to falling apart. Therefore one impor-tant challenge for future research and policy in Europe is to study the conse-quences of different welfare mixes and to balance community solidarity with social welfare services.

Empty neighbourhoods versus lively neighbourhoods during working hours were discussed, putting an emphasis on the difference it makes in terms of security, care, social cohesion and neighbourhood relationships.

This leads to:

- Intergenerational policies and community support imple-mented at an EU level: Family should be considered as an inter-generational unit. Networks of extended family solidarity should be encouraged.
- Housing, environment and community development: Given the increase of the costs of housing, there may be a need for thor-ough and comprehensive urban planning that includes:

  - Analysis of how close families live, work and go to school.

- Housing and neighbourhood planning.
- Public spaces (such as playgrounds).
- Public and private transportation.
- Neighbourhood networks.
- Proximity of care institutions, etc.

One could also look into new housing opportunities and spaces to accommodate several generations that address the needs of families in their daily lives (for example, houses for three generations of the same family). This could help augment social cohesion in neighbourhoods.

- Family associations: In the scenarios, we witnessed increased participation of local, public/private, paid/volunteer, organised/informal communities. Policy should further encourage family-related associations and organisations that can help families.
- Ageing and social cohesion: The idea of a "skills market" as a social innovation came out of the intergenerational and community discussion. The group imagined local offices that match job offers with job seekers and their qualifications. This skill market might be based on paid as well as unpaid work. Besides fulfilling the need for recognition of the elderly after retirement, this could help foster diversity in urban areas, help create a climate of trust, and help share community and family support in a variety of different ways:

  - Increased recognition of volunteering.
  - Care receivers become care-givers.
  - Be supplemental to a professional care job rather than replacing it.
  - Enable creation and re-establishment of community connections, thereby preventing isolation.

3.4.2 Importance of sufficient time for families

Another strong commonality across the 16 narratives of family life in 2035 is the aspect of time. The wellbeing of the families appeared to be related to how much time they spent together as a family. Lack of time often generated stress, tensions, more difficult family relationships, endless negotiations, health troubles, etc.

The narratives showed that in similar environments, the personal choices of parents to allow themselves more or less time for their family played an important role in the wellbeing of each family member, in providing stability for the family, and in the number of children that families had. It was observed that both parents working full-time often made it difficult to find enough time for family matters, regardless of the family form and the economic status of the family.

This 2035 society still shows that it is made up of different time "zones". One might describe these as "institutional" time (school education and work) versus "family time". There seemed to be great difficulty co-ordinating and synchronising these times. Families and individuals demonstrated that they wanted to be empowered to spend time together as a family and as a couple and to maintain family relations, and to have time to care and to be able to manage family and personal tasks.

Quoting the scenarios:

> "Klara works very successfully in a private company. Her job some-times requires her to work late and to take trips abroad. Both of them are socially and politically active outside their work and family life. In the afternoons, the grandparents usually take care of the children. Since the grandparents are often away on holiday, other forms of day-care for the children are necessary. The main challenge they have to face is prioritising between their career and parental responsibilities, as well as working on the quality of their relationship. This creates a fair amount of stress within the family and is starting to endanger its wellbeing" (S1.1).

> "[Although their housing and financial circumstances are not very good], they do not want stressful jobs and are quite satisfied with their lifestyle – they have time for themselves and their good friends who support them" (S3.1).

> "Emily and Phillip are confronted with the challenge of combining two careers with childrearing. They spend a lot of their free time with their children… but the children are still on their own a lot" (S2.3).

> "Lasse devotes a lot of time to his family and enjoys it. He had to give up many leisure activities, but he does not feel bad about it, because he is happy with his situation. His colleagues are in the same situation. They value their family time" (S2.4).

"There is more freedom and less pressure in personal relationships, since the state takes care of individuals' basic needs. The pressures and stresses of insecure jobs can, to a certain extent, be cushioned by the state, which provides childcare, schooling, housing, etc. This reduces pressure on mothers and fathers, as long as they are not employed in a high-pressure private sector job. With the minimum wage or welfare benefits in case of unemployment, a minimum standard of living is possible, so there is more time for family and personal interests" (S4).

"Kristel is also pleased with this arrangement, because somebody is now there for her at home, and she does not have to be alone so much" (S4.3).

The same immigrant couple was pictured in two different scenarios. When they can spend more time together they have another child and integrate better in Scenario 1 (S1.3), than in Scenario 3, where *"They need to work extra hours in order to save and be able to pay for medical services during pregnancy, including giving birth. They have very little time together. After the child is born Roza is allowed only one month off work, her employers being generous and paying her salary as usual. As they do not want to change the carer, Roza brings the baby with her to work… The Muslim community provides childcare at moderate cost, but in the new situation they cannot afford to move out from a shared apartment with two other couples, as they had hoped to. In these circumstances they decide not have a second baby"* (S3.3).

This leads to suggestions for policies to ease the "rush hours" in the life cycle of families:

- Policy makers should consider strategies to ease or slow down the "rush hours" over the course of life, and help synchronise institutional and family times. Based on the needs and objectives of the families, time management policies and choices are needed, as well as incentives for the employers to help employees better reconcile work and family life. Obviously, the main stakeholders - the employers - have a major part to play in this, and need to be involved in the drafting and decision process.

- One could question whether European policies encouraging the dual-earner full-time household model during the whole life-course are sustainable for families in the long term.
- Scenarios often highlight the importance of new technologies that can help ease these "rush hours" or "time bottlenecks".
- A possible social innovation coming out of the scenarios that might ease the "rush hours" in the life cycle of families is "time care insurance" or a "time credit" account of several years' duration, designed for individuals to take care of other people (young and old), to be invested over the course of family life.

3.4.3 Unpaid work and care arrangements

Care arrangements are another important issue in all of the scenarios and narratives. Unpaid work is closely linked to this issue ("time for work/family"), and was very often addressed during the Future Scenarios brainstorming sessions. There is a clear need for recognition of the unpaid work (largely care work) generated within families and communities.

In all scenarios, care work had to be done by the state with public money, by the market with private money, or by families and communities for "*no money*" (therefore "unpaid work"). In Scenario 2, when the state withdrew and did not provide any institutional care, "*the state no longer provides any care whatsoever, and private insurance and market solutions for care have increased rapidly as a result. In addition, communities began early on to step in and provide care services themselves. In both cases, responsibility within the families and intergenerational help is increasing*" (S2).

Whether or not the state is involved in promoting institutional care or parental responsibility, social and gender equality was shown to have had a certain impact on each of the different 2035 scenarios.

1. Where the state withdrew completely from financing institutional care, inequalities between different social groups were higher and society was more polarised, as in Scenario 2: "*The privatisation of childcare…has resulted in the emergence of different care markets catering to different financial and cultural needs. Quality and price are directly correlated, resulting in a polarisation between highly-qualified and well-paid carers on the one hand, and poorly educated and badly paid carers on the other. Poor families and families with more than one child do not have access to high quality childcare on the market. Many*

families in this situation, who cannot draw on the support of a community, deliberately have only one child" (S2).

2. Where the state was very strongly involved, there were lower levels of societal inequality, and society was less polarised. In these scenarios, some parental tasks could be taken over by the state-paid care institutions, which intruded to some extent on privacy and parents' rights: *"... state takes the next step in trying to bring the relationship to an end: they want to take Konstantin under Public State Care services, because administrators and social assistants are afraid of the negative effect the father could have on his son"* (S4.3).

3. In Scenario 1 *"Care systems are primarily organised by local government, and a pluralistic welfare system has developed. Local authorities provide money directly to families, and families can choose how they wish to use it (either directly for care, or to be paid to others for care services). Care is more sensitive to demands and is de-commodified. In this mix of private and public systems, there is freedom to choose between familial care and outsourced care"* (S1).

In several scenarios, parents struggled particularly with sick and handicapped family members (young and old).

The link between gender and unpaid work/domestic tasks is changing, and this is generating a need for negotiation: *"Family life and the relationship between Klara and Joseph will change as they and their children get older. Permanent negotiations take place to ensure an equal share between the two of them, since gender roles have changed into equality and their roles as mother and father have to be redefined"* (S1.1).

This leads to:

*   Recognition of unpaid care work, closely linked to the above-mentioned topic of sufficient family time.
*   A policy framework which enables families and communities to carry out care work in an environment of equality, ensuring the right balance between state involvement and parents' and communities' care responsibilities in public care policies.
*   Monitoring the impact of gender equality policies for effectiveness and unintended consequences.
*   Policy which considers alternative care arrangements, especially those linked to intergenerational (mainly grandparents)

and local communities: are they viable and possible? Are they desired by both the care-givers and care receivers?

- A societal challenge is how quality of public care can be assured, and how the optimal balance between public and private care can be provided by regional, national and European policy.

3.4.4 Children's perspectives: rights, best interests, and impact on wellbeing

The majority of the narratives address issues concerning those receiving care and how it is provided, but they do not necessarily take into account the point of view of the children involved. When given a voice in Scenario 1, we see the following dialogue: "*Mum, I want you at home tomorrow!... I want you mum, or daddy at least*" (S1.1). The wellbeing and mental health of some of the children in the scenarios is affected, as it is today, by the pressure they face from lack of time for themselves and with their families due to labour market expectations, a performance-driven society, family environments, and lack of affection.

We also see situations where parents' wellbeing does not match their children's wellbeing. The rights and interests of the parties involved are very different – they usually complement each other but can also conflict, for example in a divorce custody situation or in distributed family life situations. See Narrative 3.1, where the complicated family situation results in a financial strain for the parents and psychological problems for the children.

This leads to the following policy questions:

- Policymaking and research should look at what is important for the healthy development of the children and what is in their best interest, and not just children's rights or the parents' best interests.
    - More psychological research on children's wellbeing is necessary with regard to the variety of family forms that children live in and different care arrangements.
    - Policies need to take into account the balance between children's interests and those of their parents.
- Policies should encourage social services to empower and support families.

3.4.5 Family transitions

The families pictured in our scenarios are not static, just as today they are not static. They go through transitions and their needs and choices vary. Being able to adjust to the changes is an essential part of family wellbeing.

This leads to the need for:

- Policies that consider the life-course, including the many transitions and phases in the life of a family, taking into account the dynamic developmental processes of families (where families face both selected and unexpected events which have consequences). Families must not be considered static entities.
- Policies that favour building environments where parents are able to create and select conditions that sustain parental and child wellbeing over their life-course, and at different stages of transition.
- Policies that support actions helping couples prepare for the transition to parenthood. Policies can also support actions that acquaint potential parents with their parental responsibilities and raise awareness of the child's development and needs.

The scenarios reveal dynamic changes taking place within families as they cope with changing external conditions and the dynamic challenges faced by family members. Research focusing on families often concludes that specific family forms are disadvantaged, that different family configurations should cope with differing hardships. Many of these discussions do not assume explicitly that the family is a dynamic entity, but indirectly assume that family forms are static. Research should be focused more strongly on the causes and consequences of family dynamics.

Regarding the causes:

- Structural and ideational factors should be considered.
  - Among structural factors, institutional arrangement, family-related policies, labour market, housing, the unequal distribution of resources need to be considered.
  - Regarding the ideational factors, value orientations and attitudes, happiness could be considered.

- Considering consequences: both the material and immaterial aspects should be understood. These are: material situation, time pressures, satisfaction, mental health, stable family relationships, etc.
- When considering the dynamics of families, several kinds of dyadic relationships in the family reality should be stressed. Namely, research should focus not only on the dynamic character of partnerships (partnering, marriage, separation or divorce, re-partnering, the quality of partnerships) but also on the dynamics of childhood, parenthood, grandparenthood, network-dyads, and their changing meanings.
- The well-known life-course transitions (leaving home, leaving education, getting a first job, partnership formation, the birth of a child, divorce, unemployment and employment, retirement, becoming widowed, etc.) should be also integrated into family dynamics.
- Other possible social innovations are mediation and counselling centres, which support families and their needs during certain intended and unexpected family transitions. Policies could encourage development of such centres.
- Implement pilot programmes to evaluate the specific needs of families, employers and economic stakeholders. Family transitions call for adaptive rather than lifelong employment policies.
- Research should seek to understand the above-mentioned transitions, and the vulnerability of the individuals and families involved in these transitions. It is also important to understand if these transitions are intended or unintended, and in which cases communities or/and governments could and should help families.
- Research should also investigate how to reach families with special needs. What are new concepts to support families and prevent and resolve conflict?
- Comparative research, using quantitative and qualitative methods, may reveal the causes and consequences of family dynamics.

3.4.6 Family mainstreaming and individualisation

The central question in *Europe's family strategy* is the impact on families of all European policies. Moreover, whether a proposed policy is local, national or European, the effect of any policy on families should be studied.

The level and the form of social security rights provided in a society are seen to have an impact on family forms and the family cohesion. In scenarios where the current trend towards individualisation of social security rights has continued, and basic social security is provided by the state as an individual right of every family member, we are seeing more societal individualisation and less family cohesion. As there are more possible life choices for the individual in such societies, people could loosen the bonds they have with their families, or their local, religious or ethnic communities, and find independent paths to personal self-fulfilment – but they could also focus on strong family bonds as being relevant for their individual life.

Where there is little or no social security, family bonds and local, ethnic or religious solidarity are crucial for 'staying alive'. Hence, family and local bonds were strong in those situations, but often not a matter of choice which might influence the quality of these relationships.

Individualisation of social rights was thus seen to have ambiguous consequences. On the one hand, individualised social rights fostered social mobility, life choices and possibilities, and may have improved the wellbeing of family members. On the other hand, there was the risk that policies aiming only at the individual endangered family bonds and solidarity. There is an argument, therefore, that the family should be considered as a unit and not only as the sum of several individuals, as illustrated in Scenario 4. "*European policies focus on the individual… There is no family-oriented policy… These policies are not directed at families or designed to support families as an institution, but to support special groups (e.g. children, people with disabilities, older people, etc.)*".

Below is another example of the impact of the individualisation of social security rights. Whether or not such rights are tied to employment status makes a difference to the wellbeing of the family. The second narrative of Scenario 1 compares two family situations, in which two types of benefits are described: benefits attached to the individuals of the family regardless of employment status versus benefits attached to employment status of the parent. The outcomes for the families are drastically different:

Lily: "*Public funding is attached to the child, so whether Andrea works full-time or part-time does not affect what is available for her children*".

Cecilia: "*…benefits are all publicly funded, but tied to her working full-time… She is waiting for test results for a serious illness and is worried about what might happen to her benefits when she stops working. In the event of a serious illness, she would need to switch to private care, which might cause financial problems. She has worked* *all her life and while she has not had to spend money on care, which was publicly funded, she has had to pay high taxes. She does not have significant savings, a situation which might become difficult if she has to give up work. She worries about her children and her mother, who may need to switch to welfare day-care".*

This leads to the need for:

- Research on the consequences on the family, as a unit, of possible choices made within the family for care arrangements, and on consequences for all family forms.
- A study of "How family support should be provided?"
  - Should support be means-tested or universal?
  - What about tax policy? Taxation policy, considering the family as a unit and taking into account family size (number of dependents).

This leads to family mainstreaming as the framework of all policies:

1. Covers all different types of policies that may impact families: employment, law, education, migration, etc.
2. Addresses the family group, addresses individuals as people living in a family. Policies for special groups or targeted policies for family members as individuals are not enough.
3. Includes elderly members of the family.
4. Considers all family forms.
5. Looks at families as agents and assets and not as problems.
6. Engages them in all aspects, asking what families really want: a bottom-up process in policy-making.
7. Clarifies the true objectives of policy.
8. Includes on-going measurement of family wellbeing (i.e. included in GDP).

3.4.7 Impact of technological advance on families

As part of our 16 narratives, the emergence of new technologies plays an undeniable role in shaping the wellbeing of family members, even if it is not the leading factor. Although technological advances have not been a specific focus in working out these scenarios, the authors involved in the exercises remain convinced that technologies yet to be invented will have an impact on the wellbeing of families. The following is an outline of a few of the technological advances discussed.

Surveillance techniques: in Scenario 4, there are many mentions of non-stop surveillance systems and techniques (for example, chips implanted in children under the age of 14 with medical and other identifying information that allow for constant surveillance). Some of the types of impact: *"Parents who do not constantly know the whereabouts of their children are considered negligent"* (S4.1) and *"childhood depression* [is] *diagnosed in a large number of children…There are a substantial number of children at school with such illnesses, which are thought to arise from the constant surveillance and general lack of privacy"* (S4.1). On the other hand, surveillance techniques also support families in their care task, for example with television control or "assisted living technologies" for elder people (S1 and S3.2).

Virtual schooling is brought to the fore in several narratives: *"virtual schooling has become standard. Children in different towns, regions and countries can all be present in the same virtual classroom. For this reason, every social or cultural group can offer schooling in accordance with their beliefs and values. Access is either for community members only or, for the commercially-operated schools, regulated by fees. Very expensive face-to-face private schooling (e.g. Eton, Harrow, Salem, etc.,) is still available and serves the purpose of creating international elite networks. Consequently, social, ethnic or cultural groups ensure that their children are educated and raised the way they choose, and this has led to high diversity and de-standardisation of educational systems and increasing social segregation. The social and financial position of families therefore affects the educational development of their children, and educational inequality between social groups is increasing"* (S2).

Virtual relationships: narrative 2 in Scenario 4 is primarily based on a virtual relationship between parents and their left-behind children. What impact do they have on the "users" and how different are they from face-to-face relationships?

Communication tools and customisable media: houses have "3D media rooms" where the walls are the screens (S2.4); video-conferencing is highly developed in some scenarios and nearly omnipresent in others, enabling more working at home and reduced business travel. *"People no longer need to write or type, because voice recognition technology transfers the spoken word straight into documents, so a good pre-school needs to train for reading* *and voice skills, so that Vincent is ready for school work"* (S3.1). There are customisable TV programmes where the viewer is "virtually transferred" in the programme and becomes the hero of the programme.

This leads to:

- Policy should hold forums to discuss using technological development to support families, studying their intended and unintended consequences on all family members including the children.

3.5 Summary and conclusions

In summary, this intensive Future Scenarios exercise, based on the Foresight Approach that involved the active participation of over 30 stakeholders, helped highlight crucial policy issues and research questions that have a major impact on family life today and in the future.

Intergenerational and community solidarity played an important role in our narratives. Similarly, allowing sufficient time for families was a constant factor. Other major topics that affect the wellbeing of our families in 2035 were care arrangements and whether unpaid care work is recognised or not. The group often discussed the fact that children's and adolescents' perspectives are not always taken into account, as well as the balance between their rights and best interests and their parents'. Family transitions over the lifecourse was another topic that constantly called for research and policy attention. Given that our scenarios take place in 2035, technological advances are embedded in the day-to-day life of our characters. They also affect the wellbeing of families in many different ways. Individualisation of social security rights is a topic requiring careful attention. Across the different discussions, family mainstreaming was an underlying element which calls for a European family strategy.

- Across the different discussions, we identified some potential social innovations, such as:
  - The "skills market" (exchange of support) as a factor in social cohesion
  - "Time care insurance"
  - Mediation and counselling centres

In creating the Future Scenarios we could not describe people other than those who try to do the best they can most of the time, making choices and trying to be happy, "doing families" in their own way. What we describe in 25 years' time are people very much like ourselves today.

Whilst the families we analysed were very different in form, type and style, overall these families seem to struggle with the same everyday challenges that families face today. What we could see in the future is that the complexity of the world is not expected to decrease by 2035 – on the contrary, in some of the narratives it has significantly increased, at least as far as we can tell.

Family bonds remained a crucial field for the wellbeing of the individuals in all the narratives. The complexity of each family's environment changes and affects its members, but most families were valued by their members because, as a safety net, they reduced uncertainty and provided a framework for mutual support in the complexity of their environment. These "essentials" need to be addressed first and foremost when doing research or formulating family policies.

Families need the support of local, national and European policies to raise children. Policies should help them to have the number of children they desire, assist them when they face difficulties, and allow them to lead their lives according to their choices while respecting their obligations. Ensuring this would enable families to have and raise children who will become the responsible citizens of tomorrow's Europe.

Because of the importance of the wellbeing of families for the future of Europe, policy makers should make this a key priority. The EU2020 strategy and most European treaties are centred on the economy: we are calling for more attention to the families who are producing the economic agents of the future.

3.6 Annex - Living arrangements and family forms

3.6.1 Scenario 1

Family Form 1: Double income family with two children – living in an urban area

Names:	Joseph (40)
	Klara (35)
	Philip (6)
	Agnes (3)
Country:	Germany
Area:	Urban, Berlin

Joseph (40) and Klara (35) live in Berlin, Germany. They have two children, Philip (7), who goes to a public primary school, and Agnes (3), who attends a private kindergarten until noon each day. They own a house, and Joseph's parents live nearby. Klara's parents live in a different part of Berlin, and it takes around 45 minutes by public transport to get to them. All of the grandparents have already retired and are still very active and not dependent on help or care.

Joseph works as a teacher and returns home around four o'clock every day. Klara works very successfully in a private company. Her job sometimes requires her to work late and to take trips abroad. Both of them are socially and politically active outside their work and family life. In the afternoons, the grandparents usually take care of the children. Since the grandparents are often away on holiday, other forms of day-care for the children are necessary.

The main challenge they have to face is prioritising between their career and parental responsibilities, as well as working on the quality of their relationship. This creates a fair amount of stress within the family and is starting to endanger its wellbeing.

Flexibility is their main resource for coping with daily family life, accompanied by frequent and necessary discussions on shared responsibilities. Their main strategy for solving this problem is to strive for an ever more equal sharing of domestic work and parental responsibility, though this is accompanied by increasing ambivalence about whether this is working out or not. The parents have to consider how much time the children want them to be at home, if they see their parents as happy or not, and what they will learn in terms of work-life balance from their parents. The children basically enjoy living in this changing care environment.

To picture these tensions, listen to a typical crisis on Sunday evening:

> Klara: "Hey Joseph, have you forgotten? Tomorrow I have to fly to Lisbon for an important meeting for FAMILYPLATFORM, you know I can't miss it, I have an important presentation..."
> Joseph: "Don't say it again, Klara, I know, but Philip is looking a bit ill this evening, and I already stayed at home two weeks ago when Agnes was ill. It can't always be my responsibility..."
> Klara: "And why don't your parents help us? They're always on their own!"
> Joseph: "Why are you so unfair? They do support us in many ways. And why don't we ask your parents? We cannot go on this way! My job is important as well. In the last three months I had to stay at home four times, but I do want to be a good teacher and I have my

pupils every day in my classroom. Now it really is up to you – you are the mother!"
Klara: "Well, let's find a solution. How about finding a babysitter for tomorrow? Don't look at me that way. OK. Why don't you call the kindergarten hot-line for caring emergencies?"
Joseph: "No, they can't. The teacher stayed at home on Friday and she's pregnant. It's the fourth time this year we've asked for their support. They already look at me in a way that I feel they are wondering if we are still a family and good parents! – Philip, go to bed, you're ill."
Philip: "Mum, I want you at home tomorrow! No, I want you Mum, or Daddy at least."

Klara and Joseph have several options for dealing with this crisis:

- They can involve the other grandparents more.
- They could co-operate with a migrant woman with a valid work permit living nearby, whom they could employ when they need her support.
- They could negotiate new conditions in the partnership in terms of domestic work and childcare – both take their share in a symmetrical way.
- The conflict cannot be resolved, and within a couple of years they get a divorce.

Family life and the relationship between Klara and Joseph will change as they and their children get older. Permanent negotiations take place to ensure an equal share between the two of them, since gender roles have changed into equality and their roles as mother and father have to be redefined. New technologies will support the care task of families (e.g. television control for elder people).

Basically, this society generates wellbeing in the family because of equalities at different levels. Families can be very happy and enriched by the different opportunities and freedom to choose, but this situation can also turn into a risk factor if the family is not able to handle permanent negotiation. Families need a wide range of resources (internal and external), services and help, in order to reconcile work and family life.

Family Form 2: One parent family with two children – in a rural area

Names: Andrea (34)
           Lily (14)
           Michael (4)
Country: Austria
Area: Rural

Andrea is an Austrian single mother in her thirties. She lives with two children in a rural area in Austria. She has no family nearby and has a steady job which she wishes to make full-time but cannot because of care obligations. Her daughter Lily is in her early teens and at school. Michael, her second child, is in kindergarten and needs to be picked up at lunchtime each day. Michael has a non-life threatening but chronic orthopaedic problem that requires regular visits to specialists. The children's father is absent from the household, so he does not contribute in terms of care duties, but he does contribute financially, albeit only occasionally. Andrea thinks that social networks in rural areas are better than in a city and likes the quality of life she has where she lives. There is a network of help around her, where informal social arrangements exist between neighbours to help care for the children at certain times.

Andrea's family is facing a number of challenges. She can live on her income, has a house and a job, so there are no pressing problems, unless there is a sudden new demand on her budget. She needs to have a car, as there is no public transport available for going to work. Andrea is happy with her network, but occasionally misses a more active social life, which she cannot achieve. On an emotional level, she feels rather lonely at certain times of the day, but nothing persistent.

She fears job insecurity with the borders being opened up. She does not have specific skills and is worried. She thus wants to access lifelong learning programmes to increase her employability and also wishes to gain digital skills to help deal with a potential job loss and to open up her horizons.

She has an ethnically diverse network for her children and for herself, which she enjoys. She is ambitious for her children and herself, and to meet many of these needs she should perhaps be located in a more urban setting. There are better opportunities in a city, she feels, and one day she wishes to move, when Michael is in secondary school and she can afford it. She has an informal social network to help her with care, and there are some families in her network with fathers helping with care. She is, however, the mother of a sick child who needs care, so she is dependent on her neighbours to a great extent. Increasingly, care is becoming a problem. Her father is soon to move in with her, as he can no longer live by himself. While she needs her parents' support for her children, they both need constant care as well, particularly as one of them has Alzheimer's disease. A day nurse visits her father for a few hours in the mornings.

Andrea has adopted most of the principles of simple living. She tries to grow a variety of food and vegetables in her garden, and much of the rest of the village does the same. Lily is getting tired of the life around her. She is 14 and had a father up until four years ago. She has to travel to school in town every day. She has no friends at home. In terms of lifestyle choices, a conflict is repeatedly coming to the surface: she wants to move out, and in order to do so needs to have the same financial and social opportunities as her peers. Her mother does not want to move. Lily has been exploring EU-funded boarding schools that offer integrated learning systems including going out and exploring cities. Since Lily grew up with Moroccan friends and a Moroccan day-care mother, and has been learning French for many years now, she wants to move to France. After discussions at home with her mother, it is decided that Lily will move to a school in France.

Andrea's father is happy. His disease has not led him to be financially dependent on anyone, as his care and medication are financed by the state. His daughter's house has assisted living technologies, and the architecture is such that he will have his own space, while being supported. He also enjoys the company of his grandson. Decreasing stress levels with feelings of financial and emotional independence have helped him fight his disease.

Andrea is satisfied not just with her own lifestyle choices and support network, which help her cope, but also that she is being assisted financially to care for her father as she wishes, without restraining her daughter, who has been able to make a choice for herself. The rural health centre organises regular weekly sessions with specialist doctors, making it easy for her to get Michael treated.

Public funding is attached to the child, so whether Andrea works full-time or part-time does not affect what is available for her children. So Lily can make her own choices. As long as she stays at home, Andrea can also stay home to care for her full-time. Parental leave is available for fathers and mothers (one year for the mother and one year during the first 12 years of the child's life for the father). Andrea could also choose to stay at home to take care of Michael herself. The elderly enjoy similar advantages. Andrea's benefits are tied to both parents, so even if the children had a single father, he would enjoy similar benefits.

Andrea often thinks of her friend Cecilia, who is also a mother in another European country. Cecilia's benefits are all publicly funded, but tied to her working full-time. Cecilia does work full-time and can manage financially. However, she has to put her sick mother in an old people's home, and her small child is in a day-care centre all day. She needs to pick up her children from school and day-care and then visit her mother in her centre. She does not have a proper social network owing to the fact she works full-time. She is waiting for test results for a serious illness and is worried about what might happen to her benefits when she stops working. In the event of a serious illness, she would need to switch to private care, which might cause financial problems. She has worked all her life and while she has not had to spend money on care, which was publicly funded, she has had to pay high taxes. She does not have significant savings, a situation which might become difficult if she has to give up work. She worries about her children and her mother, who may need to switch to welfare day-care.

Family Form 3: Family with migration background

Names:	Azimbek (26)
	Roza (22)
	Myriam
	Tarek
Country:	Germany
Area:	Urban

Azimbek is in his mid-twenties, lives in Germany and comes from Kyrgyzstan. He is a first-generation migrant and has vocational qualifications as a mechanic. He started out with a temporary work visa. A reception centre helped him find housing (to avoid residential segregation) and assisted with language classes. After he had got all his papers together, he applied for family reunion, and six months later his wife Roza (22) arrived. She completed secondary education and can speak English well. She is entitled to work – in 2035 policies are fairly non-restrictive.

Azimbek works at night in a local bus station servicing public transport. The couple's dream was to increase their income, have children and one day own a house. Three days after she arrived, Roza got a job in a sheltered home for the elderly. She cleaned apartments from 09:00 to 17:00. Later on in the day, they went to a language course together in the hope that it would help them get better jobs.

After one year, they decided to have a child. Roza got 12 months' maternity leave, but in the second half was offered an intensive language course, with the child being cared for in a crèche. When the baby, Myriam, was one year old, Roza got a job as a kindergarten helper. She earned a little more money, and the crèche for Myriam was nearby. Azimbek was trying to get away from night shifts and switched to a five-year contract in the car industry. This meant longer commuting, but they still spent more time together than when he was working nights.

Soon afterwards, their second child was born. Having a child born in the receiving county entitled them to apply for citizenship and bring Roza's widowed mother from Kyrgyzstan. This new situation made Roza and Azimbek's lives easier, especially in terms of childcare, shopping and so on. But Roza's mother felt isolated and wished to take language courses herself. At that point, she was in her early 50s, well integrated into society. She soon found a new partner.

Meanwhile, Azimbek lost his job because of the economic downturn. They weighed up the possibility of going back to Kyrgyzstan, but since they were receiving support in terms of cash benefits and retraining they decided to stay. Azimbek managed to set up a little firm of his own, and Roza supported him by doing the administrative work.

A few years later, the couple faced the issue of choosing a school for their younger child. The older girl was put in a public secular school, but they could now afford to send the younger boy to a private Muslim school with a moderate tuition fee. In the long run they will both go into public tertiary education.

Roza's mother falls ill, and the couple is faced with the need to provide care for her, but that is relatively easy to arrange.

Family Form 4: Gay couple with two adopted children

Names:	Juan
	Abo
	1 child (6)
	2 child (4)
Country:	Spain
Area:	Urban

Juan and Abo are a gay couple. Juan works in a consulting company for clients throughout Europe. He can arrange his work from home with video conferences, but also travels a lot. Abo drives a minibus for money and has flexible hours.

They both wish to adopt a child, but have been given low priority on the waiting list. They decide to adopt two black brothers, aged six and four, from an institution. Both of the children are survivors from a sunken boat carrying illegal immigrants. Both partners are entitled to three months of adoption leave that can be taken at the same time.

The older boy has a psychological problem and needs a lot of attention and therapy. Abo therefore decides to take a year of homecare leave. Fortunately, the school is very supportive in providing support classes for African children and targeted counselling. The parents' association, which Juan and Abo joined, also organises counselling for parents in similar situations, teaching them about culture and providing basic language classes.

3.6.2 Scenario 2

Family Form 1: Family with migration background, two children and grandparents

Names:	Kimbacala (30)
	Gladys (29)
	Junior (3)
	Mia (3)
Country:	Germany
Area:	Urban

As a result of the total privatisation of all welfare services combined with a complete ban on migration into the EU, the society Kimbacala and Gladys are living in is highly segregated – different groups live alongside each other with very little intermingling. Membership of a group is based on common values, beliefs and culture, and determines the future unfolding of the individual life-course as well as the role and meaning of the family for its individual members and the different communities.

Kimbacala and Gladys are married and have three year old twins, Junior (male) and Mia (female). They live in an Angolan community in Berlin. Both the family itself and the community are the primary providers of welfare services. Kimbacala and Gladys' parents migrated to the EU from Angola before this became legally impossible in 2011. Gladys' parents arrived as teenagers in the mid-1990s and attended a public school for several years. Kimbacala's parents were slightly older and started to work on arriving in the EU.

Kimbacala and Gladys' parents met, married and started having children from the mid-2000s onwards. Shortly after they married, Gladys' parents set up in business as Ethnic Wedding Planners, catering to the needs of the Angolan community. Sometimes wealthy members of the market-oriented social group buy an Angolan wedding as a status symbol, leading to some mixing between groups.

Shortly after Kimbacala and Gladys were born, the state withdrew more and more from the public provision of welfare, turning it over to the market. Kimbacala and Gladys did visit public day-care and a publicly funded primary school for a few years, but due to the increasing privatisation of education their final years of schooling were spent in a community school. In this school, well educated community members teach for some hours a week as part of their contribution to the community. As a result of community schooling, Kimbacala and Gladys have an average level of education. They met at a cultural event, fell in love and started living together. Gladys' mother planned the wedding.

Kimbacala and Gladys work long hours in the family business set up by her parents, who are in their mid-50s and are still working in the business. Childcare is largely organised in the family (siblings and cousins rather than grandparents) and the community. A rotation system was set up, in which a group of parents take turns to look after each other's children, one day a week for each family.

The common background of having migrated from Angola is what holds the community together. It is a cultural resource, expressed in festivals and a certain style of dress or decoration. There is still a strong (emotional) connection to Angola.

Kimbacala and Gladys' parents are more liberal because they have much better knowledge of what they've left behind. They can view Angola with greater objectivity and appreciate the positive aspects of living in Europe. Kimbacala and Gladys have a more idealised view of their parents' homeland and try to conform to their image of it. The community supports this view and tries to reconstruct the homeland, making its members rather rigid in enforcing what they consider to be the norms and values of their country of origin.

There is a great deal of pressure within the community on Kimbacala and Gladys to conform. The business is rather vulnerable, because it depends on the goodwill of the community. Gladys' parents currently enjoy good standing in the community because they have improved their social status (and increased their wealth) through hard work (i.e. social mobility within the community is possible). They are likely to retire in the next ten years, leaving their business to Kimbacala and Gladys, who have to prove themselves in the eyes of the community. If the community shrinks, changes its marriage values or takes a dislike to the family, their business will fail. It is therefore in Kimbacala and Gladys' interest to support the traditional marriage ceremony and be perceived as good members of the community.

Kimbacala, Gladys and their parents all contribute actively to the community (e.g. caring for other people's children, giving money to the poor, donating food and decorations for festivals, providing training for young people). This could be a source of pressure, especially on Gladys, since they may be able to afford to buy childcare but would risk damaging their standing in the community (and therefore their business) if they did so.

Technological advances play an important role in education. Not only well educated community members who live together geographically teach in the community school but virtual schooling is possible for all members of a community, regardless of where they live. Communication between communities in different parts of the world is in their own language, so real-life schooling or other teaching modules are used to teach the language of the country of residence. All children in this community will grow up at least bilingual if not multi-lingual. Language mixing is likely to be widespread in everyday interaction in the community.

Although Kimbacala and Gladys work and share housework, there is still a gendered division of labour both inside the home and at work. The different types of work are evaluated differently (his is more physically demanding and more "important", hers is "easier") which translates into inequality. Each has more of a say over the upbringing of the child of their own sex. Even at the age of three the twins are treated differently, with the boy receiving more positive attention and praise. His sister is often treated as an appendage. This differential treatment will continue in the community school, which is a future source of tension between the parents and their children. Whereas the son will be aware of how rigid and restraining the community is (and may choose not to take over the family business), when the daughter grows up she will reject the community completely and seek out an "alternative" community where anyone can stay as long as they make a contribution to the self-sufficient running of that community.

Family Form 2: Patchwork Family

Names:	Maria (43)
	Erik (45)
	Bosse (12)
	Lisa (10)
	Simon (1)
	Annika (13)
	John (10)
Country:	Sweden
Area:	Rural

Maria is living with her new partner Erik. They have a child, Simon, who is one year old. Maria has two other children, Lisa and Bosse, from a previous relationship with another man. Lisa and Bosse live with their father. Maria is a ping-pong mother: every second week, she leaves her new partner and son Simon to stay with her ex-partner and their children. At first she wanted to take the children with her, but joint custody is a rule, and Bosse and Lisa did not want to leave their familiar surroundings, friends, etc.

Erik also has two children from a previous relationship. They are called Annika (13) and John (10) and live with Maria and Erik. However, every second week, when Maria is staying with her older children, Erik's ex-wife comes to live with him, Simon, Annika and John. She has an extra room in their house. This makes it easier for the children to go to a "traditional school" and not to do virtual schooling. The children do not see their siblings, who do not live with them very often, but once a year they go on holiday together to their other grandparents' house. They are adjusting to the situation, even though there are feelings of jealousy and other complications arising from this kind of living arrangement, but they are aware of the consequences of their decisions and try to cope with them.

Bosse, Lisa and Annika go to a "traditional school" in a real school building with other pupils and teachers. These schools are said to be better, but also very expensive. Maria and Erik can therefore only afford part-time education for their children. They would like them to go to school full-time and study all the subjects, e.g. learning different languages, but even now they have to struggle to pay for the education of their children.

John can't go to real school with his siblings, because he has been disabled since birth. That's why he stays at home and goes to virtual school. A carer comes every day, and the medicine is very expensive. Simon also stays at home and "goes" to virtual voluntary pre-school for a few hours every day.

The grandparents (Erik's parents) come to Maria and Erik's house every day to take care of the children and do the housework while Maria and Erik are at work. Their pensions are not enough to live on, so Maria and Erik pay them a small salary.

Maria and Erik live in a village in a big house with a garden. The grandparents live close by in a very small house, which they moved into after they gave the bigger house to Maria and Erik. They spend very little money on consumer goods, as do the children, because the services they buy are so expensive (education, care, and health-care). Most of the things they need for everyday life they exchange in online forums with other families who no longer need them.

They travel in little sky trains – every family owns a little train with which they can clink into the rails and type a destination into the onboard computer. The trains are powered by geothermal energy, as is almost everything today apart from the power generated by other renewable sources: sun, wind and water. All houses have solar cells on the roof.

Because of the lack of public care systems, having children is very important for parents to ensure that they are cared for in old age, in case they themselves cannot afford it. Investments in children and respecting children's interests are seen as very important, but only some communities have policies and institutions for these aims. Maria and Erik do not live in such a community, but they try to manage as best they can, for example, by becoming ping-pong parents.

Family Form 3: Dual earner with two children

Names:	Emily
	Phillip
	Lucia
	Romeo
Country:	Italy
Area:	Urban

Emily and Phillip live in a big city in Italy and both work as managers. They do an equal share of the housework and paid work. When Emily was in her mid-30s, in her view she was too old to have children, so the couple decided to adopt. They were lucky and were able to adopt a newborn child from another ethnic group in their country, a baby from a third-generation Colombian teenager mother.

Soon after having adopted the baby, Lucia, Emily became pregnant. With two young children Emily and Phillip are faced with the challenge of combining two careers with childrearing. Phillip has a good income, and it will not be necessary for both of them to continue working. But Emily does not want to give up her job: she would like to continue with her career. So they need all-day childcare.

With the help of an agency they find a full-time nanny. The nanny is from a migrant background and is not very well educated. With two children and a nanny they decide to buy a bigger house with room for everyone. Since the house is of a high technical standard and is expensive, they have to get a mortgage on it. Emily and Phillip spend a lot of their free time with their children. Phillip usually picks up something for dinner on his way home from work, so Emily can spend more time with the children. Both of the children are in good health.

When Lucia and her brother, Romeo, are five and four, the parents decide that the children need more social contact with their peers. They fire the nanny and hire a housekeeper, whom they find on the black market. Their idea is to save money by employing someone illegally so as to be able to afford private day-care for the children. The housekeeper has no driving licence, so Emily and Phillip have to organise taking the children to day-care and picking them up again.

To help them better organise their family life, Phillip and Emily decide to work different hours: Emily works in the early morning hours and Phillip in the late evening. This of course reduces the time they can spend with each other.

When Emily's father falls ill and requires full-time care, Emily and Phillip have to find a private solution, since there is no public health-care and all services have been privatised. The private solution is very expensive, which puts a financial strain on the family. Phillip works longer hours to compensate, but there is an increasing imbalance between paid work and housework: more housework and childcare for Emily. She criticises Phillip a lot for their unequal shares. Phillip feels more and more stressed and is in fear of losing his job.

Their relationship is deteriorating. Because of their values and their financial situation, a divorce is not an option for them. There is very little social support in society, and they find no time to repair their relationship. So they continue with the relationship for different reasons, but not because of their feelings for each other.

Emily decides to change her work life: she will stay at home and take care of the children and her father so as to save money. To earn money she has decided to care for three other children. They also move into a smaller house to save money. Emily is very unhappy with the sandwich situation.

Emily's father dies when the children are 12 and 13 years old and are going to a good school. Emily starts work again. The children have many activities in the afternoon, but they also stay on their own at home and watch TV or surf the internet. The parents have several ways of restricting internet use, but the children's media skills are better than their parents', and they get access to forbidden areas and content. Children are highly regulated and scheduled in this society, but there is not enough supervision by adults.

Emily gets a very good job offer, and the parents have discussions about their careers. They decide that Emily should take the new job and commute. The children stay with their father, and the household gets fitted out with the latest technical equipment, but the children are still on their own a lot. The family has just a few social contacts and lacks information on support services. In researching the internet they find some sub-optimal or very expensive services: for example, they spend a lot of money on health insurance. Lucia and Romeo continue their education at a good but expensive university. Phillip lives largely alone and has a secret affair. As previously mentioned, divorce is not possible because of their values.

Lucia's marriage is very costly for her parents. The high expenditure on the children and insurances has meant that Emily and Phillip cannot save any money for their retirement. They invested in a system that is not good and will not provide help for them. At the end of their working lives, Phillip and Emily have very little money, and their children live far away in relationships that are not good. The children cannot support their parents, so Emily and Phillip have to work until they are 77.

Family Form 4: Single Father with two children

Names:	Lasse (45)	
	Svea (12)	
	Lars (3)	
	Grandparents (both around 70)	
Country:	Sweden	
Area:	Urban	

Lasse lives in the suburbs of the capital city of Sweden, Stockholm. The relationship between him and his ex-wife failed. She fell in love with another man, and they ended up getting divorced. They were dual earners, and he is now the sole provider for his two children. Lasse is disappointed that his ex-wife does not want much contact with him or their children. The family is Lutheran.

To help him out and because of a desire to stop working full-time, Lasse's parents moved into the same street as Lasse and his family. He bought the house for his parents, who are both in their seventies. The grandmother quit her job, following a very demanding career, and has now moved closer to Lasse, who is her only son. She is very active in the local community, and has become very close to Svea. The grandfather is a rather distant figure, but works part-time at the community centre.

Svea is 12 years old and goes to school 20 kilometres away. The school was very carefully selected by her father. It is a small school with 15 students per class. The curriculum is fairly traditional, with a focus on sports, outdoor activities, and religious learning. School hours are from 9am till 3pm.

The father insists on taking care of Lars (aged three), even though he has a lot of responsibility at work. He works in the banking industry and is well-off, but is able to take care of Lars and devote a lot of time to him (generally in the afternoons). Lars goes to a Lutheran community pre-school a few kilometres away for a few hours a day (run by parents themselves, where older highly-educated people also work). Lars is usually picked up by his grandmother and has lunch at home.

In the house, there is a kitchen where the family cooks in a traditional way (i.e. they prepare meals by themselves instead of getting them cooked automatically). Lasse cooks a lot and enjoys doing this with his kids. Their meals are still taken together around a table. Lars usually makes a mess and enjoys dinner time, but Svea, who will soon be a teenager, is becoming less keen on this and being more and more difficult.

There is a media room in the house with 3D television (the four walls represent the screen). The family does health/fitness activities together. For example, father and daughter often play football together and hike in the surrounding hills. Svea is registered in a scouts group. The family typically stays in its neighbourhood, but they also have a summer house where they spend leisure time.

Svea is not allowed to go to the trendy local virtual clubs, nor does she go to the local school which caters to children from the strong local community of the less well-off, who are predominantly Muslim. In general, there is little mixing between these communities. Communities are socially segregated and in other parts of the city are geographically segregated. Lasse pays a tax to the local community which offers, for example, its own security and school system, but he is not involved in local decision-making. The father deliberately decided not to use those services as much as others. Despite this, his situation is actually considered slightly "different" by his Lutheran friends – most of these Lutheran friends live closer to each other and mix far less with immigrant communities on an everyday basis (shops, neighbours, some local services).

Lasse devotes a lot of time to his family and enjoys it. He had to give up many leisure activities, but he does not feel bad about it, because he is happy with his situation. His colleagues are in the same situation. They value their family time.

Some families near them go every week to church, which is a few kilometres away. Lasse and his parents go occasionally, but generally attend community-related activities rather than religious services. They are surrounded by a strong Muslim community, but they actually do not spend much time with them and have no particular views on them either. Svea has friends living in the same street, and her father and the parents of other Lutheran families at the church are far from happy about this. The grandmother, however, feels her son is a bit isolated in general, not mixing very much with members of the local or other communities. This extends to her thoughts about living in a migrant area – she feels her grand-daughter is being isolated from her peers living in the local area and worries that this will create problems when she becomes a teenager. This is not surprising considering that the grandparents have both become quite involved in the local community through working at local community centres. This fits a bit uneasily with the father's attempts to shield his family from the local community.

The family is used to regular travel, although business travel has now become less common due to new technologies. They have a good stable relationship, even though there are frequent squabbles. They respect each other. Svea begins to feel a bit frustrated, she is a teenager. Even though Lasse tries to take good care of his family, he is a bit aloof. On the other hand, there is the grandmother who has a very close and open relationship with her grandchildren, especially Svea. The grandfather enjoys family time, but is less involved in general. He used to work a lot to give his child a good education and did not spend much time with Lasse. That is also one reason why Lasse wants to spend time with his kids.

They know a few of their neighbours and help each other. Despite the chosen isolation, the grandparents' involvement in the local community has improved relations between the family as a whole and the predominantly Muslim local services and community. Health-care is generally good, because the family can afford it. The grandmother recently had new lungs implanted.

Lasse got to know a Lutheran single mother of one child in Italy who also has a good relationship with her child. She is involved in the local community there. Does Lasse want to get married again?

3.6.3 Scenario 3

Family Form 1: Family with four children – living in urban area

Names:	Katharina (40)
	George (40)
	Aaron (14)
	Anna (10)
	Calvin (5)
	Leo (4)
Country:	Netherlands
Area:	Urban, suburb of Amsterdam

Katharina and George are not married and live in the suburbs of Amsterdam with their four children. All of their grandparents live quite far away from them. Both parents have temporary jobs interspersed with regular periods of unemployment. George was born and raised in the Netherlands. Katharina comes from eastern Europe and has neither Dutch citizenship nor legal status in the Netherlands. Consequently, only the father can sign official documents for the children.

Access to citizenship is linked to legal employment and paying taxes. Both of the parents face a dilemma: either they can carry on working illegally and earn more money, or they can enter the formal labour market with lower incomes but on legal terms. Later on they can apply for some of the benefits, such as higher quality education.

The family has a very tight-knit social network with their neighbourhood, but their housing and financial circumstances are not very good. It is cold in winter and the roof is covered with tarpaulins to stop the rain getting in. The motorway is nearby and is very noisy. They have legal access to water and electricity. Their neighbourhood network cannot help them financially, so they get some financial assistance from voluntary organisations and/or church charities.

They do not want stressful jobs and are quite satisfied with their lifestyle – they have time for themselves and their good friends who support them. Again, they are facing the same dilemma as with work: do they remain in an unconventional situation, with good quality of life, or do they have to change because of their exposure to income problems, problems for the children. Should they try to legalise their situation?

The couple's relationship is very strong (no divorce likely in the future), they choose their lifestyle, they have time for each other and their children, they have a lot of support from friends, and they do manage their problems, even if they are unable to solve them.

The two elder children, Aaron and Anna, are in a public school. The school has criminal sub-cultures and Aaron, the oldest son, has started drinking alcohol at school, as well as smoking and taking soft drugs. The parents are concerned about his situation. To support the child, a sort of "family peer group conference" has been organised by a social worker from the community together with a voluntary organisation. By organising round table discussions, the voluntary organisation is looking for solutions for families, not just for families with specific problems but also as a preventive measure for other families.

Anna is a good student and she could go to a better school, but her parents cannot afford to send her there. They heard during a recent parents' evening at the school about the possibility of applying to a foundation for some grants for their daughter. Unfortunately, more information and support for the parents is needed, because they cannot handle the bureaucracy and the procedures to apply for such a grant.

The younger children, Clara and Leo, are not in kindergarten, because their parents would have to pay for it. The younger ones are living "on the street" and are cared for by the informal neighbourhood care system. Clara, the youngest daughter, has asthma, which has actually turned into a chronic illness. Since the public health system only offers some medicine and no general treatment, the parents can only help and support her with short periods of medication. They have no easy access to the health-care system and would need more information, since some private founda-tions offer financial support, but they do not know how to contact them. More information on the illness and ways of supporting Clara can be found on the internet, but since the family does not have a computer they cannot access that information easily unless they find time to sit down in an internet café.

Family Form 2: Blended family with four children – living in urban area

Names:	Kathy (46)
	John (50) – Kathy's second husband
	Mark (52) – Kathy's first husband
	Anna (44) – John's first wife
	Lucinda (17) – Kathy and Mark's daughter
	Martha (10) – Kathy and Mark's daughter
	Thomas (12) – John's son from first marriage
	Vincent (2) – Kathy and John's son
Country:	England
Area:	Urban, London

Kathy and John live in London and have a two year old son, Vincent. Vincent was born to a surrogate mother because Kathy, who has a good job, could not possibly afford to take any leave from work. In this "contract society", Kathy decided to go for an intellectual, British surrogate mother. The surrogacy was paid for by Kathy's company.

Both Kathy and John are in their second marriage. Kathy was first married to Mark: he is Danish and still lives in Denmark with their daughter Lucinda. Lucinda travels to London quite often to see her mother, Kathy. Mark is a not very successful musician with financial problems, which was the reason for the break-up of his relationship with Kathy.

John was first married to Anna, and together they had a son called Thomas who lives with Kathy and John. Anna also lives in London, and Thomas spends every other week at her place.

Anna is increasingly uncomfortable with the societal conditions around her. She lives in a co-operative community, which provides some social cohesion. She is happy that her "work" not only brings in money, but also produces some kind of commonality and brings people together. She runs a food shop with a network of people. She gets in surplus from elsewhere and sells enough for it to be just about profitable. The people working with her do so on a voluntary basis and are not paid. She does not need a lot of money, except for half of Thomas' costs. Her network has doctors who often contribute voluntary care for those in the community. In other words, Anna has built up a network of people around her.

Kathy and John have a number of problems. Financially, their children's frequent travels are causing problems. Psychologically, the situation is not easy at all: Thomas doesn't enjoy having to move from one place to the other while his step-sister, Martha, can stay in the same place all the time; she lives permanently with Kathy and John. The sisters Lucinda and Martha are separated and miss each other. Lucinda wants to move to London, where she can be close to Kathy. So Kathy is considering buying British citizenship for Lucinda, giving her legal status in Britain. Then Lucinda would have the freedom to choose if she stays in Denmark or in Britain.

With technology that enables surveillance and with pressured lifestyles, and with both Kathy and John at work, Martha and Thomas have mobile phones which can tell their parents where they are. This creates difficulties at home. Martha dislikes it immensely.

Kathy and John are looking for a good pre-school for Vincent. People no longer need to write or type, because voice recognition technology transfers the spoken word straight into documents, so a good pre-school needs to train for reading and voice skills, so that Vincent is ready for school work.

John is a highly qualified lawyer. He belongs to a local church group: he is involved in the choir, in which he has many friends. Kathy and John's resources are stretched to the limit because of the constant cash outflow to manage the multiple travelling, children, and therapy fronts. They spend what they bring in and are not financially prepared for a sudden contingency.

They live in a society which is characterised by social unrest, caused mostly by social deprivation. Police resources have long been stretched to their limits in a society of continuous conflict. Kathy and John have nothing put aside financially, and suffered a strong financial blow after an incident of violence. They did not suffer personally, but are worried. They want to move into a gated community. However, they are also worried that the many rules and imposed values in a gated community will be bad for their children. In any case, they must wait for Vincent to be of school age before they can make such a move, because he is not allowed into a gated community before a certain age.

Family Form 3: Family with migration background

Names: Azimbek (26)
          Roza (22)
          1 child (6)
Country: Germany
Area: Urban

Azimbek works at night in a local bus station servicing buses. He lives in a crowded house with other immigrant men. He fell down some stairs, broke his leg and now cannot work. He has no money saved for further treatment and has no-one to look after him. The help comes from a local Muslim community, helping with daily life, but not improving his health. Roza comes over from Kyrgyzstan and works long hours caring for an older person in a private home, earning just enough to move into better housing. Azembek gets a new, less physically demanding job as a night porter in a hotel, but with low pay.

Roza gets pregnant, but does not have a legal contract and hence no health insurance. They need to work extra hours in order to save and be able to pay for medical services during pregnancy, including giving birth. They have very little time together. After the child is born Roza is allowed only one month off work, her employers being generous and paying her salary as usual. As they do not want to change the carer, Roza brings the baby with her to work.

Roza's mother became a widow, and because so much work needed to be done, wanted to come from Kyrgyzstan to Germany, but this is not possible as immigration policies are targeted at reunion of nuclear families and at bringing in young, legally employed workers.

Roza's patient dies and she finds a new job at a supermarket. This is a legal contract, but she brings home less money and cannot take her child with her. The Muslim community provides childcare at moderate cost, but in the new situation they cannot afford to move out from a shared apartment with two other couples, as they had hoped to. In these circumstances they decide not have a second baby. The child has begun to learn German at Roza's patient's place, but forgets it all because all the other children at the childcare facility are immigrants who speak other languages.

Roza and Azimbek try to speak German at home, but this is challenging and does not help much. When she turns six, the girl goes to a public school and is the best pupil in the immigrant class. It is an open question whether this is enough to succeed in the long run.

Family Form 4: Gay couple with one adopted child

Names:	Jean
	Abo
	1 child (6)
Country:	Spain
Area:	Urban

Jean and Abo are a gay couple wishing to become parents. For that reason they contact a private international adoption agency, and have to travel to India on several occasions.

In the absence of public counselling they are not well prepared for adopting a child. With their financial background they only can afford to adopt a child of school age – a six year old child. Legislation in their country does not provide any form of adoption leave, so they have to put a substantial amount of money aside if they want to stay home with the child for the first few weeks after adoption. The child has psychological problems, and the attendance of the parents is necessary, but homecare leave is not available for Jean and Abo.

The couple faces a dilemma: either one partner has to quit work completely, or they have to place the child in an expensive private school with counselling services. They decided on the second option. The child gets better, but the financial stress is adversely affecting their relationship. Needing to work more, Abo and Jean spend less and less time together. Their lives become more and more disparate. Will their relationship survive?

3.6.4 Scenario 4

Family Form 1: Two mothers with five children

Names:	Wendy (36)
	Mandy (37)
	Björn (15)
	Vanessa (12)
	Miko (6)
	Stig (6)
	Nelson (2)
Country:	Scandinavia
Area:	Rural

The family comprises two women with five children. Their stable relationship is based on mutual support and care. Wendy's child, Björn, was the unexpected result of an "experimental phase" at university. Mandy is an only child and believes it's important for children to grow up with lots of brothers and sisters. She is the mother of three of the children (Vanessa, Stig and Miko) and is altruistic, which is why she wanted to adopt the refugee baby (Nelson) and give him a good home. She trained and works as a nursery nurse.

The two women met when Wendy brought Björn to the day-care centre where Mandy works. He was six months old at the time (public childcare entitlement starts on expiry of parental leave six months after the birth). At some point, the two women decided to move in together to share childcare, etc. Altogether another four children followed (three through natural or artificial insemination), including an African child, Nelson, who was adopted. There is a pan-European adoption system for child refugees shipwrecked trying to get into Fortress Europe. There is thorough screening of parents for such adoptions, training by a unit of the European Border Agency, and regular follow-up screenings. It remains to be seen whether adopted children will want to return to where they were originally born, because the scheme is quite recent – though it remains a point of contention and concern.

The children are between three and 15 years old and all are either in public day-care or school. Wendy is a freelance graphic designer in a non-standard, insecure job. She acquires most of her work via the internet, where she is part of a large professional network. She works irregular hours (sometimes not at all) and has an irregular income.

Mandy works in a day-care centre and is therefore employed in the public sector. Her job is not very well-paid. She inherited a country house (to fill with children and cats and one old dog), which means that they do not have to pay rent. Nevertheless, with five children and their low incomes it is a struggle to make ends meet. The pre-school children's clothes are either handed down from oldest to youngest or come from swaps organised in the village or at the day-care centre. The mums have a garden and grow their own vegetables as well as keeping chickens, although this isn't really done for the small amount of food it yields – more because of the mums' values and the environment it creates for the children.

Technological developments are highly relevant for the work of both mothers. Whereas Wendy organises and lives her professional life on the internet, Mandy is constantly under surveillance during her work in the day-care centre. The centre is full of closed circuit television cameras (CCTV), so parents can always look in on their children via the internet; questions of privacy in such a setting have long since been subsumed in the desire to monitor and ensure the safety of children: parents who do not constantly know the whereabouts of their children are considered negligent. The city streets are also full of CCTV, although they are still less common in the countryside. All cars are tracked by satellite – under the pretext of preventing car thefts (though this is also for crime detection purposes, enforcing speed limits, and for calculating vehicle taxes). Everyone has to carry an ID card that can be read remotely by scanners. Children under 14 (age of criminal responsibility) have had a chip implanted just under the skin, with medical and other identifying information. The authorities would like the implant to stay there for longer, but have not yet managed to pass the necessary legislation. Parents, however, can use the internet and the chip to locate their children. With this technology it is possible to know where everyone is all of the time.

The family has social links to many people in the village, including older "volunteer grandparents", who occasionally do things with one or more of the children on an informal basis, such as taking them swimming or fishing.

Björn is starting his last year at school and is faced with the choice of either academic or vocational further education. Unfortunately, his mums cannot afford to keep him at school, because after 16 all education is private. The city school is very large, with 1,000 to 2,000 pupils. He decides to leave school and do an apprenticeship before joining the European Border Guards to finance his university degree.

Together with his mother, Björn is now using the internet to find his biological father. They've located him on Facebook 10.0, but are somewhat apprehensive about how to approach him. Björn wants to find out what sort of a person his father is, and Wendy is also curious to see what he has turned into.

Vanessa has been diagnosed as suffering from depression. There is a long waiting list for treatment because of the high level of child depression and limited resources, so it takes several months for her to see the specialist. The treatment or therapy is targeted at the individual, not the family, so Vanessa visits a child psychologist outside of the home. There is a child health centre next door to the school, with paediatricians and child psychologists who care for children at the school. This kind of situation is fairly normal, with childhood depression diagnosed in a large number of children and recognised as being similar to depression in later life. There are a substantial number of children at school with such illnesses, which are thought to arise from the constant surveillance and general lack of privacy. Due to its high prevalence, however, depression is not as stigmatised as it was in the past.

Vanessa goes to the same school as her brother, Björn, which is in the city. They get there with the free school bus but have to come straight back home because they cannot afford to stay longer and then have to pay the normal bus fare.

The twins, Miko and Stig, are in their first year at primary school in the village. They walk to school on their own. The first year at school is full of tests, screenings and assessments.

Nelson is just becoming aware that he's different. His mothers are very supportive and his older brother, Björn, is very protective. Nelson has had some racist comments directed at him in the local village, but is still too young to understand them.

Family Form 2: Children living without their parents

Names:	Pol (16)
	Sophie (11)
	Gaby (4)
Country:	Romania
Area:	Rural

Pol, Sophie and Gaby are living in Romania, which is a member of the United States of Europe. In this federation, there are the "first-class countries" and then their respective economic "colonies".

Alexander, the children's father, emigrated to Poland and works as a construction worker. He is married to Josephine, who is also a migrant worker living in Germany and working as a nanny. They live 100 kilometres apart, but meet in a city halfway between the two every other week. Alexander has been in Poland for ten years. Josephine left nine years ago and then came back four years ago to deliver her third child. They used to live on a farm. The children had to stay behind to "keep" the farm. Their grandparents live in the same village on their own farm and usually visit every day. The parents have created a network of carers for their children. The whole community helps them (former neighbours, teachers). The family is not the only one to have gone abroad. Several other families are in the same situation. That is why they understand each other, and there is strong solidarity.

No adults live with the children. Pol has a lot of responsibility and is the main carer for his two sisters. He no longer goes to school, having just graduated. He is doing a vocational degree in agriculture and works on the farm (he feeds the animals, harvests the corn, does the yard work). Before then it was Alexander's brother who looked after the farm.

Every day the children have breakfast together with their parents via communication technologies. The parents have also installed a monitoring system so they know, for example, exactly who visits the house, if something is burning in the kitchen, if a water pipe is leaking, and what is in the food cupboards and fridge. The parents control the weekly food menu, so they make sure the children eat healthily. According to a financial agreement between families, dinner is provided by the neighbours.

Parents and children communicate frequently, using basic communication technology, which includes 3D cameras. Gaby loves the evening bedtime story with her mother, while she tells it to the children of her host family. Actually, Sophie and Gaby have become virtual friends with the children of their mother's host family. After school, Gaby also enjoys the cartoon, in which she is the heroine.

Josephine's job consists of taking care of twin babies and a school child. The twins were diagnosed with a serious genetic disorder, but thanks to pre-implant therapy they were cured. The host parents have to travel a lot (even though travel has become very expensive). She implements robotic programming systems in city infrastructures. He is a physiotherapist for actors.

Because energy is so expensive, the number of cars has drastically decreased across Europe. Solar-powered public transport is widely available; however, the "cyclo" is a growing mode of transport. It is actually because of long-distance transport being so expensive that Josephine and Alexander rarely go back home. People across Europe typically have a healthy way of life as a consequence of the energy crisis. Even though pollution levels have decreased in general, there are many UV radiation alerts, during which people have to stay indoors.

The father and mother come back every autumn to help with the "big work" on the farm. Alexander comes back more often, for a few weeks at a time, when he does not have work in Poland.

The father instructs his son on what to do on the farm. Pol is very frustrated with his parents being away, but he deals with it because he has a strong sense of responsibility. In their society, it is very important for women to have a good education (provided for free by the state). When Gaby is sick, the parents ask the grandparents to take care of her.

Climate change has had a huge impact on their family life. It has changed migrant workers' seasonal work. The short-term goal of the parents is to alternate seasonal work between the two of them so as to be able to spend more time with their children. When he gets married, Pol will probably bring his bride home (to the farm). Gaby and Sophie share a room. The father is a bit afraid to come back home forever because he is not used to living under the same roof as his family. Josephine longs to come back forever. The parents have no close friends, because they are not there often enough to form close friendships.

Even if Gaby receives lots of presents to compensate for the absence of her parents, the additional money earned by her father and mother is used to pay for farm tools and energy infrastructure. Energy has become outrageously expensive, and the parents are earning money to buy the infrastructure for their farm to become "energy self-sufficient". Land is very precious, because it represents a source of energy. Water has also become a rare commodity. Once the Ramonov family have their energy-generating windmills installed, the family will feel safer and will be able to retire easily, because they will be able to sell the energy.

They are also saving for a better education for their daughters. All facilities (school, shops, sports activities, etc.) are available in the city close by. At school, Gaby is learning to sing the European anthem and shares virtual classes with other four year olds from Spain. Sophie and Gaby are also learning to speak several European languages at school.

Luckily, Pol is there to do the physical work at home while the girls are out. There are tools to clean the house automatically, so he does not have to do that much housework.

The family pays a lot of tax to Europe (it is at the same level in each country). Tax revenue is redistributed among the Member States.

Family Form 3: Single Parent

Names:	Linda (45)
	Kristel (10)
Country:	Norway
Area:	Urban

By the year 2035, climate change and dwindling resources have caused Europe to become a closed fortress. A "mega-state" was established, to control migration and all resources, including population. Other parts of the world were severely affected by climate change. Most of Africa, parts of Asia and South America became deserts, and people can no longer live there. Although itself heavily affected by climate change, Europe is still quite a good place to live because it was able to establish a Supranational Organisation (which developed from the current EU). Technological developments (funded by this SNO) enabled sea water to be desalinated. Water is currently the most important natural resource.

This is a society dominated by women; they occupy most of the leading positions in government, science, business, and society in general. Men occupy mostly subordinate positions, but there are a few of them in key positions. This was the result of many years of different outcomes in education. Women slowly started to occupy leading positions because of their higher educational attainment. It started in science, later in business and politics.

Linda (45) is a very well educated woman, working as a professor at the university in a city where there is a big research centre trying to improve the desalination process. Because of the higher position of women and the lack of men with higher education and a similar social position, it is hard for her to find an equal partner. This gap has led to a marriage or partnership squeeze for well-off women. During her education and career, Linda has also moved a lot around Europe, which has made forming a stable relationship hard. This led her to be alone a great deal.

Various reproductive technologies, including genetic cloning, have made it easier for the state to take an increased role in controlling reproductive issues (supporting certain groups of people in having children with certain characteristics, which are needed from the point of view of society). The state chooses whom to support, based on their genetic material and intelligence. Linda is a perfect candidate and has decided to have a child.

The state has offices where you can request insemination and medical treatment. Eventually, she got pregnant and gave birth to a daughter. She wanted to have a daughter (a younger version of herself) because the position of men in society is not so good. It was possible to select the gender of the child as well as many other characteristics. So when she applied, she wanted her daughter to be intelligent and good-looking. The girl was named Kristel. They live together in a city close to the sea in the Northern part of Europe (formerly known as Norway).

Linda does not want to have a man by her side and intends to make this child her own project. She expects to be a good mother with one real strong relationship because she has no other committed relationships in her life. Additionally, motherhood has a very high status, because only a few people are formally supported in having children on the basis of their own decision. But the development of the child takes an unexpected turn. Kristel was three when the private lessons in languages started, and new activities were added as she got older. Despite her taking lessons in music, arts, horse riding and languages from an early age, she is not very good at anything and makes no effort.

To make things more complicated, the child does not even do well at the public school, which is obligatory for everyone. This is frustrating for Linda because of all the unfulfilled dreams she had for Kristel, and the relationship becomes worse and colder. They do not interact much, because Kristel still takes part in various activities in children's facilities (school, private lessons, etc.,) while Linda is working a lot in order to further her career and to earn the money they need for maintaining their lifestyle. Although they live in an urban area which is culturally rich and offers many opportunities, Kristel lives very reclusively, without friends, without special interests, playing hologram games, watching TV and communicating with a girl from abroad, whom she cannot meet in person. So this is her best friend and she is hungry for another relationship, but unable to find one and build it up. She spends her free time in her mother's stylish apartment, being lazy. Linda's frustration is growing: she finds motherhood a burden.

By the time Kristel turns ten, it is clear that she will remain a loser in her mother's eyes. At this point Linda meets a man she is physically attracted to and has a passionate affair, but at first it is only a physical relationship. He is poorly educated, his social status is low, and his attitudes concerning education and childrearing are rather traditional, so they do not have a lot in common. However, the man is keen to have a "real" family and is ready to build a stable relationship with Linda and her daughter.

Linda has to make a choice. There are three possible scenarios for this family:

First: she does not want to let the man form any parent-like relationship with Kristel, because this would make her feel that she has been a complete failure. So she ends the relationship, or starts treating it as purely sexual. Kristel is under less pressure, but still lacking close personal relationships. Because of the reduced pressure, she is able to take part in activities which she likes, but which are not designed to further her career (e.g. cookery courses). As she gets a bit older she can express her own wish to go to a different school, for example the boarding school where the only childhood friend she has is going, and since the connection with the mother is not so vital for the mother anymore, she is allowed to do so.

Second: the mother sees the possibilities in a relationship with this man. He will manage the household and care for the child, so she will be freer than before. He takes his chance because it is an opportunity for social mobility and to live fatherhood, which does not happen that often to a man of his position in this kind of society. Additionally, he will have a higher level of social insurance and the comfort of a high technical standard in the household. Kristel is also pleased with this arrangement, because somebody is now there for her at home, and she does not have to be alone so much.

Third: the mother gets fed up after trying to live with the man because of different cultural ideals, and ends the relationship. But since the man has played a parental role for Kristel during the relationship, and she accepted him as a father figure, they continue to have a good relationship after the break-up. This makes it easier for Kristel to cope with all the pressures of doing well in education because of the acceptance and recognition of the "father", which is unconditional (compared to the high hopes of her mother).

Family Form 4: Love-related partnership (cohabiting) with one child

Names: Mia (mother) (35)
              Daniel (father) (30)
              Konstantin (son) (7)
              Grandparents (70)
Country: Germany
Area: Urban

Europe's borders are closed to the rest of the world, but within them it is a mega-state, and everyone speaks the same language. Many men work as border guards protecting the borders, or are housekeepers, or do other badly paid jobs, because they are mostly less well educated than women. Childcare enjoys high social prestige, and if men do this, they have the opportunity to achieve higher standing in society.

People cannot fall below a specific income level, since equality is assured by the welfare state, but on the other hand, wealthy citizens cannot accumulate as much money as they want, because they have to pay high taxes for guaranteeing the equal level of social services. This tax system is very transparent, everybody knows how much each citizen pays in taxes, and so wealthy people are under social pressure.

During the time of open borders to countries outside Europe, Mia met Daniel on holiday in India, and they fell in love. Just before the borders were closed, he came to Europe as a legal refugee, because India was flooded as a consequence of climate change. They have one child (Konstantin), who is seven years old and has already started compulsory school. Mia is now pregnant again. Her parents, who are about 70, are both living in her house. They worry a bit about their daughter's decision to have such a "risky" partnership with a man from Outside; they would have preferred Mia to have a European partner.

Mia is a university professor and has a very good income, enough for all the family. Daniel cares for the grandparents, even though caring for the grandfather is very time-consuming, so they do not need public services. Because of Daniel's status as a migrant, the family is isolated, since Europeans do not like people from Outside. Mia also has a younger brother, who is working as a soldier protecting European borders. He does not approve of his sister's relationship.

Daniel can go out, but he has no friends. Like all legal migrants who came to Europe before the borders were closed, he is allowed to work in a low-paid job, for example, as a housekeeper or cleaner. Thus, at the beginning of the relationship, Daniel was working casually in the black market as a housekeeper. After Konstantin was born, Daniel and Mia decided that he should care for Konstantin, for two main reasons: first, because Mia is working full-time and secondly, as a kind of strategy, since caring for the child could enhance Daniel's prestige and social recognition. They think that pursuing this strategy will gain them increased acceptance from their neighbours and Mia's parents.

Daniel achieved no social recognition, however, and the plan failed: Europeans are still taking into account the fact that he is a migrant, which counts more than the fact that he is caring for his son.

Konstantin is a special child, compared to other children of his age: he speaks Hindi as his first language, has a greater capacity for reasoning and a higher emotional intelligence and greater skills than his classmates, just because of his family situation and the time his parents spend with him. At school, he is often alone; the other children avoid him, because he is half from Outside, but also because they are a little bit jealous. His teachers also have trouble dealing with him, because they do not have the training. Konstantin is not a very good pupil, because he already knows almost everything that is taught. The school curricula are very rudimentary, because public education does not include more than the necessary minimum, which means there is no education in arts, music or culture. There is no information about the Outside, except that people from there have a worse way of life and could harm Europe.

Mia is forced to choose between giving up her highly paid job or breaking up with Daniel because he is from outside Europe. She quits her job and goes to work full-time in a low-paid job. Because they can no longer afford to care for the grandparents at home, they have to send them to a public home.

The state takes the next step in trying to bring the relationship to an end: they want to take Konstantin under Public State Care Services, because Administrators and Social Assistants are afraid of the negative effect the father could have on his son. They do not trust him – he could be a terrorist or a secret agent.

Mia and Daniel therefore decide to get married as a symbol of their love. Although it is formally permitted, they cannot find anyone to conduct the ceremony; first because he is from Outside and secondly because marriage has lost its importance, so only very few people are permitted to conduct the ceremony.

Then, after a check-up, Mia's gynaecologist predicts the unborn child will be born disabled. In Europe, children with disabilities are seen as not having high productive potential compared to others, so in general should be aborted. But Mia and Daniel want to have this baby, regardless of whether he or she is disabled or not. With these developments and the death of the grandfather in public care, they decide to escape to Asia with the help of Mia's younger brother.

3.7 References

- Beier, L., Hofäcker, D., Marchese, E., Rupp, M. (2010). *Family Stuctures & Family Forms – An Overview of Major Trends and Developments. Working Report.* Bamberg. Available from: *http://hdl.handle.net/2003/27689.*
- Belletti, F., Rebuzzini, L. (2010). *Local Politics – Programmes and Best Practice Models. Working Report.* Milan. Available from: *http://hdl.handle. net/2003/27694.*
- Blaskó, Z., Herche, V. (2010). *Patterns and Trends of Family Management in the European Union. Working Report.* Budapest. Available from: *http://hdl. handle.net/2003/27695.*
- Blum, S., Rille-Pfeiffer, C. (2010). *Major Trends of State Family Policies in Europe. Working Report.* Vienna. Available from: *http://hdl.handle. net/2003/27692.*
- Gordon, T. J., Helmer, O. (1964). *Report on a Long-Range Forecasting Study.* Rand Corporation, Santa Monica/ California. Available from: *http://www. rand.org/pubs/papers/2005/P2982.pdf* [accessed 14.06.2011]
- Kuhlmann, S. (2002). *Foresight and Technology Assessment as Complementing Evaluation Tools.* In: Fahrenkrog G., Polt, W., Rojo, J., Tübke, A., Zinöcker, K. (eds.) *RTD Evaluation Toolbox. Assessing the Socio-Economic Impact of RTD-Policies.* EUR 20382 EN. Sevilla: European Commission, 192-199.
- Kuronen, M. (ed.) (2010). *Research on Families and Family policies in Europe. State of the Art.* Final Work Package 1 report. Jyväskylä. Available from: *http://hdl.handle.net/2003/27686.*
- Kuronen, M., Jokinen, K., Kröger, T. (2010). *Social Care and Social Services. Working Report.* Jyväskylä. Available from: *http://hdl.handle. net/2003/27696.*
- Leccardi, C., Perego, M. (2010). *Family Developmental Processes. Working Report.* Milan. Available from: *http://hdl.handle.net/2003/27690.*
- Livingstone, S., Das, R. (2010). *Media, Communication and Information Technologies in the European Family. Working Report.* London. Available from: *http://hdl.handle.net/2003/27699.*
- Lutz, W., Sanderson, W., Scherbov, S. (2008). *IIASA's 2007 Probabilistic World Population Projections.* IIASA Word Population Program Online Data Base of Results 2008. Available from: *http://www.iiasa.ac.at/Research/ POP/proj07/index.html?sb=5* [accessed 10.3.2011].

- Reiska, E., Saar, E., Viilmann, K. (2010). *Family and Living Environment. Working Report.* Tallinn. Available from: *http://hdl.handle.net/2003/27693.*
- Turoff, M., Linstone, H. A. (eds.) (2002). *The Delphi Method. Techniques and Applications.* Available from: *http://is.njit.edu/pubs/delphibook/index.html* [accessed 10.3.2011].
- Wall, K., Leitão, M., Ramos, V. (2010). *Social Inequality and Diversity of Families. Working Report.* Lisbon. Available from: *http://hdl.handle. net/2003/27698.*

Chapter 4: Research Agenda on Families and Family Wellbeing for Europe

Marina Rupp, Loreen Beier, Anna Dechant & Christian Haag
(with the support of Dirk Hofäcker and Lena Friedrich)

4.1 Introduction

The Research Agenda is based on those elements of FAMILYPLATFORM which emerged from extensive earlier discussion: the state of the art, the critical review and the foresight scenarios. Before outlining the agenda itself, we will summarise the most important societal trends and political challenges.

4.1.1 Main societal trends

There are general and important trends that impact on all family-related fields. These are mutually interdependent.

The first of these are the effects of globalisation combined with individualisation. Globalisation has led to rising plurality and increased demand for flexibility. In addition, globalisation has also resulted in increasing uncertainty (especially in relation to employment and workplace), and a high degree of interconnectedness through new information technologies. These tendencies are in part responsible for the growing gap between those who can deal with the demands of globalisation and those who cannot afford to be mobile or flexible (e.g. families with many children, lone-parents, the elderly, adult children who have care responsibilities for their older parents, and people with low educational attainment levels). This produces new forms of inequality and a higher risk of social exclusion and financial deprivation.

The second big challenge is demographic change. It comprises delayed timing of family formation and decreased fertility rates below the level needed to sustain the population. At the same time, there is societal ageing as a result of higher life expectancy. Both contribute to a changing age-dependency ratio, which in turn has an impact on social security systems.

Other significant developments include increasing levels of educational attainment, and higher labour market participation of women. These are very much connected to demographic development, and also shape gender roles. Thus, for each of the following research areas it is important to bear in mind that there are huge gender disparities.

These major trends impact in a variety of ways on families, leading in particular to rising uncertainty in many areas of life. A high proportion of young people already benefit from long-term education – but the target of the EU is to raise the percentage of tertiary education up to at least 40 per cent (European Commission, 2010). This may lead to longer phases of education, affecting families in different ways: young people struggle to plan long-term, and it is difficult for them to achieve financial independence, so they experience a lack of (material) security. This often leads to less stable relationships and to delayed family formation (Klijzing, 2005; Mills/Blossfeld, 2005). The question here is how to make potential young parents feel secure enough to have children, or to have more children. Today, many young people have to go through several important transitions - especially starting a career and starting a family - in a short period of time. This leads to the so-called rush hour of life, especially for women, who are still the main care-givers.

Difficulties in reconciling work and family are one reason for the lower fertility of highly educated women. As one aim of the Agenda 'Europe 2020' as well as of the 'Lisbon Agenda' is to achieve a higher participation of women in the labour market, measures to ease the burden of work are also required. This is important not only for mothers but also for fathers, who should not only be breadwinners, but also parents with appropriate rights and duties. Family-friendly working conditions, care facilities, and the whole environment (such as housing, infrastructure and basic needs) are important for gender equality – especially as far as the division of paid and unpaid work is concerned.

Demographic change has made care a significant issue for the EU and all Member States. In general, it is the family which has to deal with the growing demand for care, especially care for elderly family members. But because women are spending more time at work, there is less time available for providing adequate care.

Mobility of European workers is a major goal of the European Union. Most migrants, however, come from non-EU countries and often have different cultural backgrounds. Immigration raises the issue of how the migrant population can be successfully integrated into the host society. Integration does not mean assimilation, but rather the acceptance of cultural variety. Thus, in terms of social security, there is a need for integration concepts which ensure the wellbeing of migrant families and diminish social exclusion.

In general, the increasing use of new information technologies and media-related opportunities and risks creates demands for new forms of education and skills, as well as new opportunities and new forms of inequality.

4.1.2 Key recent policy issues

A number of policy responses are required for dealing with the consequences of the societal trends described above. Existing measures may need to be adjusted, but new concepts are also needed. The following suggestions reflect the main topics discussed by FAMILYPLATFORM.

Regarding *care*, it is necessary to consider different areas: childcare and care of elderly or disabled people. How can policies provide appropriate and comprehensive care services and provisions to support families? It is important to take into account the viewpoints of the care-receivers as well as of the care-givers, so as to better integrate different policies influencing care arrangements. For childcare, it is important to create policies that help parents realise their preferred arrangements – with a combination of care provisions, high quality external childcare, leave schemes, adequate working time arrangements, self-determined flexibility in working hours and financial support. It is very important to create these possibilities equally for both women and men (Wall *et al.*, 2010c; Blum/Rille-Pfeiffer, 2010a; Kuronen *et al.*, 2010a). Care of elderly or disabled persons may take place within or outside the family, and both situations need special attention. First, a family leave scheme and remuneration would facilitate care-giving. To relieve family carers, a high quality system of external care, investments in retirement housing and palliative care are needed. It is important to bear in mind that care-receivers are a rather heterogeneous group. Care-givers and their special needs must also be taken into account. Women provide care more often for all relatives and accept that their commitment will mean loss of income, career opportunities and future pensions. Care personnel are to a large degree female, and some have a migration background. Therefore, policies providing social protection for carers (regardless of whether they are family members or external helpers) are necessary (Wall *et al.*, 2010c). Overall, it is necessary to integrate all the different policies that influence care.

Another major political challenge relates to '*doing family*'. As female participation in the labour market has changed and continues to change the effects on how household chores are shared increasingly need to be taken into account. Management of families has become more complicated and ambitious, as less time is spent within the home and different timetables have to be organised. Doing family is connected to how families divide or reconcile paid and unpaid work, and this in turn is related to gender equality, as most unpaid care work is done by women. As gender equality is one of the goals of the European Union, policies should address this problem, for example by means of labour market regulations (e.g. legislation on part-time work, flexible working hours, well-paid leave schemes, life-long learning) and incentives for companies (e.g. promotion of a family-friendly certificate) (Wall *et al.*, 2010c; Blaskó/Herche, 2010a; Kuronen *et al.*, 2010a). It seems to be especially important to encourage men to participate more in unpaid work.

It is very important to bear in mind that due to increased dynamics in family life and increased freedom of choice there is a *growing variety of family forms* besides the so-called standard nuclear family, for example single-parent families, same sex families, step-families, patchwork families, and others. Each of them has special needs and issues. Policies not only have to be responsive to this, but have to respect the different living arrangements and support all of them to avoid inequalities (Beier *et al.*, 2010a; Wall *et al.*, 2010c). Special attention when designing policies has to be drawn to families of minorities – taking into account ethnic background and family size.

This is true for all stages of the life-course *and to all transitions in family life*, so policies have to react to the pace of change, and should facilitate it. Significant transitions in the life-course are those to adulthood and to parenthood (Leccardi/Perego, 2010a). With regard to the former, policies (e.g. education or employment) and institutional settings need to be reconsidered. This is important because the transition to adulthood influences processes of family formation. More policies supporting young adults in starting a family are needed, because the timing of family formation is related to whether a couple decide to have children, and the average number of children they may end up having. In turn this has important effects on demographic change. Higher and longer parental benefits seem to be one good means of opening up possibilities for couples to have more children, and therefore increase fertility rates. Essentially, integrated transition policies are necessary (Stauber, 2010) to ease the transition to parenthood and family formation, while there is a societal demand for higher levels of educational attainment over longer periods of time.

Spatial mobility is an important issue in Europe, as its citizens have the right to move freely from one Member State to another in order to take up employment and settle. Additionally, there is a significant flow of migration from non-EU countries, and there are different forms of migration: long- or short-term migration, within a country or beyond borders, commuting, circular migration, seasonal migration and other forms of movement. Migrants and mobile people are a very heterogeneous group and need differentiated legislation. Up til now, policies have treated people as individuals who are not embedded in a social context. Regardless of whether migration is voluntary or involuntary, questions of integration and tolerance arise. It is obvious that there are differences between various immigrant groups as far as participation in the host society is concerned, for example with regard to the educational attainment of children or social exclusion of the family (Wall *et al.*, 2010a). Policies have to cope with this problem to support the wellbeing of the whole family and especially of children.

Inequality and material deprivation are important issues not only for migrant or mobile families but for all, because there is growing polarisation between families with very low and very high incomes. In particular, child poverty has to be avoided to ensure the wellbeing of children (Wall *et al.*, 2010a). Income deprivation is an important starting-point, but the resulting loss of dignity, lack of suitable housing, education, health services, nutrition and other relevant opportunities in society have to be kept in focus as well. As material deprivation is often 'inherited', policies supporting all generations are essential. Financial help is necessary but not sufficient, as it has to be accompanied by empowerment.

Families are not always a secure place. *Violence* can occur in different forms: psychological, economic, physical and sexual. It is to be found between partners, parents and their dependent children or elderly parents and their adult children. It is often assumed that victims are female, but men are affected as well (Wall *et al.*, 2010a). Therefore, policies have to provide help for all victims, regardless of their gender and age. Common standards should be implemented in all European countries, including the following: violence in marriage or against children should be proscribed, and there should be no sexual obligations within marriage. Additionally, services that provide help, refuge or counselling to victims have to be extended.

Media and new technologies bring opportunities and risks to families. New information and communication technologies, such as the internet, allow people to stay in contact with relatives and friends living far away. On the other hand, they also entail a number of risks. Parents are often ignorant of the dangers or do not know how to protect their children, as they have not grown up with these technologies themselves. Here, help for the parents, more information and family (life) education are necessary. In addition, the question of availability of relevant media has to be discussed. Another aspect related to media is the representation of families and family life and the question of how this affects the attitudes, values and behaviour of (young) people (Livingstone/Das, 2010).

Family (life) education in general is needed to help parents guide and educate their children. Therefore, access to services supporting parents should be ensured, and projects should be promoted in order to empower parents. A sustainable strategy for family education is accompanied by financial support and empowerment of parents in educational and social aspects (Wall *et al.*, 2010c).

Security, uncertainty and social policy are relevant policy issues and closely related to poverty, migration and care. Therefore, it would seem to be important to promote more policy evaluation and benchmarking strategies in order to have a reliable basis for creating new policies and for an international exchange of good practice (Blum/Rille-Pfeiffer, 2010a).

One way of ensuring that all of these topics are properly considered from the viewpoint of families is *family mainstreaming*. With an international plan of action, the family dimension could be integrated into overall policy making. This would lead to reconsideration of all policy fields with regard to how they affect families and the different family members: men, women, children and the elderly, in all stages of the life-course. In addition to scientific data, relevant information for policy making can be brought into the decision-making process by family organisations, which understand the needs and wishes of the families they represent, as well as the feasibility of different political strategies.

4.2 Main research areas and methodological issues

A couple of central thematic areas emerged from the various working packages of FAMILYPLATFORM. These constitute a challenge not only for future European societies but also for family research and family politics. The central thematic areas were care, life-course and transitions (including family forms and structures), doing family (including family roles and gender as well as work-life balance), monitoring and evaluation of social policies, mobility and migration, demographic change, violence, financial deprivation, media and environment, security or insecurity and social policy, family (life) education, family relationships, and minorities.

These points are arranged according to the priorities set out by the members of the Consortium and Advisory Board, and it should be made clear that they overlap somewhat. The following key research areas will be analysed and discussed in greater detail below: *family policies, care, life-course and transitions, doing family*, and *migration and mobility*, with a focus on concrete research questions and subjects. The remaining topics will be discussed more briefly.

Whilst *demographic change* is a key aspect and should always be borne in mind, it is far more the driving force behind future development and, as the corresponding Existential Field report has shown, it is also relatively well researched. The area of *family structures* is included in the *life-course research field*. *Family relationships* are discussed under *doing family*.

Equality of men and women as well as attainment of equal opportunities are central objectives of the European Union. Setting targets addresses a multitude of aspects such as the division of duties within the family, the division of participation in the workplace, social security, burdens, the concept of roles, etc. We have not, however, devoted a separate chapter to this aspect, as this would have led to needless reformulation of the research issues. The issue of equality should, however, be borne in mind in all that follows.

The *wellbeing of children* is an extremely important cross-cutting theme. A central objective of family policy is to ensure that all the lives of young children are full of possibilities. On the whole, conditions for children's wellbeing are transmitted through the family. Family development, structure and resources are therefore always to be seen from the point of view of the wellbeing of children. In this context, a clear change in perspective has developed in recent times: the family is conceived of from the child's point of view, not only nationally but at the EU level too, and measures and regulations are increasingly being oriented towards the wellbeing of children. This aspect should therefore be kept at the forefront of the discussion.

4.3 General methodological remarks

Discussion during FAMILYPLATFORM raised many questions and demands on the topics and methods for future research. Due to their crosscutting characteristics and in order to avoid redundancies, a short overview of existing statistics at the EU level, general methodological issues, and data requirements are first outlined.

Official European statistical data available to family researchers

The main advantage in using official European statistics is that data provided by Eurostat are harmonised, representative and comparable throughout the entire Union as far as possible. The current legal framework enables access to anonymous Eurostat microdata for scientific purposes. Family researchers have access to the following resources at the European level, including Eurostat:

- EU-LFS (Labour Force Survey);
- EU-SILC (European Union Statistics on Income and Living Conditions);
- ECHP (European Community Household Panel), running from 1994-2001;
- Eurobarometer Surveys.

SHARE (Survey of Health, Ageing and Retirement in Europe), ISSP (International Social Survey Programme), GGP/GGS (Generations and Gender Programme/Generations and Gender Survey), EVS (European Values Study) and ESS (European Social Survey) are other examples which will not be further discussed.

The European Union Labour Force Survey (EU-LFS) is a quarterly, large-sample survey providing results for the population in private households in the EU, the European Free Trade Association (EFTA) and EU Candidate Countries. The survey is carried out more or less at the same time, using the same questionnaire, common classifications, and a single method of recording in all countries. In the context of family research, some of the core variables, particularly the EU-LFS 2005 *ad-hoc* module "Reconciliation between work and family life", are of special interest.

European Union Statistics on Income and Living Conditions (EU-SILC) is an instrument which aims to collect timely and comparable cross-sectional and longitudinal multidimensional microdata on income, poverty, social exclusion and living conditions. The EU15, Estonia, Norway and Iceland data were first collected in 2004; the ten new Member States (with the exception of Estonia) started in 2005; and the instrument was implemented in Bulgaria, Romania, Turkey and Switzerland from 2007. Social exclusion, housing condition information and some data on income are collected at the household level, while labour, education, health, and income information are collected at a very detailed level. Answers to questions on individuals' satisfaction with life are also obtained.

Each Standard Eurobarometer consists of approximately 1,000 face-to-face interviews per Member State, except Germany (2,000), Luxembourg (500), United Kingdom (1,300 including 300 in Northern Ireland). In addition, Special Eurobarometer extensively addresses special topics, such as family issues or gender roles. Flash Eurobarometers are *ad hoc* thematic telephone interviews that enable the Commission to obtain results relatively quickly on focussed and specific target groups. The qualitative Eurobarometer studies investigate the in-depth motivations, feelings and reactions of selected social groups towards a given subject or concept by listening and analysing their way of expressing themselves in discussion groups, or with non-directive interviews. For example, in 2010 a qualitative Eurobarometer study survey on children's rights was conducted. The study was carried out amongst young people in all 27 Member States of the EU and consisted of 170 focus groups.

From the point of view of family researchers, official European statistical data suffer from the drawback that they ignore family relations and partnerships that extend beyond the household. Thus the data provided at the comparable EU level is not profound and differentiated enough to give answers to many research questions developed by FAMILYPLATFORM (e.g. legal family relationships, rare family forms, etc.). Information is particularly sparse for more recent Member States of the EU. Research in the new Member States should therefore be encouraged, and candidate countries need to be better integrated, in order to avoid a similar situation in the future. Current all-encompassing research should be extended and deepened. The most important point in any research is to assess the advantages and disadvantages of each methodological approach.

New topics and common indicators in basic data

A wider range of basic statistics at the European level is needed for all families, but particularly for rarer family forms, where basic data about their size and socio-demographic background is lacking. While some family forms like lone-parents, married couples with children and consensual unions with children can be differentiated, important data on aspects such as biological, legal and social parenthood are not collected. Moreover, the number of same sex couples in official European surveys is usually too small for detailed analysis, thus larger samples (e.g. through oversampling) are needed to obtain valid information. Information is also lacking with regard to the development of family forms and transitions between them in respect of national and cultural differences. Here, one suggestion has been to oversample rare family forms in national and international surveys. This is necessary if advanced methods are to be applied and if the most important research questions are to be answered, for example finding out what kinds of inequalities different family forms face. To achieve these aims, it is also necessary to overcome the prevalent household concept and to collect data at the individual, family and network level, in order to get more insight into relationships and support networks. We need possibilities of systematically analysing differences according to qualifications, social class and regional structure, and it is therefore necessary to develop common indicators. Basic statistics at an EU level should cover all age groups sufficiently in order to ensure that all cohorts are analysed. This is important because it enables conclusions to be drawn on the process of (social) change across the generations and over time.

To give an overview of migration flows, we need data to be collected at the EU level, not solely at the national level. A new institution monitoring European migration processes would be very helpful. This might identify different kinds of movements as well as trends in trans-nationalism and their background. These studies will need to be repeated over time with the same respondents to discover developments and trends, identify the effects of policies at the national and EU level, and differentiate between the effects of age, cohorts and time (e.g. different decades). Information has to be sufficiently detailed to allow differentiation at sub-national or regional levels.

At present, the level of information on these issues differs significantly between countries. The use of common indicators should therefore be made a precondition for the funding of studies at an EU or national level.

As mentioned above, it is necessary to discuss existing indicators and find new ones to describe the situations of families and countries more precisely, addressing aspects such as wellbeing, financial deprivation, education, and different forms of inequality – not only income-based indicators or GDP (for suggestions see Stiglitz *et al.*, 2009); these have to be differentiated for family forms and family networks. Additionally, existing typologies of welfare regimes have to be reconsidered, especially with respect to new Member States and Candidate Countries. So we need a pool of advanced method-ological approaches that is seen as a common standard. To realise these aims, it would be helpful to have an institution providing them; this might be achieved by extending the remit of Eurostat.

Each member of a family has their own position, roles, relationships to others, and perspective on the family system. Research should cover every aspect, especially when asking people what their needs and demands are and what kind of support would be helpful. Adequate and differentiated indicators of wellbeing are needed in order to describe the reality of families more precisely, for example to describe dissolution, the family as a network, or intergenerational relationships (Fleurbaey, 2008).

Methodological approaches

Current research is mainly static. As the dynamics of family life increase, lack of information on development processes becomes more of a problem. Hence the need for more adequate measurements increases as well. There is a need for scientific institutions at the national and EU levels to cope with the challenge of accelerated change. However, it is clear that it is not possible to cover all research questions in this way. Ideally, strategies would contain large-scale representative data-sets accompanied by smaller in-depth studies. In order to improve our general understanding of the family, and to get an insight into its dynamics, we need more differentiated qualitative research, for example, by addressing transition processes in family life. This research should include the positions of both men and women, take into account the viewpoint of children, and focus on decisions and their causes, using a process-related approach.

Possible examples of the above are qualitative studies in specific regions, for instance those with high or low fertility. Retrospective narrative inquiries, which differentiate between male and female points of view, and between couples and single people would contribute to a better understanding of postponement. Longitudinal studies addressing transitions and their effects, for example the impact of divorce on children's wellbeing, or the effects of different models of parenting, would help to gain insight into important family-life processes and their impacts (Thomson/Holland, 2003; Thomson *et al.*, 2003; Smith, 2003). Longitudinal studies should also be carried out in order to accompany migrants in their immigration and integration processes.

To get deeper insight into these fields, it is necessary to decide on the appropriate qualitative methods and limit different scientific approaches (Denzin/Lincoln, 2007; Flinders/Mills, 1993). Exploratory studies would sharpen our perceptions of rare family forms and their living conditions, as well as of changes caused by transition, with a special focus on the children's perspective (Neale/Flowerdew, 2003). In addition, the need for and usefulness of longitudinal studies should be properly discussed: on the one hand they are ideal for understanding processes and their causes and impacts, but on the other the costs are high, and it takes longer to obtain useful results. There is therefore a risk that the results of such studies may not reflect social and political changes.

The extent to which it may be possible to engage in secondary-level analysis in order to make greater use of qualitative and especially qualitative longitudinal data should be further promoted and applied. Some European countries (e.g. Austria, Denmark and Spain) already have qualitative data archives. Even if secondary analysis of qualitative material is uncommon, it can nevertheless be a good way of obtaining answers to research questions using existing in-depth research, rather than generating new data (for a critical discussion see Gillies/Edwards, 2005; Kelder, 2005; Thorne, 1999).

Target groups

Discussions in FAMILYPLATFORM reiterated the need to incorporate children's points of view, which is indeed often missing in sociological research. We must therefore expand the scope of our work to incorporate research done by other disciplines, especially psychology and pedagogy, so that it embraces a broader range of topics and also addresses younger children (Langsted, 2002) and more thematic dimensions. It is important to improve our understanding of the living conditions of children today, and their wishes and experiences. This would help us to design better care solutions and improve societal and legal frameworks and social systems, thereby contributing to the wellbeing of children and their families. Adolescents and the elderly are additional target groups which should be researched to a far greater extent than they are today, and with more innovative approaches (Steele *et al.*, 2007).

Research on social innovations also needs to be improved. For many of the challenges discussed, we still have few ideas about how to solve them

– for example, what future care arrangements will look like. Here we have to search for new models and to accompany them with research. This point leads to the call for more evaluation of demonstration projects in many areas: e.g. work-life balance, care arrangements and custody arrangements. We also need scientific monitoring of innovative projects in empowerment and family education.

To sum up, there is a significant need for more advanced research methods. This means creating new indicators that afford better insight into the wellbeing of families in Europe, in line with the suggestions already made in the Stiglitz Report (2009). There is a need for greater variety and creativity in research methods, and for light to be shed on under-researched areas, where it is hard to get even basic information, such as violence or in research questions which have rarely been looked into up to now, such as family empowerment. Here we need to find ways of gaining access to particular target groups such as victims, perpetrators or minorities, and to decide on what types of research we will carry out. How can new media, like the internet, be used for research and what alternatives do we have for different kinds of inquiries? What other indicators can we use to draw conclusions, for example court files and medical records?

4.4 Family policies

Right from the start of FAMILYPLATFORM, family policies were seen as an important research area, and they are increasingly seen as a major policy field in many European countries (Blum/Rille-Pfeiffer, 2010b). In this context the main question is *"what governments do and why"* (Blum/Rille-Pfeiffer, 2010a: 66). To answer this, it is necessary to get deeper insight into policy development and decision-making processes and the norms and rules governing them, and to assess the influence of those who develop policy and make decisions including researchers and NGOs. To improve family policies, European countries should learn from each other, regarding the outcomes of different strategies applied in the many different frameworks which exist at national and sub-national levels.

The demands of family policy research are a major challenge, because of the enormous heterogeneity of the European countries, in the same way that family policies vary in their degree of institutionalisation (Blum/Rille-Pfeiffer, 2010b; Bahle/Maucher, 1998). Additionally, they address a great variety of topics and aspects which are handled not only within so-called family policies but also in different policy fields (economy, education, etc.). In this context, we can see how policy decisions made in different policy fields have unintended impacts on families. Family policies in Europe have different normative backgrounds and historical developments alongside different models, ideals and cultures of support (Mühling, 2009). Thus, the first requirement of research is to provide an overview. The second is for it to evaluate and test different family policies throughout Europe. And thirdly, it has to incorporate the views of families and their representatives.

Welfare regimes

In the social sciences, efforts often are made to classify countries according to their different social systems. Classification reduces complexity, provides a better overview and a clearer structure and generally makes it easier to explain differences to non-scientists. It can also help to detect impacts of different policy strategies and to find similar problems and possibly solutions which can be transferred from one country to another.

Well-known attempts to classify countries according to significant aspects of their welfare regimes are those of Esping-Andersen (1990) or Lewis and Ostner (1994). Esping-Andersen (1990) analysed the social protection of citizens against market risks and social stratification. He concluded that there were three types of welfare regimes: liberal, social-democratic and conservative. This simplistic classification failed to adequately describe the situations in all European countries. It was broadened by introducing two additional types: the Southern European cluster (Ferrera, 1996) and the Eastern European 'post-socialist' countries (Hofmeister *et al.*, 2006). The Esping-Andersen classification has been practically tested, well established and frequently confirmed for different policy fields. Critical points are that it is highly policy-centred, does not take gender inequality and family aspects into account properly and does not consider social change adequately.

Other efforts to classify countries have been attempted, for example the Family policy typology of Lewis and Ostner (1994). Here, assignment depends on the family model and the division of labour between the spouses. This classification ranges from a strong male-breadwinner to a weak male-breadwinner model. This classification is more sensitive to family aspects, but it does not fit for the plurality of family forms. Additionally, there is little systematic reference to data and it was only applied to a limited number of countries. As these two examples show, it is difficult to combine different effects of policies and social-economic backgrounds into a small number of categories. In fact, it is nearly impossible to include all countries into such a classification because social change is taking place too rapidly and it is as yet unknown where it will end (e.g. Eastern European countries). There are risks of over-simplifying realities in order to place all countries into a scheme and then overlooking special cases as a result; country-typology outliers have to be taken into account. Furthermore, classifications are neither case-sensitive nor field-sensitive, and often ignore the issue of time-dependency. Thus demands for a classificatory system have to be carefully balanced with the risks of imposing such a system on cases where it does not fit.

Monitoring of European family policies

Against this background, a basic challenge is to provide an overview of family-related frameworks, laws and rules throughout European nations. Some promising steps towards such an overview have been taken (e.g. the European Observatory on National Family Policies and the Observatory on the Social Situation and Demography), but they have led neither to ongoing monitoring nor to a visible trend of conversion of national politics. Because of time lags and conceptual differences between these initiatives, they are of only partial use to us. But we should use the existing work as far as possible when moving forwards in order to get an idea of what has been done in the past. At present the main source of information on family policies is the MISSOC tables, covering all areas of social protection in EU Member States. While the 'Council of Europe Family Policy Database' and 'OECD Family Database' address important aspects, they do not include all EU Member States. With regard to regional aspects, welfare systems and family policies in several Member States are significantly under-researched. This is especially true of the new Member States, but also of some older Member States such as Denmark, Ireland and Portugal (Blum/Rille-Pfeiffer, 2010a: 62).

To understand how family policy structures impact on family policies, the first step is to monitor – at every level. It is therefore necessary to build up a reliable overview of existing mechanisms and measures relating to family policies. Firstly, research has to be carried out on all members of EU27, to obtain an overview of their present status with regard to family policies. Ideally, this should also cover Candidate States. Secondly, we should find out what each Member State's intentions are and how they relate to EU goals. Thirdly, we should look at how cultural background factors such as attitudes and norms influence the development of national policies. There are different ideals of family (e.g. nuclear vs. plural family) and different traditions in dealing with family-related political issues, e.g. pro-natalism, gender equality objectives, and to reflect the impacts of different models of motherhood and fatherhood. Future research has to take into account the fact that relevant policy strategies (e.g. provision of childcare facilities) should not be located at a national level but at communal or regional level. For this reason greater differentiation will be required.

With regard to different social security schemes in EU countries, we have to reconsider existing typologies, and analyse different types of normative background with regard to different institutional frameworks in the EU (for example, whether or not there is a government department responsible for family affairs, or whether responsibility is spread across various departments).

As political strategies change at different rates, comparative and longitudinal studies should examine the differing effects of stable and changing family policy regimes. In this context there is a need to understand how different institutional family policy structures (e.g. whether or not there is a specialised government department) influence the outcome of policies.

Monitoring does not simply mean summarising existing policy strategies. All relevant policy areas have to be taken into account, and discrepancies between different policy fields need to be analysed. Because family affairs touch upon every area of politics and society, other relevant policy fields - for example health and occupational policies - have to be taken into account too. It is a major challenge to define what measures should be researched and where the inner circle of family policies ends. Therefore, we need a common definition of what constitutes family policy (Bahle/Maucher, 1998). To the extent that many policy measures concern the management of families and their resources, a broad view of the political framework is required, including employment and educational policies and the organisation of welfare systems.

To measure the impact of different policies more precisely, consensual criteria (common indicators on family forms, relationships, financial deprivation and education) and new categories of political interventions and mechanisms are needed. This would in turn enable the comparison of means and their effects. As comparisons are mostly made with macro-quantitative methods, there is a need for additional research of a smaller and qualitative design. This can be used to sharpen "the view for historical development" (Bahle/Pfenning, 2000:3), as well as to examine specific details. Comparative evaluation and testing can highlight the differing impacts of various policy strategies, for example tax or cash benefits, different leave schemes or care provisions.

A special form of monitoring is provided by calculation models. These show the diverse regulations (e.g. remuneration replacements such as parental pay) and make it possible to record their effects on various family constellations. Calculation models help to assess the effects of changing measures, for example, on the material situation, thereby uncovering related structures of inequality, as well as any unintended effects. Models of this kind have already been used at national level, for example, on the consequences of increased tax allowances for children, and they could be introduced in a similar way at the EU level. This would enable us to monitor which measures are advantageous to particular family forms and which are not (for example, standard marital status tax relief as opposed to individual taxation), and thereby to study their effects on social inequality. These models could also be used to evaluate to what extent people adjust their behaviour to different measures; it is necessary to focus not only on material effects, but also on equal opportunities and gender equality effects.

In order to make family policies more sustainable, policies should take future trends into account. Research in this field is already being done in some spheres, for example demographic development. Existing approaches to modelling future trends could be extended to other research fields by doing more surveys on future perspectives.

Evaluation of policies

Evaluation of policies has been called for in almost every political field and research area, as well as at every level. It is true that we have very few evaluations of national policies and even fewer at the EU level. Even existing scientific knowledge and empirical data only rarely finds its way into legislation, especially regarding the outcome of political bargaining processes. We know some reasons for these trends:

- Information is often lacking or unreliable, and often does not cover representative groups, areas, or nations;
- Results, findings or interpretations vary according to different theoretical, methodological, cultural, normative or regional backgrounds;
- Results which are 'bought', i.e. where public authorities such as ministries fund or finance research.

Major problems arise not only because of a lack of data and from differences of interpretation, but also because of the different ways in which policy makers handle available information. When calling for more research and more evaluation, the priority must be dealing with this problem. If this is not dealt with, evaluation work will have no impact on policies.

The first step to more effective evaluation is to carefully select the persons or organisations entrusted with the work and to define how policy makers deal with the results. Regular exchange of information and establishment of mixed institutions consisting of researchers, stakeholders and politicians would help to remove barriers such as different languages and to ensure transfer of knowledge.

The second step is to think about the longer term (Wall et al., 2010c): people seldom wait for changes in legislation or support to make their plans and choices (e.g. family formation). They often react with a time lag; some-times they make a small change in order to obtain some additional advan-tages and avoid disadvantages or because they are unsure about a new situation. Additionally, we have to face the fact that people - and especially younger people - do not ask what the concrete rights and legal outcomes of their decisions will be. For example, only a few people study the legal implications of marriage and divorce before they become engaged. When we ask for evaluation, we have to ask for prior knowledge of rights and enti-tlements. And we have to ask whether policies help to make the citizens concerned better informed or not. Furthermore, people sometimes do not realise which areas of life will be affected by a change of policy – for example, allowances and custody after divorce. Evaluation of policy has to take into account how much people know and how long it will take before people react on disseminated information. Hence, it is clear that some changes in policies have a longer-term impact, while others influence people's behav-iours quite quickly. Evaluation has to take these aspects properly into account and to explain why these differences arise. To clarify the impacts of family policies, changes in other political fields have to be kept in mind, as the impacts of policies in one field could be diminished or thwarted by those in other political domains. To deal with this problem, cross-cutting effects should therefore be examined with care.

Thirdly, when an evaluation of political mechanisms and their (longitu-dinal) effects is sought, policies have to be valid for a longer period. This is also true for associated political fields. Serious evaluation must be able to rely on a stable legal framework for the necessary duration of its studies, and it is important to start the evaluation process before a new measure comes into force to have the possibility of identifying the effects of the new legislation.

Fourthly, and as mentioned above, evaluation has to take into account the variety of social situations (family forms and phases, social groups, etc.) because outcomes and scientific recommendations will vary accordingly.

Fifthly, regular benchmarking of 'family-friendliness' indices should be introduced at a European level (Blum/Rille-Pfeiffer, 2010b) in order to demonstrate how family-friendly different nations or regions are.

Because calls for evaluation come up against these major challenges, they often seem to struggle with reality. But if all relevant actors are aware of the problems, possible solutions should be easier to find: these could take the form of smaller steps, such as concentrating on small target groups or narrower policy areas, learning by doing, and finding better ways to interact.

Before evaluations are carried out, decisions are required (in the light of policy priorities) on what type of evaluation is preferred[1]:

[1]   On formative and summative evaluation see Sager, 2009; Wholey, 1996; Chambers, 1994.

- *Formative evaluation* involves guiding new or renewed policies (or strategies). It might be carried out on a small or large demonstration project, and might relate to one or more special social groups. All such aspects, intentions, target groups and expected outcomes have to be explained and recorded in detail. The next step is to choose the methodological method(s) best suited to the research question. This includes who (persons, groups, etc.) should be covered by investigation or other methods. Preliminary indicators of positive effects have to be discussed with specific reference to previously defined objectives. A group of relevant actors may then decide to implement changes in the project (or not), and the next evaluation cycle begins. The process of formative evaluation can be carried out several times over a fixed period, at the end of which final conclusions can be drawn and implemented. Formative evaluation allows us to react fairly quickly, despite the risk of over- or underestimating effects because of short observation periods. It is much more appropriate for smaller, limited strategies, rather than for broader policies.
- *Summative evaluation* tests outputs. This kind of evaluation examines stated policy objectives and tries to find measurements which tell us whether the objectives have been reached and what other effects have been observed. One significant problem in family policy is that objectives may not be very precisely defined, and ways of fulfilling those objectives are not always clear. What is to be done, for example, to provide support to children in large families? Give those families more money, lower their taxes, or provide free access to education or care facilities? Outcomes will vary according to the kind of support we choose to provide. Lower taxation, for example, might not be feltsubjectively at all, or there is no subjective relation to the number of children in a family. More prior research is required, as mentioned above, to understand the possible effects on different social groups. This example also highlights the fact that research and policy making have to interact from the outset in order to develop a precise outline of intended effects and determine what measures should be implemented in order to achieve them. This might be one way to get reliable information. Time frames also have to be taken into account. Summative evaluation is time-consuming, because it needs to keep track of possible effects. This is a disadvantage, because policy is not able to react quickly to unintended effects (Weiss, 1999). The benefit of this method is that the results are clearer and more reliable.

Listening to families and their representatives

In addition to these research strategies, it is also necessary to listen regularly to the voices of the persons affected by policies, including family members and their representatives, in order to be able to bring their needs and wishes into the process of policy formation. This could be done by means of direct representative data collection, but again, every family member should be taken into account.

Another way of structuring policies in a family-friendly way is to bring in experts and representatives from family associations from different backgrounds into the policy making process. Several methods can be used to incorporate expert or specialist knowledge, for example, Delphi and group discussions, as well as qualitative interviews. In general, knowledge transfer between research and policy making should be improved through continual exchange. Thus, there is a need to understand how different types of family organisations can contribute to policy making and there is a lack of knowledge about the relationships between government and NGOs (Blum/Rille-Pfeiffer, 2010a).

Means and models of participation at all political levels (communal, regional, national and EU) need to be developed, especially for the inclusion of family associations. In some fields, such as family education, politicians and organisations often work hand in hand. In others we find a lack of participation. So we need greater insights into how to achieve effective participation and ensure that families are heard. Research is needed on how to organise such processes and devise methods of gathering the knowledge of people working for families on a day-to-day basis. One way of doing this might be to explore those fields where there already is effective participation (for example, at the communal level), in addition to finding and testing new methods of participation.

While family policy should rely on forecasting research, it should not be limited to "the power of the factual" (Schubert/Blum, 2010 in: Blum/Rille-Pfeiffer, 2010a: 64); the feasibility of policies has to be researched too. Effective consultation on policy would be fostered by a "platform between politics and research, which builds on a sustainable basis and a bottom-up, pluralist approach" (Blum/Rille-Pfeiffer, 2010a: 64).

Family education

Due to the already mentioned societal changes, over the past few years, family education has become an important field, both for policy making and for NGOs. Family education is planned and organised at various levels (Rupp, 2003), sometimes as part of national policy, but more often as part of local policies. As a result, many different institutions are involved. It is difficult to get an overview of education strategies and activities, but extremely important that we do so. At the same time the requirements on parents in relation to the upbringing and education of children have greatly increased (Rupp *et al.*, 2010). One example is the importance of promoting children's school performance. A fact that has to be recognised here is that families demonstrate a large degree of diversity. Individual family biographies differ, particularly in terms of their educational background and (financial) resources (Wissenschaftlicher Beirat für Familienfragen, 2005). One central as well as action-oriented concern is what kinds of support each family needs. This may depend on their specific setting or the transition they are in, and how they can make best use of this support.

Up until now there has been little empirical evidence or data on the suitability of support strategies, and their acceptance by specific types of families. It is crucial to include the family-specific, demand-oriented point of view derived from a sensitive approach when developing criteria and contents for family information. This means that initial exploratory studies should be carried out to evaluate the necessary differentiation among the population. Thereafter, standardised measures can be used to obtain data from a larger sample.

4.5 Care

Care emerged as the topic of greatest concern amongst participants of FAMILYPLATFORM. Care relations involve different actors from within families and from external care providers. They have recipients with a wide range of individual needs and resources, and are influenced by different regulations and policy schemes. Care is usually seen as practical help for the frail elderly or for the upbringing of children, but it is necessary to broaden this view. It can also be regarded as general assistance, as in providing an environment to live in and grow, as well as the place where general wellbeing is fostered. Therefore, it is important to extend the focus to adjunct areas, beyond obvious physical care activities.

Changing norms and role models have led to different understandings of family, work, and life responsibilities. Global developments continue to influence families, which have to cope with high uncertainty on labour markets, and adjustments to national economies in response to global crises. Additionally, due to changes in family formation and the increasing diversity of family forms, the traditional nuclear family is no longer the only family model in Europe. All these aspects have profound effects when considering how to best to manage care needs.

Against this background, an evaluation of existing care arrangements for the EU27 in a comparative study would be a first step for the Research Agenda. The goal here should be to extend existing knowledge (e.g. Anttonen/Sipilä, 1996; Bettio/Plantenga, 2004) in a comparative design for all Member States. In fact, there is increasing demand for home care (Kuronen *et al.*, 2010b), yet each of the Member States supports a different model. It is necessary to evaluate to what extent welfare states push families towards providing care and to what extent they support them, e.g. providing services, insurance or allowances. A distinction between childcare, elderly care, assistance for those who are (temporarily) ill or otherwise in need of assistance, or caring for persons with disabilities must be made, as each group has special needs and resources. For a better understanding of the various aspects of care, it is necessary to consider the differentiation between care recipients and care providers.

Once the basic knowledge on care arrangements and care schemes throughout Europe has been gathered, its use in comparative studies may provide information for successful future plans as well as the optimisation of existing care schemes of EU countries. Such a comparison would have to consider the various aspects and crosscutting topics that are directly and indirectly linked to the actual provision of care, e.g. the gender gap or changes in relationships. As the ideal care situation is likely to differ from the practical solution, considering the circumstances and resources at hand, we need insights into decision-making processes, with special emphasis on the impact that state policies can exert on families to provide care. It is also necessary to investigate what support families need to fulfil these duties.

Childcare provisions vary between Member States, as do care arrangements, especially with regard to the extent of institutionalised care offered, used and accepted. Private childcare is mostly provided by mothers, although fathers are (very) slowly taking on a greater role here, thus effects on income, career and social security, therefore, are mainly on the female side. Further research on social systems and cultural background is required, using a life-course approach. It is important to focus on how women and men can manage employment, childrearing and care, and still find time for themselves for personal development and rest.

Gender is a factor that needs special attention. This is due to the greater involvement of women in family work and rising female employment (Kuronen *et al.*, 2010b), as women have a greater burden of obligations which need to be negotiated. Given this gender imbalance, the question arises of why there is such a great lack of male care-givers, in the private as well as in the institutional sector, and how this situation can be changed. One factor might be changing the image and prestige of care work so as to achieve higher levels of gender equality.

In this context, employment is a major dimension, as other obligations often have to be negotiated in accordance with the specific employment arrangements in each family. The reconcilability of care and paid work depends on the flexibility of employers and of the social security system in general. It is necessary to think about the value to be placed on care provided by family members who give up paid work, or reduce their working hours to care for dependent relatives.

The family network plays an important role in the provision of care (Kuronen *et al.*, 2010b), but the societal changes mentioned above have also affected the availability of care. Future studies need to take family networks into consideration, not only as a resource for care-giving, but also as a general background variable determining the environment of any given family.

The viewpoint of care recipients is an aspect of major importance that must not be neglected. It is essential to consider care ethics, to make the intrusion into privacy as acceptable as possible and to respect care recipients' wishes. Therefore, a major point of interest should lie in the wishes, views and needs of the different groups of care-recipients in order to hear their voices. Qualitative and quantitative research into different groups of care recipients is required.

It should be determined, for children at various different ages, how satisfied they are with their care arrangements and whether they prefer alternatives. This is information which has not been gathered on a comparative level before, and must be differentiated according to social backgrounds and types of care arrangements against the backdrop of the social systems.

Due to the increasing number of frail elderly people, care deficits are likely to develop (Hoffer, 2010). In order to ensure an adequate mix of types of care provision, the elderly should be asked what kinds of care relations they prefer. This also includes investigating decision-making processes and their considerations for care-givers. People with disabilities often need long-term assistance. The focus should therefore be on a lifespan approach.

Longitudinal measures are particularly important in connection with children and care recipients who are ill or disabled, in order to capture the effects of care relations for the future course of their lives. For the elderly, a longitudinal setting would allow for a comprehensive view of the last phase of life, in order to create sustainable support. Only with profound knowledge of the process of physical and mental deterioration and related care needs can policies be devised to support both care recipients and family members who are also care providers.

At a later stage, exploring innovations in care could give us valuable information on how to reshape care relations within families and in co-ordination with families and professional care providers. This is true for elderly care as well as for other care solutions, particularly childcare. Among new forms of care relations, migrant workers (who are most prominent in care for the elderly) pose new challenges. Many questions surrounding the legal status of migrant carers, the affordability of care services in general and the quality of the care provided remain unanswered, including who cares for the families of carers.

Furthermore, there is a need for information on the extent to which technological innovations are of assistance to care-givers and how far care recipients can regain independence through the use of technological appliances or innovative programmes like home assistance networks.

The last major area for future research on care is future political strategies for care arrangements in general. Based on the knowledge of desired care relations, policies can be adjusted to remove obstacles and support care-givers. At the same time, the financial and economic considerations of providers of care need to be taken into account. Policy-making needs to recognise the specific environments of families, which are likely to differ not only from state to state but also according to social classes and groups.

4.6 Life-course and transitions

Family life changes over the life-course. Needs and interests are therefore not stable but shifting. Although the life-course approach (Mayer, 1987; Elder, 1978) has become more important in the social sciences, there is a lack of research that uses it. At the EU level we find comparative data mostly at the individual or household level (e.g. Eurostat, Eurobarometer, EU-SILC, SHARE and GGS).

Transitions in life-course and in family life have become more difficult, and some have become more frequent (Leccardi/Perego, 2010b). For example, the divorce rate in the whole of the EU has risen markedly in the last decade (Beier et al., 2010a), and forms of transition to parenthood have become more diverse. Whether this takes place with or without a marriage certificate, alone or as a couple, as natural or social parents makes a difference in the context of the demands placed on starting a family, social security and possible risks, as well as with regard to intergenerational relationships (Stauber, 2010). The greater diversity of family life also raises the issue of inequality of opportunities, especially for children. Research on families' wellbeing should accordingly follow the life-course and focus on these more diverse forms of transition.

Transition to parenthood

Some data on the transition to parenthood is available for Europe as a whole, for example, the age of the first-born child, the desire for children and attitudes to childcare and employment (Leccardi/Perego, 2010a). However, little is known about the interplay between the development of these patterns and policy measures (Philipov *et al.*, 2009; Gauthier, 2007). There is also a lack of longitudinal studies on (potentially) relevant factors and observations on trends and changes (Stauber, 2010). An up-to-date confirmation and evaluation of existing studies is essential if measures are to be taken to encourage couples to start a family and obstacles are to be removed. Then again, differences between countries and the differing cultural and socio-political backgrounds must be taken into account (Blum/Rille-Pfeiffer, 2010a). In order to carry out such research properly, a longitudinal design involving both partners would be ideal, but it would be easier to ask representative couples who are at the end of their fertile phase questions about their decision-making process.

Scientific research into decisions on family formation and the resultant (different) family forms is necessary to assess the impact of national social policies and attitudinal trends, as well as to compare the various measures in place in Europe. To achieve this, survey data relating to the various target groups is needed, ideally for all European Member States. Additionally, qualitative research is needed to get deeper insight into motivations, attitudes and the anticipated impact of parenthood, as well as the interaction of the partners.

Dissolution, separation, divorce and reorganisation

The decrease in the stability of relationships is a major cause of family dynamics and multiplicity of present family forms (Leccardi/Perego, 2010a; Beier *et al.*, 2010a). The available basic data is not sufficient to identify the different legal relationships between parents and children or the number of preceding marriages. FAMILYPLATFORM stresses the need for in-depth studies which go beyond existing basic data into the field of separation and divorce (Wall *et al.*, 2010c). With regard to research into causes, the main factors in partnership break-ups and the role of the children in this context have to be identified, as well as differences between the sexes concerning expectations, reasons and decision-making processes. Additionally, research has to look for differences arising from different normative, cultural and ethnic backgrounds, as well as from different legal frameworks.

Another requirement is to develop intervention studies in order to generate ways of stabilising family relationships. The wellbeing of children is the relevant focus in this context. Care and custody arrangements and particularly their impact on parent-child relationships have to be researched in detail and also from the children's point of view. A very important question is the development of family relationships after separation and how and when children can be involved in decision-making processes. These questions can be approached by interviewing experts and asking older children. The material situation of post-divorce families and their development over time, in the context of the applicable legal frameworks, are also relevant topics.

Variety of family forms

The increased variety of family forms is based on greater tolerance of non-traditional family forms in most EU Member States (Beier *et al.*, 2010a, Wall *et al.*, 2010c), but the actual number of, for example, cohabiting couples with children, varies significantly between European countries (Beier *et al.*, 2010a) according to their legal status in each of them. Research should therefore focus on family forms in situations where their legal status may be expected to place them at a disadvantage, examining in particular the effects of such situations on the wellbeing of the children.

Step- and patchwork families are complex entities whose specific structural demands stand out against the background of a lack of common family history (Beier *et al.*, 2010a). We know little about how members of such families deal with this situation and how difficult it is for them not only to establish a family identity but also to be involved in extended family networks. Specific research could help to explain what family members go through and what forms of support they need. Of great importance here is which legal forms of relationships are involved (and possible), particularly between parents and their stepchildren, because these have lasting consequences for the security of the children.

The variety of family forms implies different support needs. Thus we have to obtain more information about the living conditions of pair-headed-, single-parents-, homosexual-, teen-mothers, patchwork and migrant-families, married and unmarried parents as well as families from ethnic and religious minorities. As mentioned above, it would be helpful to oversample these family forms in common surveys in order to have reliable data for differentiation and comparison.

Family phases

Demands on the family change according to the age of the children in it. This also means there are changes in parental tasks and the resources they need (Van Dongen, 2009). Hitherto, research and most family policy measures have paid insufficient attention to these facts. We need to learn more about the shifting challenges of parenting, variations in the division of labour within the family and between family and professional employment. In this context, sources of instability in the phases of family development should also be taken into account.

For example, the scope for employment of mothers usually increases with the age of the children (OECD, 2010a; Pettit/Hock, 2009). This means that it is even more difficult to deal with the need for support which occurs later on in life – for example with regard to problems at school. A specific study to deal with these transitions could contribute to providing institutions and family policy measures with the necessary resources, thereby making family life easier and more attractive. Comparison of the different institutional frameworks of European nations is needed as background information. In this context, we need to ask what the perceived needs of parents are. Can they take leave and organise their working lives accordingly, and what factors prevent them from taking advantage of these opportunities?

Transition to large families

Although there is an intimate connection between demographic development and the reduction in the size of 'large' families (Beier *et al.*, 2010a), research has focused little on this question. The point is to examine which mechanisms, considerations and attitudes played a role with regard to the decision to have (or not have) a large family (Eggen/Rupp, 2006). The existence of different gender roles and parenthood concepts need to be taken into account. Research is needed to determine the influence of these concepts and of families' financial and social circumstances on fertility decisions. A fascinating question, for example, is how young people experience growing up in a large family and what consequences this experience has on their own family ideals. Furthermore, Eurobarometer could design a survey on the acceptance of large families in different European societies or Member States. Such a survey could provide key information on whether and how the image of these families could be improved. The knowledge of representatives of large families should also be incorporated, for example through expert surveys.

Families and relationships of the elderly and the transition to the fourth age

Establishing and structuring of relationships has become significantly more important for older generations. Alongside questions of how the elderly find a partner and establish a relationship, it is also important (from the point of view of sociology of the family) to understand how intergenerational relationships develop as a result. Qualitative and quantitative data are needed for a better insight of these aspects.

With regard to the later stages of life, questions concerning the needs of the elderly, and what resources are available to them, have grown in importance. The Survey for Health, Ageing and Retirement in Europe (SHARE) shows that children are prepared to step in for their parents (Schmidt *et al.*, 2009). Questions arise, however, as to whether and to what extent they are able to do so, and what arrangements will be chosen and are affordable.

4.7 'Doing family'

Managing daily life within the family in modern societies has become a major task and creative challenge. Even though the division of work between members of the family is no longer (fully) predetermined according to gender roles, the European objective of gender equality is still a long way off (European Commission, 2009). Therefore, doing family focuses on the processes of managing daily life within the family. This becomes more important as more women and mothers take part in employment. Doing family touches not only on the aim of gender equality, but also on the higher quota of women's labour force participation. Doing family means matching different demands from various parts of society with their own processes. If we take leisure activities, educational pursuits (e.g. music lessons) and further social duties (e.g. looking after elderly family members) into consideration, it becomes clear that family members are involved in many different tasks, and that they also follow diverse routines which are not easily harmonised.

As a rule, women have much more responsibility for management of household and care tasks than men (Blaskó/Herche, 2010a). Even though the fulfilment of gender roles is vital for the personal gender identity of the partners (Bielby/Bielby, 1989), it can lead to dissatisfaction, overload, conflicts and frustration (Baxter, 2000). A satisfying arrangement is important for the stability of partnerships and therefore directly and indirectly for the growth and development of the children as well.

The status of previous research in this area is very heterogenic: the participation of men and women and fathers and mothers in the workplace is relatively well documented and up-to-date (e.g. European Commission, 2011). There is however a significant lack of research on the distribution of labour within the family (Blaskó/Herche, 2010b). Firstly, the available data is not up-to-date at the EU level, which is especially critical given the labour market trends mentioned above. Secondly, there is no comparable information for all Member States of the EU27. Thirdly, differing concepts and measures of unpaid work, household labour and childcare make comparisons difficult. Fourthly, not much of the data has been obtained using the more labour-intensive diary method (Bonke, 2005). On the whole, the available information is based on estimated time spent, which in the current discussion over methodology is seen to be less valid than the diary method. Often task participation rankings or data on responsibility are used. These make it difficult to obtain differentiated results, particularly according to age, number of children, and family form.

Another problem arises from different approaches to and concepts of measuring unpaid work, housework and childcare. Various task areas need to be identified and defined empirically. Professional work including overtime, time taken in commuting to work, training, participating in special events, etc. must be differentiated and recorded precisely. In order to be able to work out how different tasks and duties are to be reconciled (or not), it is also necessary to clarify at what times each task and activity is performed during the day (Schulz/Grunow, 2007). However, it is necessary to check which methods of data collection are appropriate on a significantly broader base. For this we need specialist research into the relative strengths of the various concepts governing the collection of data on the distribution of household tasks and investment of time within the family. To date, the diary method has provided us with concrete records, and beside that estimated time units and task participation indexes are available and used more often. Additionally, we find that tasks are categorised differently and different tasks were collected. There is accordingly an urgent need for unity and comparability. The amount of time invested in childcare is barely recorded, and is therefore difficult to compare. One exception worth mentioning here is the harmonised European time use studies, which use a diary method and allow free reporting of activities in five main dimensions. More concepts need to be developed in order to differentiate between the diverse tasks according to the ages and the needs of children. Furthermore, it should be noted that tasks in the household and childcare are often carried out at the same time. This situation must be borne in mind in order to record possible signs of overloading factors on the one hand, and to identify gender-related differences on the other. It is also necessary to capture the subjective experience of overload and stress in a differentiated way. Further methodical development is vital here too, and is seen as an interdisciplinary task.

The question of how unpaid work in general can be sufficiently differenti-ated poses a further challenge. Advanced research methods will be required for a better appreciation of the importance of these tasks. This brings us to the question of how current social security systems influence gender roles and the distribution of labour. The views of men and fathers in particular need to be explored to gain a better understanding of how work is currently divided.

Changes in gender roles have made reconciling work and family a key issue. It is also of immense importance for the wellbeing of family members. There are significant differences between countries in support levels in this area (e.g. regulations on parental leave and opportunities for part-time work; OECD, 2010b; OECD, 2010c) and it is not easy to make a general evaluation of the situation. There is a need for further comparative research on atti-tudes to work-life balance and concrete measures to achieve it on the part of employers. This data would have to be examined alongside particular state regulations in a more concrete way. It would also have to be confronted with the practical concerns of parents, as identified in EU-wide surveys.

Doing family changes over the life-course of the family, and the internal distribution of labour has to be adjusted accordingly. It is often forgotten that new challenges and structures arise when children get older, and that it is not only the early years which are demanding for parents. For this reason the life-course perspective needs to be kept in mind during research on this area.

The change in gender roles derives mainly from the attitudes and the ability of women to structure their own lives (Lück, 2009). But what about changes on the male side? In the end, a family-centred reorganisation of labour would be a potential solution to the work-life balance problem. The fact that research on fathers is still in its early stages means that it urgently needs to be further developed. In this area, research is faced with numerous challenges. Alongside the need for Europe-wide collection of data on the attitudes, practices and perceived restrictions of fathers, it is necessary to take into account innovative models of work-life balance (e.g. support at the company or public sector level, but also in the form of socio-political conditions). At the same time, we also have to study the mother's viewpoint in detail. With respect to workplaces, how can flex-ibility in living arrangements be achieved, and be positively evaluated and structured according to individual needs? In addition to an overview of best practice models, surveys of employers concerning their attitudes and prejudices are important.

Comparable basic data should be collected in all European countries and standardised as far as possible. Ideally, this should be carried out on a regular basis. In-depth and explorative studies should be made available which differentiate according to socio-cultural and socio-economic background, as well as for different selected regions (e.g. strongly traditional or modern-egalitarian areas).

Children's points of view

The distribution of labour within the family from the child's point of view is another important topic for future research. Here there are two main questions: one concerns the contribution of children to managing everyday life, and the other concerns children's desired solutions. This should be studied Europe-wide and be differentiated according to ethnic, regional and socio-economic conditions.

Here, it is a question of which tasks children carry out at which age and to what extent, and which areas of responsibility they are entrusted with. Answers could be arrived at by implementing quantitative studies of parents and children which are accompanied by qualitative approaches. It is more difficult to clarify the question of which psycho-social responsibilities children are burdened with. To address these topics, it is presumably necessary to start with exploratory studies, in order to probe the field and further develop the questions which need to be asked, as well as the methods for evaluating them (for example by following the example of research into divorce).

Family relationships

With regard to family relationships, a number of different aspects need to be addressed: first there is the relationship between parents, as well as between parents and their children, and with respect to the age of the children, we have to differentiate between infants and adults. Addressing parent-child relationships, we mostly focus on the specific responsibility of parents to ensure children's wellbeing. Hence, we need deeper insights into the nature, structures and formation of relationships and their vulnerability. This can only be derived from interdisciplinary studies of longer duration, and innovative approaches. Even though there has been research into evidence of attachment between children and parents, there is still a need for more longitudinal studies that give a deeper insight into the effects of parental absence and how to solve this problem (e.g. fostering or adoption). Another basic theme is parental absence for a longer time (e.g. single-parent families). Rare forms of parenting, like foster and adoptive families, also have to be taken into account. Additionally, we need information to understand how parents and children could be assisted in these situations, and their long-term wellbeing ensured.

Another relevant aspect is the relationships between adult children and their elderly parents. Despite the fact that real support is very important for the wellbeing of all members of the (wider) family, we need to see emotional ties as being far more relevant. Another point of interest is the relationship to and interaction between grandparents and grandchildren, both of which are shaped by rising employment rates and mobility. Research on emotional relations as well as care giving, education and financial support is therefore important.

4.8 Migration and mobility

Mobility and migration are research areas of increasing importance. Defined as a permanent change of residence or at least change of residence for a relatively long duration, *migration flows* have to be differentiated from *mobility*. An important characteristic is that the centre of the migrant's life shifts from one place to another. We refer to migration as international migration if national borders are crossed during the migration process. Additionally, we have to distinguish between migration flows amongst European Member States and migration flows from third country nationals into the European Union, as there are huge legislative differences between them (Wall *et al.*, 2010a).

Many OECD and EU countries have tried to attract highly qualified people from abroad as the economy becomes increasingly knowledge-intense and needs more human capital than their own educational system can provide. They try to achieve a 'brain-gain' strategy, which from the perspective of the country of origin is often a 'brain drain'. Inequality between countries is produced if migration flows only in one direction. This problem is aggravated by policies which encourage and retain only those with high potential to immigrate. It is often forgotten that they may have families in the country of origin, and migrating is unattractive for them. Hence, it is also often assumed that those who return to their countries of origin are individuals on their own, but they may find a partner and start a family in the country of destination and not want to go back, as this would affect their family life. In both cases, more research into the effects of brain drain/ brain gain on social inequality and family life is necessary. Comparisons of data on immigration flows are difficult, however, as there are different measurements and concepts in the various different countries, e.g. of what an immigrant is (Dumont/Lemaître, 2008; Kupiszewska/Nowok, 2008). The 'regulation on Community statistics on migration and international protection' (EC Regulation No 862/2007) (European Union, 2007) is a step towards the harmonisation of measures, but the data collected is not sufficient and detailed enough to answer most of the important research questions.

Against this background of increasing migration, it is essential to analyse social and demographic data on the extent and structure of immigration processes into and within the EU, as well as the origin, destination and motivation of migrants. At a European level, it is unknown if certain intra-European migration systems can be observed. On the basis of an institutionalised and harmonised reporting system on intra-European migration, it would be possible to number and describe different mobile groups and analyse their staying or returning behaviour.

Research on migration of third country nationals is closely linked to their integration processes. Irrespective of the differences in existing definitions, the concept of integration is mainly understood as the process of incorporating new populations into the existing social structures of the host society. From the point of view of sociology, integration means a long-lasting, dynamic and complex process neutralising social exclusion or separation. Often the goal is for immigrants to take part in all societal spheres of the host society, to the same extent as native inhabitants. They are normally expected to learn the local language and to know and behave according to the law. According to Heckmann (2001), the receiving society has to meet certain conditions if it is to provide a good foundation for this process. Assimilation, diversity or exclusion are potential alternatives to integration. Research on the latter focuses on the individual level; there are gaps especially with regard to seeing the family as a research object, but also with regard to subjective aspects of integration.

Migrant families and their special needs have not been an explicit object of many European research studies so far (Wall *et al.*, 2010a). Therefore, it is unknown if migrant's family structures (e.g. family forms or intergenerational relationships) resemble those of their home country or those of the receiving country. Furthermore, it is not known how these structures change with the migrant generations (and the aspect of socialisation in dissimilar societies for different generations) if there are differences between migrant groups or between EU countries. Bi-national families seem to be even less researched, though their potential cultural differences may be important for family life and family decisions. A Europe-wide survey is required to research these aspects. At the very least, the largest migrant groups in each European country should be interviewed. A panel study design would allow the investigation of changes over time. This would be especially helpful for comparing the different needs of newly immigrated families, as well as their needs after they have lived in a country for a longer period of time and are more integrated into the receiving society.

Besides migrant families as explicit research objects, the impact of family resources as intervening factors in successful integration have so far not been studied in detail. Existing studies are mostly quantitative. For the future, it is important to focus on the causal relationships between the various factors and processes of integration. These aspects could best be covered by qualitative identification of family resources which promote or hinder integration processes of the individual family members. Financial, emotional or moral support could be essential factors for their wellbeing. Little is known about hindering factors, for example, ethnic or religious attitudes, and the underlying mechanisms here have to be studied in order to be able to react to them.

From the receiving society's point of view, questions about the acceptance of migrants and differences from non-national group to group need to be answered. It is necessary to analyse how immigrants from different countries of origin are seen in the 27 Member States, and how this is affected by various factors like the proportion of non-nationals, media representation or historical aspects such as colonialism. This knowledge would help to improve the image of non-natives and to create a helpful climate for integration.

For people migrating from one European Member State to another, education is a very important factor motivating mobility. Higher education in particular is a strong motivational factor here. The European Union supports international lifelong learning through several programmes for different age groups: Comenius for students at school, Erasmus for higher education, Leonardo da Vinci for vocational education and training, and Grundtvig for adult education. There is no study comparing who is mobile for educational reasons in different European countries. The answer to this question is needed to assess the impact of this kind of mobility on societal inequality and whether different EU grants are able to minimise potential social differences which may affect opportunities of going abroad for education. When analysing this, it is important to take into account that there are already differences between European countries with regard to the participation of different social, ethnic or gender groups in educational systems. Thus the question arises whether or not these inequalities can be mitigated by EU programmes.

Integration is just one concept used to describe the relationship between migrant minorities and the receiving society. Another - relatively new - concept is transnationalism: migration processes, especially international labour migration, are regarded as a normal component of the life-course in a globalised world. These go along with the evolution of transnational systems, which build a specific culture which includes aspects from the country of origin and from the destination country. It is unknown if migrant groups living a transnational way of life are to be identified, and if they are found, how their families are affected by this way of living. Research projects should begin with analyses of migration processes to establish which countries are strongly connected by international circular migration. These insights would help to determine whether it is possible to identify a transnational system, and how this could be characterised if there is one. On this basis, migrants should be characterised according to their socio-economic status, family situation and subjective feeling of belonging.

Mobility, especially work-related mobility, is another topic of great importance for family life. Spatial mobility and its impact on families have been studied to some extent (Schneider/Meil, 2008; Schneider/Collet, 2010), but there is a lack of research on the importance of mobility for the occupational careers of employees from a European comparative perspective. This is especially true with regard to the effects of this kind of mobility on the careers of spouses and variations in this domain between European countries. These questions could be studied by analysing existing national longitudinal data sets on individual occupational careers. In addition, it would be interesting to investigate their impacts on partnership and family life, especially with regard to family formation and relationship stability. As appropriate and comparable data sets are not available for all European Member States, new data should be collected to answer these questions.

4.9 Inequalities and insecurities

From the outset, inequality and family diversity were major topics of importance to FAMILYPLATFORM. And because inequality is one of the so-called cross-cutting aspects, some particularly important themes should be mentioned here.

Inequality and financial deprivation

It was shown during the discussions that inequality has many facets, with material deprivation being a major concern. The risks to children are significantly higher than for the average of all Europeans (Wall *et al.*, 2010a). More information is needed in order to improve our social systems and the wellbeing of families and their children. As with many other research topics, a need for more cross-national comparative basic data has been called for, using more than just 'income' indicators.

Research that focuses on income-poverty falls short in (at least) two dimensions (Wall *et al.*, 2010a): it helps in identifying the poor, but it fails to give information about the process and the experiences of people who are at risk or belong to poor groups. Firstly, families could suffer from one of many forms of deprivation, for example, educational deprivation, illiteracy, lack of social acceptance, etc. This disconnects them and their children from societal participation in very important areas. Secondly, it diminishes their chances for future development and success, particularly for children. Major efforts should be devoted to modelling new measurements and scales in order to strengthen these 'weak' indicators of social deprivation. To go ahead with these ideas, more explorative and qualitative studies are necessary. Thirdly, we need studies which are able to model movements in and out of poverty, while taking several important dimensions into account (see above), and their impact on these developments.

Violence

There are several significant aspects to family violence and violence in the wider social environment. Violence *per se* is defined in different ways across different nations and cultures (OECD, 2010d), and varying types of sanctions accompany these definitions. Closely related are variations in the individual risk of experiencing violence. The probability that men, women, children, elderly, disabled people, as well as people in institutional or private care and individuals from certain social groups or living in specific areas, etc. will be victims of violent acts varies (see Wall *et al.*, 2010b, Hagemann-White *et al.*, 2008).

Because family violence is often a taboo subject, it is rather difficult to gain access to research this; it is estimated that many cases go unreported. Hence, the demands for research, especially for comparative research, are high. First of all, a common and standardised definition of violence must be developed in order to analyse basic data from public sources, police statistics, case statistics from courts and district attorneys' offices, as well as information from institutions for victim counselling. It is very likely that this definition needs to be gendered (see Wall *et al.*, 2010b). Efforts to generate knowledge have been undertaken in social science research on violence, as well as in the struggle against violence, and particularly violence against women. Thus attempts have been made to conduct representative surveys (e.g. in Germany: GIG-NET, for Violence against men see BMFSFJ, 2004) or to draw up first findings at a European level (see Eurobarometer, 2010).

The next research challenge is access to victims and perpetrators of violence, in order to estimate the prevalence of certain types of violence affecting not only children and men but also the elderly and the disabled, which are particularly difficult target groups. A sensitive approach is needed, and new methods must be sought to investigate abuse in care relationships. Qualitative methods should be used to shed light on the relationship between victims and perpetrators. In terms of potential offenders and areas which involve a higher risk of violence, research gaps and differences also have to be dealt with. The aim is to make progress on risk assessment and prevention. However, research into the impact of violence on health, both physical and mental, should be included and incorporated in this type of work. Information on the experience of being a victim and being a perpetrator would also be valuable (see Martinez *et al.*, 2007).

For comparable studies, common methods and indicators must be developed and adopted when designing studies in the area of domestic violence. Advanced forms of measurement, comparative methods and a consensus on such standards, in accordance with a common definition of violence, are crucial. There is a pressing need for more exact descriptions of the situation of specific target groups (e.g. family members or people with different social backgrounds, ethnic minorities), and these could be achieved by more qualitative analysis (see Hagemann-White *et al.*, 2008; Martinez *et al.*, 2007).

It is particularly important to improve our knowledge and methods with regard to children as victims of violence and also regarding child offenders. Appropriate age-specific measures and indirect indicators to assess abuse, negligence and psychological violence need to be developed (e.g. studies of case statistics from the courts and police statistics).

Minorities

There were frequent calls within FAMILYPLATFORM to intensify research on certain social groups. This is particularly the case for ethnic or national minorities, e.g. the Roma, who are most numerous in Romania, Spain, the former territory of Yugoslavia, Bulgaria and Hungary[2], the Basques in France and Spain, or the Sorbs in Germany (Malloy, 2005; European Commission, 2004). These groups often suffer from social and financial inequalities, including a lower standing in society and fewer opportunities in the labour market, as well as a higher risk of unemployment (Turton, 1999). In this context, it would be important to evaluate how successful different national policies have been in integrating these groups into society as equal citizens by simultaneously recognising their background and traditions. In doing so, it would be interesting to see how different political strategies affect their wellbeing and how they see their role in society.

In addition to national minorities, special family constellations are also seen as marginal groups, e.g. homosexual parents (Rupp, 2009). Research should focus on their living conditions and on whether they are treated fairly and equally and given social recognition. In this respect, it was

---

[2] The Roma are not seen as national minorities in all the countries mentioned.

suggested that the relationship between policy and practice should be evaluated. However, there is insufficient information on the potential differences between policies on homosexual and heterosexual couples and their consequences on family life, particularly for the status of the children.

Living environment and housing

The living environment has a considerable impact on the wellbeing of families, as do access to nature, the quality of the neighbourhood, and housing. "Good living conditions [...] are not equally distributed across Europe or within different social groups" (Reiska et al., 2010a: 84). Poorer people, foreign minorities or people from rural areas tend to have more problems with low quality housing. Thus, inequalities concerning the quality of life reflect the economic and social situation (Reiska et al., 2010b; Bolte et al., 2009). It would be important to know if a 'bad' living environment has a greater or lesser negative impact on future opportunities than it does on living conditions, such as family situation, financial situation or health in general (Reiska et al., 2010a).

One important connection to the living environment in the context of social inequalities is the process of transition to parenthood and its possible influence on childbearing decisions. Furthermore, it is important to know if and in which ways families have worse chances of living in adequate environment compared to people living without children.

Adequate data is only available with regard to housing, which is covered by several larger databases (with the exception of the distribution of homeless people or those living in emergency shelters). However, detailed research studies do not cover the whole EU. Moreover, there is no statistical information on safety and crime at the EU level – though this is available for OECD countries. In this regard, it was also suggested that the different related categories of "housing, neighbourhood and closer natural environment as a whole" should be taken into account (Reiska et al., 2010a: 85).

In other respects, the most useful data available is that covering the EU15 Member States. Data for new and future Member States is rather sparse, so it was suggested that candidate states' data should be included, in order to avoid this problem in the future. Furthermore, there is a need for pertinent, comprehensive, comparable and country-specific research with regard to detailed projects on living environments and neighbourhood. For this purpose, it is important to try to harmonise conceptual definitions, particularly bearing in mind the rather subjective nature of the residents' satisfaction (Reiska et al., 2010a).

4.10 Media

Some form of media, such as TV or the telephone, is available in almost every family and is closely connected with the organisation of everyday life. The use of media structures family life, places demands on resources, and increases demand for new skills which are not common to everybody (Livingstone/Das, 2010). So the availability of media causes great social differences and contributes to social inequality, and it also has effects on family management and relationships, a topic which is a field in itself.

In general, there is a deficit in comparative research on the use of media in families. There is a need for more in-depth data, so that scientists may draw conclusions and provide policy-makers with appropriate recommendations for action. Such data needs to be collected for differentiable age groups and has to be specific enough to show up differences in media literacy and consumption between social classes, ethnicities and different cultures. Research must cover not only media consumption but also the environment in which this takes place - enabling and hindering (particular forms of) consumption - as well as the entire media diets of families and each of their members.

Research has to distinguish between two types of influence: the first is how trends in media development and dissemination shape family life and behaviour. Thus, we should examine the development of communication, the frequency with which it is used by family members (and others), information flows, and risks combined with the new opportunities. Looking at the flow of communication from the other direction, it is essential to understand which trends in family life influence the development and demand for (particular forms of) media. Here the question arises of which social groups are at the forefront of new trends and which families are excluded.

In general, future research in the broad area of the media needs to pool questions and outcomes of a greater range of disciplines such as psychology, sociology and communication sciences, in order to learn from the knowledge already gained and to specify future research questions.

4.11 Summary

Central societal trends are globalisation, demographic change, developments in gender roles and the processes of education and employment, as well as increasing multiplicity and dynamics in family life. They lead to new demands on the framework of families and family policy. In order to achieve such political aims as sustainable growth and gender equality, it is necessary to recognise the relationship between policy measures, societal conditions and decisions taken at both the individual and the family level. This is essential to remove obstacles, as well as to be able to provide the necessary support. The present Research Agenda is one step in the direction of making this knowledge available, because it outlines future research needs and poses central questions.

The family is a very broad subject, and the Research Agenda concentrates on central aspects that were discussed by FAMILYPLATFORM. It is organised into key research areas and provides essential methodological advice for future research. The report draws mainly from the European level, as detailed insights into the status of research have already been provided in the description of the Existential Fields[3]. Because of the huge size of the topic of family, and factors which influence it, decisions had to be made which narrowed down the content of the Research Agenda. The main areas of research, worked out by the members of the Consortium of the FAMILYPLATFORM in conjunction with the Advisory Board, and discussed at length, were as follows: the monitoring and evaluation of policy measures and strategies; the area of care; family studies, which are oriented to the life-course and various family forms; the area of doing family; and the challenges which occur as a result of migration and mobility. Many other themes were discussed, and are included here as research areas in shortened form. The roadmap for future family research in Europe is divided into five main areas and a number of subsidiary areas, including violence, insecurity, deprivation, environment, media, family education and minorities.

In summary, it can be said that more and better differentiated official statistics are required. Furthermore, for particular subjects, like transitions within the family biography, it is necessary to set up longitudinal studies. In order to gain deeper insight into motivation and decision-making processes, qualitative and innovative methods are required. It would be generally advisable to establish mixed methods to assess the complex areas of research in order to pool various sources of information (e.g. initial surveys, secondary analyses, expert interviews, case study analyses, etc.). New media (e.g. the internet) and methods of research and access to the target groups were demanded in various areas, for example, with regard to the study of violence, as well as the media themselves.

A further demand concerned the need for advanced indicators, able to adequately measure material situation, 'wellbeing', and for quantifying unpaid work. Principally, the implementation of common and standardised indicators in Europe-wide research is as essential as the inclusion of all the Member States and the expansion of research to include candidate countries. It would behelpful to establish a co-ordinating body which drives

---

[3] See *http://www.familyplatform.eu.*

this development forward and monitors compliance with these standards. Although a lot of research into the family has been and is being done, there is still a great deal to be achieved. This is especially true with regard to the further development of research into family policy in Europe.

4.12 References

- Anttonen, A. & Sipilä, J. (1996). *European Social Care Services: Is It Possible To Identify Models?* Journal of European Social Policy 6 (2): 87-100. DOI: 10.1177/095892879600600201.
- Bahle, T. & Maucher, M. (1998). *Developing a family policy database for Europe. Working Paper.* Mannheim: Mannheimer Zentrum für Europäische Sozialforschung. Available from: *http://www.mzes.uni-mannheim.de/publications/wp/wp1-27.pdf* [accessed 17.02.2011].
- Bahle, T. & Pfenning, A. (2000). *Introduction.* In: Pfenning, A. & Bahle, T. (eds.) *Families and Family Policies in Europe: Comparative Perspectives.* Frankfurt am Main: Peter Lang, 1-11.
- Baxter, J. (2000). *The Joys and Justice of Housework.* Sociology 34 (4): 609-631. DOI:10.1177/S0038038500000389.
- Beier, L., Hofäcker, D., Marchese, E. & Rupp, M. (2010a). *FAMILYPLATFORM Existential Field 1: Family Structures & Family Forms – An Overview of Major Trends and Developments. Working Report.* Available from: *http://hdl.handle.net/2003/27689.*
- Bettio, F. & Plantenga, J. (2004). *Comparing care regimes in Europe.* Feminist Economics 10 (1): 85-113. DOI: 10.1080/1354570042000198245.
- Bielby, W. & Bielby, D. (1989). *Family ties: Balancing commitments to work and family in dual-earner households.* American Sociological Review 54 (5): 776-789.
- Blaskó, S. & Herche, V. (2010a). *FAMILYPLATFORM Existential Field 5: Patterns and Trends of Family Management in the European Union. Working Report.* Available from: *http://hdl.handle.net/2003/27695.*
- Blaskó, S. & Herche, V. (2010b). *FAMILYPLATFORM Existential Field 5: Patterns and Trends of Family Management in the European Union. Working Report – Summary.* Available from: *http://www.familyplatform. eu/en/doc/91/EF5_Summary_Family_Management.pdf* [accessed 10.02.2011].
- Blum, S. & Rille-Pfeiffer, C. (2010a). *FAMILYPLATFORM Existential Field 3: Major Trends of State Family Policies in Europe. Working Report.* Available from: *http://hdl.handle.net/2003/27692.*
- Blum, S. & Rille-Pfeiffer, C. (2010b). *FAMILYPLATFORM Existential Field 3: Major Trends of State Family Policies in Europe. Working Report – Summary.*

Available from: *http://www.familyplatform.eu/en/doc/95/EF3_Summary_ State_Family_Policies.pdf* [accessed 10.02.2011].

- BMFSFJ (Federal Ministry for Family Affairs, Seniors Citizens, Women and Youth) (2004). *Violence against men. Men's experiences of interpersonal violence in Germany – Results of the pilot study.* Available from: *http:// www.bmfsfj.de/RedaktionBMFSFJ/Broschuerenstelle/Pdf-Anlagen/M_C3_ A4nnerstudie-englisch-Gewalt-gegen-M_C3_A4nner,property=pdf,bereich =bmfsfj,sprache=en,rwb=true.pdf* [accessed 18.02.2011].
- Bolte, G., Tamburlini, G. & Kohlhuber, M. (2010). *Environmental inequalities among children in Europe – evaluation of scientific evidence and policy implications.* European Journal of Public Health 20 (1): 14-20. DOI: 10.1093/eurpub/ckp213. Available from: *http://eurpub.oxfordjournals.org/ content/20/1/14.full.pdf?ijkey=AlTQSRrwXiDzkcL&keytype=ref* [accessed 17.02.2011].
- Bonke, J. (2005). *Paid Work and Unpaid Work: Diary Information Versus Questionnaire Information.* Social Indicators Research 70 (3): 349-368. DOI: 10.1007/s11205-004-1547-6.
- Chambers, F. (1994). *Removing confusion about formative and summative evaluation: Purpose vs. time.* Evaluation and Program Planning 17 (1): 9-12. DOI: 10.1016/0149-7189(94)90017-5.
- Denzin, N. & Lincoln, Y. (eds.) (2007). *The SAGE handbook of qualitative research.* Thousand Oaks: SAGE.
- Dumont, J.-C. & Lemaître, G. (2008). *Counting foreign-born and expatriates in OECD countries: a new perspective.* In: Raymer, J. & Willekens, F. (eds.) *International Migration in Europe. Data, Models and Estimates.* Chichester: John Wiley & Sons, 11-40.
- Eggen, B. & Rupp, M. (eds.) (2006). *Kinderreiche Familien.* Wiesbaden: VS Verlag.
- Elder, G. H. Jr. (1978). *Family History and the Life Course.* In: Hareven, T. (ed.) *Transitions. The Family and the Life Course in Historical Perspective.* New York: Academic Press, 17-64.
- Eurobarometer (2010). *Domestic Violence against Women – Report. Special Eurobarometer 344.* Available from: *http://ec.europa.eu/public_opinion/ archives/ebs/ebs_344_en.pdf* [accessed 18.02.2011].
- European Commission (2004). *The Situation of Roma in an Enlarged European Union.* Luxembourg: Office for Official Publications of the European Communities. Available from: *http://www.equalrightstrust. org/ertdocumentbank/Situation%20of%20Roma%20Enlarged%20EU.pdf* [accessed 16.12.2010].
- European Commission (2009). *Report from the Commission to the Council, the European Parliament, the European Economic and Social Committee*

and the Committee of the Regions. *Equality between women and men – 2010. COM (2009) 694 final/SEC (2009) 1706*. Brussels. Available from: *http://eur-lex.europa.eu/LexUriServ/LexUriServ. do?uri=COM:2009:0694:FIN:EN:PDF* [accessed 14.02.2011].

• European Commission (2010*). Europe 2020. A European strategy for smart, sustainable and inclusive growth*. Brussels. Available from: *http://europa. eu/press_room/pdf/complet_en_barroso___007_-_europe_2020_-_en_ version.pdf* [accessed 04.01.2011].

• European Commission (2011). *European economic statistics*. Luxembourg: Publications Office of the European Union. Available from: *http://epp. eurostat.ec.europa.eu/cache/ITY_OFFPUB/KS-GK-10-001/EN/KS-GK-10- 001-EN.PDF* [accessed 23.02.2011].

• European Union (2007). *Regulation (EC) No 862/2007 of the European Parliament and of the Council of 11 July 2007 on Community statistics on migration and international protection and repealing Council Regulation (EEC) No 311/76 on the compilation of statistics on foreign workers*, 11 July 2007, No 862/2007. Available from: *http://www.unhcr.org/refworld/ docid/48abd548d.html* [accessed 17.03.2011].

• Fleurbaey, M. (2008). *Individual well-being and social welfare: Notes on the theory*. Available from: *http://www.stiglitz-sen-fitoussi.fr/documents/ individual_well-being_and_social_welfare.pdf* [accessed 14.02.2011].

• Flinders, D., & Mills, G. (eds.) (1993). *Theory and concepts in qualitative research. Perspectives from the field*. New York: Teachers College Press.

• Gauthier, A. (2007). *The impact of family policies on fertility in industrialized countries: a review of the literature*. Population Research and Policy Review 26 (3): 323-346. DOI: 10-1007/s11113-007-9033-x.

• Gillies, V. & Edwards, R. (2005). *Secondary Analysis in Exploring Family and Social Change: Addressing the Issue of Context*. Forum Qualitative Sozialforschung/Forum: Qualitative Social Research 6 (1): Art. 44. URN: urn:nbn:de:0114-fqs0501444. Available from: *http://www.qualitative- research.net/index.php/fqs/article/view/500/1077* [accessed 30.03.2010].

• Hagemann-White, C., Gloor, D., Hanmer, J., Hearn, J., Humphreys, C., Kelly, L., Logar, R., Martinez, M., May-Chahal, C., Novikova, I., Pringle, K., Puchert, R. & Schrottle, M. (2008). *Gendering Human Rights Violations: The case of interpersonal violence: Final Report 2004-2007. Project Report. CAHRV.* Brussels, Belgium. Available from: *http://www.cahrv.uni-osnabrueck.de/ reddot/CAHRV_final_report_-_complete_version_for_WEB.pdf* [accessed 10.02.2011].

• Heckmann, F. (2001). *Integrationsforschung aus europäischer Perspektive*. Zeitschrift für Bevölkerungswissenschaft 26 (3/4): 341-356.

• Hoffer, H. (2010). *Irreguläre Arbeitsmigration in der Pflege: Rechtliche und politische Argumente für das notwendige Ende einer politischen Grauzone.* In: Scheiwe, K. & Krawietz, J. (eds.) *Transnationale Sorgearbeit. Rechtliche Rahmenbedingungen und gesellschaftliche Praxis.* Wiesbaden: VS Verlag für Sozialwissenschaften.

- Kelder, J.-A. (2005). *Using Someone Else's Data: Problems, Pragmatics and Provisions.* Forum Qualitative Sozialforschung/Forum: Qualitative Social Research 6 (1): Art. 39. URN: urn:nbn:de:0114-fqs0501396. Available from: *http://www.qualitative-research.net/index.php/fqs/article/view/501/1079* [accessed 30.03.2010].
- Klijzing, E. (2005). *Globalisation and the early life course. A description of selected economic and demographic trends.* In: Blossfeld, H.-P., Klijzing, E., Mills, M. & Kurz, K. (eds.) *Globalisation, Uncertainty and Youth in Society.* London/New York: Routledge: 25-50.
- Kupiszewska, D. & Nowok, B. (2008). *Comparability of statistics on international migration flows in the European Union.* In: Raymer, J. & Willekens, F. (eds.) *International Migration in Europe. Data, Models and Estimates.* Chichester: John Wiley & Sons, 41-71.
- Kuronen, M., Jokinen, K. & Kröger, T. (2010a). *FAMILYPLATFORM Existential Field 6: Social Care and Social Services. Working Report.* Available from: *http://hdl.handle.net/2003/27696.*
- Kuronen, M., Jokinen, K. & Kröger, T. (2010b). *FAMILYPLATFORM Existential Field 6: Social Care and Social Services. Working Report – Summary.* Available from: *http://www.familyplatform.eu/en/doc/92/EF6_Summary_ Social_Care%26Social_Services.pdf* [accessed 10.02.2011].
- Langsted, O. (2002). *Looking at Quality from the Child's Perspective.* In: Ferguson, I., Lavalette, M. & Mooney, G. (eds.) *Rethinking Welfare. A critical perspective.* London: SAGE, 28-42.
- Leccardi, C. & Perego, M. (2010a). *FAMILYPLATFORM Existential Field 2: Family Developmental Processes. Working Report.* Available from: *http://hdl. handle.net/2003/27690.*
- Leccardi, C. & Perego, M. (2010b). *FAMILYPLATFORM Existential Field 2: Family Developmental Processes. Working Report – Summary.* Available from: *http://www.familyplatform.eu/en/doc/88/EF2_Summary_Family_ Developmental_Processes.pdf* [accessed 10.02.2011].
- Livingstone, S. & Das, R. (2010). *FAMILYPLATFORM Existential Field 8: Media, Communication and Information Technologies in the European Family. Working Report.* Available from: *http://hdl.handle.net/2003/27699.*
- Lück, D. (2009). *Der zögernde Abschied von Patriarchat. Der Wandel von Geschlechterrollen im internationalen Vergleich.* Berlin: edition sigma.
- Malloy, T. (2005). *National Minority Rights in Europe.* Oxford: Oxford University Press.

- Martinez, M., Schröttle, M., Condon, S., Springer-Kremser, M., May-Chahal, C., Penhale, B., Lenz, H.-J. & Brzank, P., Jaspard, M., Piispa, M., Reingardiene, J. & Hagemann-White, C. (2007). *Perspectives and standards for good practice in data collection on interpersonal violence at European level.* Available from: *http://www.cahrv.uni-osnabrueck.de/reddot/FINAL_ REPORT__29-10-2007_.pdf* [accessed 10.02.2011].
- Mayer, K. -U. (1987). *Lebenslaufforschung.* In: Voges, W. (ed.) *Methoden der Biographie und Lebenslaufforschung.* Opladen: Leske + Budrich, 51-73.
- Mills, M. & Blossfeld, H.-P. (2005). *Globalisation, uncertainty and the early life course. A theoretical framework.* In: Blossfeld, H. -P. & Klijzing, E.,Mills, M. & Kurz, K. (eds.) *Globalisation, Uncertainty and Youth in Society.* London/ New York: Routledge, 1-24.
- Mühling, T. (2009). *Familienpolitik im Europäischen Vergleich.* In: Mühling, T. & Rost, H. (eds.) *ifb-Familienreport Bayern 2009. Schwerpunkt: Familie in Europa. ifb-Materialien 6/2009.* Bamberg: Staatsinstitut für Familienforschung, 9-32. Available from: *http://www.ifb.bayern.de/ imperia/md/content/stmas/ifb/materialien/familienreport_2009_ mat_2009_6.pdf* [accessed 16.02.2011].
- Neale, B. & Flowerdew, J. (2003). *Time, texture and childhood: the contours of longitudinal qualitative research.* International Journal of Social Research Methodology 6 (3): 189-199. DOI: 10.1080/1364557032000091798.
- OECD (2010a). *LMF2.2 The distribution of working hours among couple families and adults in couple families individually, by broad hours groups, presence of children, and age of youngest child.* OECD Family Database. Available from: *http://www.oecd.org/dataoecd/1/56/43199547.pdf* [accessed 16.02.2011].
- OECD (2010b). *LMF2.4: Family-Friendly Workplace Practices. OECD Family database.* Available from: *http://www.oecd.org/dataoecd/1/52/43199600. pdf* [accessed 16.02.2011].
- OECD (2010c). *PF2.1: Key characteristics of parental leave systems. OECD Family database.* Available from: *http://www.oecd.org/ dataoecd/45/26/37864482.pdf* [accessed 16.02.2011].
- OECD (2010d). *SF3.4 Family violence. OECD Family Database.* Available from: *http://www.oecd.org/dataoecd/30/26/45583188.pdf* [accessed 17.02.2011].
- Pettit, B. & Hook, J. L. (2009). *Gendered Tradeoffs. Family, Social Policy and Economic Inequality in Twenty-One Countries.* New York: Russell Sage Foundation.
- Philipov, D., Thévenon, O., Klobas, J., Bernardi, L. & Liefbroer, A. C. (2009). *Reproductive Decision-Making in a Macro-Micro Perspective (REPRO). State-of-the-Art Review.* Available from: *http://www.oeaw.ac.at/ vid/repro/assets/docs/ed-researchpaper2009-1.pdf* [accessed 14.02.2011].

- Reiska, E., Saar, E. & Viilmann, K. (2010a). *FAMILYPLATFORM Existential Field 4: Family and Living Environment. Working Report.* Available from: *http:// hdl.handle.net/2003/27693.*
- Reiska, E., Saar, E. & Viilmann, K. (2010b). *FAMILYPLATFORM Existential Field 4: Family and Living Environment. Working Report – Summary.* Available from: *http://familyplatform.eu/en/doc/89/EF4a_Summary_Family_and_ Living_Environment.pdf* [accessed 07.02.2011].
- Rupp, M. (2003). Einführung: *Niederschwellige Familienbildung im Überblick.* In: Rupp, M. (ed.) *Niederschwellige Familienbildung, ifb-Materialien 1/2003.* Bamberg: Staatsinstitut für Familienforschung, 9-12. Available from: *http://www.ifb.bayern.de/imperia/md/content/stmas/ifb/materialien/ mat_2003_1.pdf,* [accessed 15.02.2011].
- Rupp, M. (ed.) (2009). *Die Lebenssituation von Kindern in gleichgeschlechtli- chen Lebensgemeinschaften.* Köln: Bundesanzeigerverlag.
- Rupp, M., Mengel, M. & Smolka, A. (2010). *Handbuch zur Familienbildung im Rahmen der Kinder- und Jugendhilfe in Bayern. ifb-Materialien 7/2010.* Bamberg: Staatsinstitut für Familienforschung.
- Sager, F. (2009). *Die Evaluation institutioneller Politik in der Schweiz.* In: Widmer, T., Beywl, W. & Carlo Fabian (eds.) *Evaluation. Ein systematisches Handbuch.* Wiesbaden: VS Verlag für Sozialwissenschaften, 361-370.
- Schmidt, C., Raab, M. & Ruland, M. (2009). *Intergenerationale Austauschbeziehungen im internationalen Vergleich.* In: Mühling, T. & Rost. H. (eds.) *ifb-Familienreport Bayern 2009. Schwerpunkt: Familie in Europa. ifb-Materialien 6/2009.* Bamberg: Staatsinstitut für Familienforschung, 143-166. Available from: *http://www.ifb.bayern.de/imperia/md/content/ stmas/ifb/materialien/familienreport_2009_mat_2009_6.pdf* [accessed 16.02.2011].
- Schneider, N. F. & Collet, B. (eds.) (2010). *Mobile Living Across Europe II. Causes and Consequences of Job-Related Spatial Mobility in Cross-National Comparison.* Opladen/Farmington Hills: Barbara Budrich.
- Schneider, N. F. & Meil, G. (eds.) (2008). *Mobile Living Across Europe I. Relevance and Diversity of Job-Related Spatial Mobility in Six European Countries.* Opladen/Farmington Hills: Barbara Budrich.
- Schulz, F. & Grunow, D. (2007). *Tagebuch versus Zeitschätzung. Ein Vergleich zweier unterschiedlicher Methoden zur Messung der Zeitverwendung für Hausarbeit.* Zeitschrift für Familienforschung 19 (1): 106-128.
- Smith, N. (2003). *Cross-sectional profiling and longitudinal analysis: research notes on analysis in the longitudinal qualitative study, 'Negotiation Transitions to Citizenship'.* International Journal of Social Research Methodology 6 (3): 273-277. DOI: 10.1080/1364557032000091888.
- Stauber, B. (2010). *FAMILYPLATFORM Expert Report 1: Transitions into*

Parenthood. Available from: *http://hdl.handle.net/2003/27691.*

- Steele, R., Secombe, C. & Brookes, W. (2007). *Using Wireless Sensor Networks for Aged Care: The Patient's Perspective.* Available from: *http:// epress.lib.uts.edu.au/research/bitstream/handle/10453/2910/2006010605. pdf?sequence=1* [accessed 14.02.2011].
- Stiglitz, J., Sen, A. & Fitoussi, J.-P. (2009). *The Measurement of Economic Performance and Social Progress Revisited. Reflections and Overviews.* Available from: *http://www.stiglitz-sen-fitoussi.fr/documents/overview- eng.pdf* [accessed 11.02.2011].
- Thomson, R. & Holland, J. (2003). *Hindsight, foresight and insight: the chal- lenges of longitudinal qualitative research.* International Journal of Social Re- search Methodology 6 (3): 233-244. DOI: 10.1080/1364557032000091833.
- Thomson, R., Plumridge, L. & Holland, J. (2003). *Longitudinal qualitative re- search: a developing methodology.* International Journal of Social Research Methodology 6 (3): 185-187. DOI: 10.1080/1364557032000091815.
- Thorne, S. (1999). *Secondary analysis in qualitative research: issues and implications.* In: Janice M. Morse (ed.) *Critical issues in qualitative research methods.* [Reprint]. Thousand Oaks: SAGE: 263-279.
- Turton, D. (1999). *Introduction.* In: Turton, D. & Gonzáles Ferreras, J. (eds.) *Cultural Identities and Ethnic Minorities in Europe.* Bilbao: Universidad de Deusto. Available from: *http://www.humanitariannet. deusto.es/publica/PUBLICACIONES_PDF/01%20Ethnic%20Minorities. pdf* [accessed 17.02.2011].
- Van Dongen, W. (2009). *Towards a democratic division of labour in Europe? The Combination Model as a new integrated approach to professional and family life.* Bristol: Policy Press.
- Wall, K., Leitão, M. & Ramos, V. (2010a). *FAMILYPLATFORM Existential Field 7: Social inequality and diversity of families. Working Report.* Available from: *http://hdl.handle.net/2003/27698.*
- Wall, K., Leitão, M. & Ramos, V. (2010b). *FAMILYPLATFORM Existential Field 7: Social Inequality and Diversity of families. Working report – Summary.* Available from: *http://www.familyplatform.eu/en/doc/93/EF7_Summary_ Social_Inequality_and_Diversity_of_Families.pdf* [accessed 10.02.2011].
- Wall, K., Leitão, M. & Ramos, V. (2010c). *FAMILYPLATFORM Critical Review of Research on Families and Family Policies in Europe. Conference Report (July 2010, published September 2010).* Available from: *http://hdl.handle. net/2003/27687.*
- Weiss, C. H. (1999). *The Interface between Evaluation and Public Policy (keynote speech at the conference of the European Evaluation Society in Rome).* Evaluation 5(4): 468-486. London/Thousand Oaks/New Delhi. Available from: *http://evi.sagepub.com/content/5/4/468.full.pdf+html* [accessed 17.02.2011].

- Wholey, J. (1996). *Formative and Summative Evaluation: Related Issues in Performance Measurement*. American Journal of Evaluation 17 (2): 145-149. DOI: 10.1177/109821409601700206. Available from: *http://aje.sagepub.com/content/17/2/145.full.pdf+html* [accessed 14.02.2010].
- Wissenschaftlicher Beirat für Familienfragen (2005). *Familiale Erziehungs-kompetenzen. Beziehungsklima und Erziehungsleistungen in der Familie als Problem und Aufgabe*. Weinheim/München: Juventa Verlag.

Annex 1

Participants in FAMILYPLATFORM

The Consortium

1. Technical University Dortmund (Co-ordinators)
2. State Institute for Family Research at the University of Bamberg
3. Family Research Centre, University of Jyväskylä
4. Austrian Institute for Family Studies, University of Vienna
5. Demographic Research Institute, Budapest
6. Institute of Social Sciences, University of Lisbon
7. Department of Sociology and Social research, University of Milan-Bicocca
8. Institute of International and Social Studies, Tallinn University
9. London School of Economics
10. Confederation of Family Organisations in the European Union (COFACE), Brussels
11. Forum delle Associazioni Familiari, Italy
12. Mouvement Mondial des Mères Europe, Brussels

All the following persons were involved in the work of FAMILYPLATFORM. We thank them all:

Members of the Consortium

- *Loreen Beier,* ifb - State Institute for Family Research at the University of Bamberg (Germany)
- *Zsuzsa Blaskó,* DRI - Demographic Research Institute Budapest (Hungary)
- *Sonja Blum,* ÖIF - Austrian Institute for Family Studies, University of Vienna (Austria)/ Institute of Political Science, University of Münster (Germany)
- *Francesco Belletti,* FDAF - Forum delle Associazioni Familiari (Italy)
- *Julie de Bergeyck,* MMM - Mouvement Mondial de Mères Europe (Belgium)
- *Ranjana Das,* LSE - London School of Economics and Political Science (United Kingdom)
- *Anne-Claire de Liedekerke,* MMM - Mouvement Mondial de Mères Europe (Belgium)
- *Anna Dechant,* ifb - State Institute for Family Research at the University of Bamberg (Germany)
- *Matthias Euteneuer,* TUDO - Technical University Dortmund (Germany)

- *Linden Farrer*, COFACE - Confederation of Family Organisations in the European Union (Belgium)
- *Christian Haag,* ifb - State Institute for Family Research at the University of Bamberg (Germany)
- *Leeni Hansson*, Institute for International and Social Studies, Tallinn University (Estonia)
- *Veronika Herche*, DRI - Demographic Research Institute (Hungary)
- *Dirk Hofaecker*, ifb - State Institute for Family Research at the University of Bamberg/MZES - Mannheimer Zentrum für Europäische Sozialfoschung (Germany)
- *Kimmo Jokinen*, JYU - Family Research Centre at the University of Jyväskylä (Finland)
- *Olaf Kapella*, ÖIF - Austrian Institute for Familiy Studies, University of Vienna (Austria)
- *Teppo Kröger*, Family Research Centre at the Universtiy of Jyväskylä (Finland)
- *Marjo Kuronen*, JYU - Family Research Centre at the University of Jyväskylä (Finland)
- *William Lay*, COFACE - Confederation of Family Organisations in the European Union (Belgium)
- *Carmen Leccardi*, University of Milano-Bicocca (Italy)
- *Mafalda Leitão*, ICS - Institute of Social Sciences, University of Lisbon (Portugal)
- *Sonia Livingstone*, LSE - London School of Economics and Political Science (United Kingdom)
- *Sveva Magaraggia*, University of Milan-Bicocca (Italy)
- *Elisa Marchese*, ifb - State Institute of Family Research of the University of Bamberg (Germany)
- *Petra Marciniak*, TUDO - Technical University Dortmund (Germany)
- *Dorota Pawlucka*, TUDO - Technical University Dortmund (Germany)
- *Miriam Perego*, University of Milano-Bicocca (Italy)
- *Marietta Pongracz*, DRI - Demographic Research Institute (Hungary)
- *Lorenza Rebuzzini*, FDAF - Forum delle Associazioni Familiari (Italy)
- *Epp Reiska*, IISS - Institute of International and Social Studies, Tallinn University (Estonia)
- *Christiane Rille-Pfeifer*, ÖIF - Austrian Institute for Family Studies, University of Vienna (Austria)
- *Vasco Ramos*, ICS - Institute of Social Sciences, University of Lisbon (Portugal)
- *Marina Rupp*, ifb - State Institute of Family Research at the University of Bamberg (Germany)

- *Ellu Saar*, IISS - Institute of International and Social Studies, Tallinn University (Estonia)
- *Kim-Patrick Sabla*, TUDO - Technical University Dortmund/University Vechta (Germany)
- *Zsolt Spéder*, DRI - Demographic Research Institute (Hungary)
- *Joan Stevens*, MMM - Movement Mondial de Mères Europe (Belgium)
- *Uwe Uhlendorff*, TUDO - Technical University Dortmund (Germany)
- *Karin Wall*, ICS - Institute of Social Sciences, University of Lisbon (Portugal)
- *Hayet Zeghiche,* COFACE - Confederation of Family Organisations in the European Union (Belgium)

Members of the Advisory Board

- *Katerina Cadyova*, MoLSA - Ministry of Labour and Social Affairs of the Czech Republic (Czech Republic)
- *Clem Henricson*, UEA - University of East Anglia (United Kingdom)
- *Jonas Himmelstrand*, Haro (Sweden)
- *Krzysztof Iszkowski*, European Commission - DG Employment, Social Affairs and Equal Opportunities
- *Jean Kellerhals*, UNIGE - University of Geneva (Switzerland)
- *Lydie Keprova*, MoLSA - Ministry of Labour and Social Affairs of the Czech Republic (Czech Republic)
- *Nina Parra*, BMFSFJ - German federal ministry of family affairs, senior citizens, women and youth (Germany)
- *Lea Pulkkinen*, JYU - University of Jyväskylä (Finland)
- *Gilles Seraphin*, UNAF - Union Nationale des Associations Familiales (France) *Emanuela Tassa*, European Commission - DG Employment, Social Affairs and Equal Opportunities
- *Madeleine Wallin*, Haro (Sweden)

External experts

- *Katherine Bird*, Bundesforum Familie (Germany)
- *Pierre Calame*, FPH - Fondation Charles Léopold Mayer (France)
- *Elie Faroult,* former official of the European Commission, DG Research
- *Aila-Leena Matthies,* JYU - University of Jyväskylä, Kokkola University Consortium Chydenius (Finland)
- *Barbara Stauber*, University of Tuebingen (Germany)

Stakeholder representatives and other participants

- *Hibah Aburwein*, EFOMW - European Forum of Muslim Women (Belgium)
- *Silvan Agius*, ILGA Europe - International Lesbian, Gay, Bisexual, Trans and Intersex Association (Belgium)
- *Robert Anderson*, EUROFOUND - European Foundation for the Improvement of Living and Working Conditions
- *Claudine Attias-Donfut*, CNAV - Caisse nationale d'assurance vieillesse (France)
- *Anneli Antonnen*, Tampere University (Finland)
- *Maks Banens*, MODYS - Université de Lyon (France)
- *Angelo Berbotto*, NELFA - Network of European LGBT Families Associations (United Kingdom)
- *Tijne Berg-le Clercq*, NJI - Netherlands Youth Institute (Netherlands)
- *Gabrielle Chabert*, FEF - Forum Européen des Femmes (Belgium)
- *Anne Charlier-des Touches*, FEFAF - Fédération Européenne des Femmes Actives au Foyer (Belgium)
- *Anna Maria Comito*, COFACE-handicap/Co.Fa.As.Clelia - Confederazioni di Organizzazione di Famiglie con persone disabili dell'Unione Europea (Italy)
- *Michela Costa*, COFACE - Confederation of Family Organisations in the European Union (Belgium)
- *Agata D'Addato*, Eurochild (Belgium)
- *Baudouin de Lichtervelde*, EDW - European Dignity Watch (Belgium)
- *Luk de Smet*, Gezinsbond (Belgium)
- *Eric de Wasch*, Gezinsbond (Belgium)
- *Lourdes del Fresno*, Spanish Family Forum (Spain)
- *Marién Delgado*, UNCEAR - Unión de Centros de Acción Rural (Spain)
- *Fred Deven*, Kenniscentrum WVG - Kenniscentrum Welzijn, Volksgezondheid en Gezin (Belgium)
- *Shirley Dex*, IoE - Institute of Education, University of London (United Kingdom)
- *Christiane Dienel*, Nexus - Institute for Cooperation Management and Interdisciplinary Research (Germany)
- *Elena Dobre*, Ministry of Labour, Family and Social Protection (Romania)
- *Maria do Rosário Mckinney*, FEF - Forum Européen des Femmes (Belgium)
- *Annemie Drieskens*, Gezinsbond (Belgium)
- *Nathalie d'Ursel*, NWFE - New Women for Europe (Belgium)
- *Karolina Elbanowska*, Zwiazek Duzych Rodzin "Trzy Plus" (Poland)

- *Tomász Elbanowski*, Parents Rights Association (Poland)
- *RadaElenkova*, BGRF - Bulgarian Gender Research Foundation (Bulgaria)
- *Daniel Erler*, pme Familienservice GmbH (Germany)
- *Colette Fagan*, University of Manchester (United Kingdom)
- *Jeanne Fagnani*, CNRS - Centre national de la recherche scientifique (France)
- *Lena Friedrich*, ifb - State Institute for Family Research at the University of Bamberg (Germany).
- *Ada Garriga Cots*, IFFD - International Federation for Family Development (Spain)
- *Judit Gazsi*, Ministry of Social Affairs and Labour (Hungary)
- *Gabor Gerber*, Semmelweis University Budapest (Hungary)
- *Zsuzsa Gerber*, HWA - Hungarian Women's Alliance (Hungary)
- *Anne Girault*, Femina Europa (France)
- *Margen Celene Gómez Vargas*, European Large Families Confederation (Spain)
- *Marion Gret*, NELFA - Network of European LGBT Families Associations (France)
- *Maria das Dores Guerreiro, ISCTE-IUL - Lisbon University Institute (Portugal)*
- *Hana Haskova*, SOU - Institute of Sociology of the Academy of Sciences of the Czech Republic (Czech Republic)
- *Marek Havrda*, European Commission - DG SANCO (Belgium)
- *Brian Heaphy*, University of Manchester (United Kingdom)
- *Marielle Helleputte*, FEFAF - Fédération Européenne des Femmes Actives au Foyer (Belgium)
- *Joelle Hennemanne*, FEF - Forum Européen des Femmes (Belgium)
- *Jaime Hernández*, HO - HazteOir.org (Spain)
- *Helena Hiila*, The Family Federation of Finland (Finland)
- *Maria Hildingsson*, FAFCE - Federation of Catholic Family Associations in Europe (Belgium)
- *Sven Iversen*, AGF - Arbeitsgemeinschaft der deutschen Familien-organisationen (AGF) e.V. (Germany)
- *Alexandra Jachanova-Dolezelova*, EWL - European Women's Lobby (Czech Republic)
- *Jana Jamborová*, NWFE - New Women for Europe (Belgium)
- *Lea Javornik Novak*, Ministry of Labour and Social Affairs (Republic of Slovenia)
- *Josef Jelinek*, ONZ - Obcanske sdruzeni (Czech Republic)
- *Ghislaine Julemont*, CAL - Centre d'Action Laïque (Belgium)
- *Karin Jurczyk*, DJI - German Youth Institute (Germany)

- *Renata Kaczmarska*, LSE - London School of Economics and Political Science (United Kingdom)
- *Teresa Kapela*, Związek Dużych Rodzin "Trzy Plus" (Poland)
- *Jean Kellerhals*, UNIGE - University of Genève (Switzerland)
- *Naureen Khan*, NSPCC - National Society for the Prevention of Cruelty to Children (United Kingdom)
- *Magdalena Kocik*, IPS - University of Warsaw (Poland)
- *Anna Kokko*, The Family Federation of Finland (Finland)
- *Zsuzsa Kormosné Debreceni*, NOE - National Association of Large Families (Hungary)
- *Martina Leibovici-Mühlberger*, ARGE Erziehungsberatung (Germany)
- *Arnlaug Leira*, UiO - University of Oslo (Norway)
- *Liliane Leroy*, FPS - Femmes Prévoyants Socialists (Belgium)
- *Kathrin Linz*, ISS - Institute for Social Work and Social Education (Germany)
- *Marie-Liesse Mandula*, MMM - Movement Mondial de Mères Europe (France)
- *Laszlo Marki*, European Large Families Confederation (Hungary)
- *Claude Martin*, EHESP - Ecole des hautes études en santé publique (France)
- *Michiel Matthes*, Alliance for Childhood European Network Group (Belgium)
- *Françoise Meauzé*, CNAFC - Confédération Nationale des Associations Familiales Catholiques (France)
- *Rommel Mendes-Leite*, Université de Lyon (France)
- *Cascón Mercedes*, The Family Watch (Spain)
- *Michael Meuser*, TUDO - Technical University of Dortmund (Germany)
- *Anneli Miettinen*, Population Research Institute, The Family Federation of Finland (Finland)
- *Roumjana Modeva*, Women and Mothers against Violence (Bulgaria)
- *Daniel Molinuevo*, Eurofound - European Foundation for the Improvement of Living and Working Conditions (Ireland)
- *Andreas Motel-Klingebiel*, DZA - German Centre of Gerontology (Germany)
- *João Mouta*, PFA - Parents Forever Association (Portugal)
- *Leonids Mucenieks*, ULLFA - Union of Latvian Large Families Associations (Latvia)
- *Philip Muller*, Spanish Family Forum (Spain)
- *Gordon Neufeld*, UBC - University of British Columbia (Canada)
- *John Hebo Nielsen*, Joint Council of Child Issues (Denmark)
- *Margaret O'Brien*, UEA - University of East Anglia (United Kingdom)

- *James O'Brien*, Westlink Consulting
- *Livia Oláh*, Stockholm University, Department of Sociology, Demography Unit (Sweden)
- *Ilona Ostner*, Georg - August University Goettingen (Germany)
- *João Peixoto*, ISEG - Instituto Superior de Economica e Gestão, Universidade Técnica de Lisboa (Portugal)
- *Heloísa Perista*, CESIS - Centre for Studies for Social Intervention (Portugal)
- *Judith Peters*, Mothers Union (Ireland)
- *Ann Phoenix*, IoE - Institute of Education, University of London (United Kingdom)
- *Elisabeth Potzinger*, KFÖ - Katholischer Familienverband Österreichs (Austria)
- *Ceridwen Roberts*, University of Oxford (United Kingdom)
- *Yves Roland-Gosselin*, COFACE - Confederation of Family Organisations in the European Union (Belgium) / CNAFC - Confédération Nationale des Associations Familiales Catholiques (France)
- *Virginie Rothey*, Femina Europa
- *Raul Sanchez*, IESP - Institute of Advanced Family Studies, International University of Catalunya (Spain)
- *Verena Schmidt*, ILO - International Labour Organisation (Hungary)
- *Alexander Schwentner*, UNICEF Austria - United Nations International Children's Emergency Fund Austria (Austria)
- *Cinzia Sechi*, ETUC - European Trade Union Confederation (Belgium)
- *José A. Simões*, UNL - New University of Lisbon (Portugal)
- *Jorma Sipilä*, Tampere University (Finland)
- *Ignacio Socias*, The Family Watch (Spain)
- *Owen Stevens*, Sientific Consultant for Movement Mondial de Mères Europe (Belgium)
- *Sabrina Stula*, Observatory for Sociopolitical Developments in Europe (Germany)
- *Dorottya Szikra*, Eötvös Loránd University (Hungary)
- *Phillippa Taylor*, Care for Europe - Christian Action Research and Education for Europe (United Kingdom)
- *Luis Tejedor López, HO* - HazteOir.org (Spain)
- *Tobias Teuscher*, ULB - Université Libre de Bruxelles (Belgium)
- *Anália Torres*, ICS - Institute of Social Sciences, University of Lisbon (Portugal)
- *Olga Tóth*, HAS - Institute of Sociology, Hungarian Academy of Sciences (Hungary)
- *Stanislav Trnovec*, Club of Large Families (Slovakia)

- *Kaija Turkki*, University of Helsinki (Finland)
- *Bettina Uhrig*, Nova - Norwegian Social Research (Norway)
- *Vera Urban*, AFG - Association of German Family Organisations e. V. (Germany)
- *Anna Maria Vella*, Cana Movement (Malta)
- *Attila Vida*, EUROCSALÁD - Association of Divorced Fathers (Hungary)
- *Florence von Erb*, MMM - Movement Mondial de Mères Europe (France)
- *Sylvie von Lowis*, MMM - Movement Mondial de Mères Europe (France)
- *Eric Widmer*, UNIGE - University of Genéve(Switzerland)
- *Anna Záborská*, European Parliament (Slovakia)

Special thanks to

- *Marc Goffart*, European Commission, DG Research
- *Ralf Jacob*, European Commission, DG Employment
- *Pierre Valette,* European Commission, DG Research

for their support and guidance of the work of the FAMILYPLATFORM.

Annex 2

Author Biographies

Loreen Beier

Loreen Beier is a sociologist (Dipl. Soz.). During her studies at the Otto-Friedrich-University of Bamberg she focused on empirical research methods and statistics, as well as demography. Since June 2008 she has been working as a research scientist at the State Institute for Family Research at the University of Bamberg (ifb). For FAMILYPLATFORM she was co-responsible for the organisation of the meeting in Bamberg and the conference in Brussels as well as for writing the Research Agenda. Her research interests include quantitative and qualitative research methods, comparative analyses of welfare states and their effects on families, especially gender issues and the compatibility of work and family, with specific attention to particular differences between the new and old Länder in Germany.

Julie de Bergeyck

Julie de Bergeyck joined MMM (Mouvement Mondial des Mères) to work on FAMILYPLATFORM as Project Manager. She is a mother of three and has a background in communications. She has worked in the internet advertising industry in Brussels, and in the US, where she spent eight years in a leading advertising agency. She recently took a three-year break to raise her third child and has volunteered for different local organisations.

Anna Dechant

Anna Dechant is a sociologist (Dipl. Soz.). During her studies at the Otto-Friedrich-University of Bamberg she focused on empirical research methods and statistics, as well as demography. Since December 2008 she has been working as a research scientist at the State Institute for Family Research at the University of Bamberg (ifb). She was involved in a DFG-funded project analysing the division of labour in couples after the transition to parenthood. For FAMILYPLATFORM she was co-responsible for the organisation of the meeting in Bamberg and the conference in Brussels as well as for writing the Research Agenda. Her research interests include quantitative and qualitative research methods, the divison of labour in couples, gender issues and the compatibility of work and family.

Matthias Euteneuer

Matthias Euteneuer is a research assistant at the Faculty of Educational Science and Sociology at the Technical University of Dortmund. For FAMILYPLATFORM he was Administrative Co-ordinator, the intermediary between the participants and the European Commission; he was also responsible for overseeing all deliverables provided by the Consortium. His general research interests cover family research, social work with families, sociology of work and qualitative research methods.

Lena Friedrich

Lena Friedrich is a sociologist (Dipl.-Soz.). She studied at the Ruprecht-Karls-University of Heidelberg and the Otto-Friedrich-University of Bamberg. She worked from 2007 till 2009 at the research department of the Federal Office of Migration and Refugees in Nuremberg. Since 2010 she has been working as a research scientist at the State Institute for Family Research at the University of Bamberg (ifb). Her major interests include sociology of migration, life course research as well as family education.

Christian Haag

Christian Haag (Dipl.-Soz.) studied sociology at the University of Bamberg and at the National University of Ireland, Galway. He specialised in empirical methods, population studies, organisational psychology and family and the life-course. In April 2010 he began teaching undergraduate students at the Chair of Sociology I at the University of Bamberg. In June 2010 Christian Haag joined the staff of the State Institute for Family Research at the University of Bamberg (ifb), where he is currently working on a project concerned with women's re-entry into the labour force after the birth of a first child under the new parental-leave act in Germany. His research interests include men and masculinities, transitions within the life course with a focus on the transition towards parenthood, research on non-heterosexual individuals, and families in the context of homosexuality. For FAMILYPLATFORM Christian Haag participated in writing the research agenda.

Dirk Hofäcker

Dirk Hofäcker, Dr. rer. pol., studied Sociology and Political Economics at the University of Bielefeld, Germany. During his years of study, he spent a research term in 1997 at the State University of St. Petersburg, Russia, financially supported by a TEMPUS/TACIS scholarship. His major research interests include comparative analyses of welfare states, the effects of welfare systems on the social structures of families and labour markets, and quantitative multivariate methods of social science data analysis.

Kimmo Jokinen

Dr. Kimmo Jokinen is professor of family studies and head of the Family Research Centre at the University of Jyväskylä. He was educated at the University of Jyväskylä, where he obtained his Ph.D degree in 1997. His research interests have focused on family studies (recently complex and challenging family relations and emotional security in these relations), cultural studies, children and the media, and school and youth.

Olaf Kapella

Olaf Kapella studied social pedagogic and works additionally as therapist and counsellor and has training in sexuality education. He is senior researcher and research co-ordinator at the Austrian Institute for Family Studies, University of Vienna, and his main research areas include: work-life-balance, gender, sexuality education, evaluation of family policies and violence.

Marjo Kuronen

Dr. Marjo Kuronen is Senior Lecturer in Social Work at the University of Jyväskylä, Department of Social Sciences and Philosophy. She was awarded her Ph.D at the University of Stirling (UK) in 1999. Her research interests include gendered practices of parenting and post-divorce family relations, feminist social work, and qualitative cross-cultural research methodology.

Mafalda Leitão

Mafalda Leitão is a sociologist (Master of Social Sciences) working as a Research Assistant at the Institute of Social Sciences (ICS) of the University of Lisbon, Portugal. She has carried out research on voluntary work and citizenship as well as on the reconciliation of work and family life. She is currently working on developments in family policies in Portugal and in the European Union. Her research interests include family policies, leave policies and parenthood, social care, gender equality and family oriented services.

Anne-Claire de Liedekerke

Anne-Claire de Liedekerke, has been president of MMM (Mouvement Mondial des Mères) Europe for nearly three years. Its mission is to represent the voice of mothers to European institutions and to raise awareness of the importance of mothers' role in the social, cultural, and economic development of our societies. She is an art historian and a mother of three grown-up children. Her family has lived in many different parts of the world, which has given her the opportunity to gain experience as a volunteer, with professional commitments. She also launched the yearly guide "Expats in Brussels", of which she was until recently Co-editor.

Vasco Ramos

Vasco Ramos is a sociologist and a junior research fellow at the ICS (Institute of Social Sciences) of the University of Lisbon since 2009. He is carrying out research on family policies and on social inequalities and family trajectories from an intergenerational perspective. His research interests include family and gender equality, social and geographical mobility and youth subcultures.

Marina Rupp

Marina Rupp, Dr. rer. pol., studied Sociology at the University of Bamberg, Germany. Her issues are wide-ranging, including questions related to pregnancy, the paths to parenthood, divorce, as well as gay and lesbian families and large families. Additionally she has been working on the relationship between law and social developments resp. family life, e.g. in the field of violence in the family and concerning rainbow families. Marina Rupp is the head of the State Institute of Family Research at the University of Bamberg (ifb) and was Scientific Co-ordinator of FAMILYPLATFORM.

Uwe Uhlendorff

Uwe Uhlendorff is professor of social pedagogy at the Faculty of Educational Science and Sociology at the Technical University of Dortmund. He was Overall Co-ordinator and the main representative of FAMILYPLATFORM, and has previously co-ordinated several other international co-operation projects and research networks including the international DFG Graduate Programme "Youth Welfare and Social Services in Transition", sponsored by German Research Foundation. He also led the pilot study "Change of Childhood and Family Images in Public Child-Rearing – Reform Processes of Social Services for Families in European Comparison", funded by the Federal Ministry for Education and Research. Prof. Uhlendorff has organised several international conferences in Dortmund, the Metropolitan University Tokyo, and Nara University of Education. He has published nine academic books and over 40 journal articles and book chapters. He was involved in numerous international conferences as moderator and he developed and led future scenario exercises, and was previously a Research Professor at the University of Tokyo (2009).

Karin Wall

Karin Wall is a sociologist and a senior research fellow at the Institute of Social Sciences (ICS) of the University of Lisbon. She has carried out national and cross-national research in the areas of sociology of the family and social policies affecting families. Her research interests include demographic trends and changing family patterns, family interactions and social networks over the life-course, gender and family, reconciliation of work and family life, family policies and social care in Europe, migrant women and families, and lone parents. Her recent publications examine the diversity of contemporary family relationships, family life from the male perspective, attitudinal change to gender and family, the politics of parental leave and care regimes in Southern Europe, and changing female migration trajectories.

Annex 3

FAMILYPLATFORM Reports and Publications

- Beier, L., Hofäcker, D., Marchese, E. & Rupp, M. (2010). *Family Structures and Family Forms in the European Union: An overview of major trends and developments.* (Existential Field Report #1). Available from: *http://hdl.handle.net/2003/27689.*
- Belletti, F. & Rebuzzini, L. (2010). *Local Politics: Programmes and best practice models.* (Existential Field Report #4a). Available from: *http://hdl.handle.net/2003/27694.*
- Blaskó, Z. & Herche, V. (2010). *Patterns and Trends of Family Management in the European Union.* (Existential Field Report #5). Available from: *http://hdl.handle.net/2003/27695.*
- Blum, S. & Rille-Pfeiffer, C. (2010). Major Trends *of State Family Policies in Europe.* (Existential Field Report #3). Available from: *http://hdl.handle.net/2003/27692.*
- Farrer, L., Rebuzzini, L., Aurora, L., de Liedekerke, A.-C., Donnelly, J. & Mandula, M.-L. (2010) *Civil Society Perspective: Family Organisations at the Local, National, European and Global Level – Three Case Studies.* Available from: *http://hdl.handle.net/2003/27703.*
- Farrer, L. & Lay, W. (2011). *Spotlights on Contemporary Family Life.* (FAMILYPLATORM – Families in Europe #2). Available from: *https://eldorado.tu-dortmund.de/handle/2003/27708.*
- Kapella, O., de Liedekerke, A.-C. & de Bergeyck, J. (2011). *Foresight Report: Facets and Preconditions of Wellbeing of Families.* Available from: *http://hdl.handle.net/2003/27688.*
- Kuronen, M. (2010). *Research on Families and Family Policies in Europe: State of the Art.* Available from: *http://hdl.handle.net/2003/27686.*
- Kuronen, M., Jokinen, K. & Kröger, T. (2010). *Social Care and Social Services.* (Existential Field Report #6a). Available from: *http://hdl.handle.net/2003/27696.*
- Leccardi, C. & Perego, M. (2010). *Family Developmental Processes.* (Existential Field Report #2a). Available from: *http://hdl.handle. net/2003/27690.*
- Livingstone, S. & Das, R. (2010). *Media, Communication and Information Technologies in the European family.* (Existential Field Report #8). Available from: *http://hdl.handle.net/2003/27699.*
- Matthies, A. L. (2010). *The Professional Standards of Care Workers – The Development of Standards for Social Work and Social Care*

Services for Families. (Existential Field Report #6a). Available from: *http://hdl.handle.net/2003/27697*.

- Reiska, E., Saar, E. & Viilmann, K. (2010). *Family and Living Environment: Economic Situation, Education Levels, Employment and Physical Living Environment*. (Existential Field Report #4a). Available from: *http://hdl.handle.net/2003/27693*.
- Rupp, M., Beier, L., Dechant, A. & Haag, C. (2011*). A Research Agenda on the Family for the European Union*. Available from: *http://hdl.handle.net/2003/28901*.
- Stauber, B. (2010). *Transitions into Parenthood*. (Existential Field Report #2b). Available from: *http://hdl.handle.net/2003/27691*.
- Stevens, J., de Bergeyck, J. & de Liedekerke, A.-C. (2010). *Realities of Mothers in Europe*. Available from: *http://hdl.handle.net/2003/27685*.
- Wall, K., Leitão, M., & Ramos, V. (2010). *Critical Review of Research on Families and Family Policies in Europe*. Available from: *http://hdl.handle.net/2003/27687*.
- Wall, K., Leitão, M., & Ramos, V. (2010). *Social Inequality and Diversity of Families*. (Existential Field Report #7). Available from: *http://hdl.handle.net/2003/27698*.